Principles and Practices of Air Pollution Control and Analysis

Principles and Practices of Air Pollution Control and Analysis

Contributors

Frederick W. Lipfert et al.

AURIS
Reference

www.aurisreference.com

Principles and Practices of Air Pollution Control and Analysis

Contributors: Frederick W. Lipfert et al.

Published by Auris Reference Limited

www.aurisreference.com

United Kingdom

Principles and Practices of Air Pollution Control and Analysis

ISBN: 978-1-78154-833-2

British Library Cataloguing in Publication Data
A CIP record for this book is available from the British Library

Printed in the United Kingdom

Exclusively distributed by CBS Publishers & Distributors Pvt. Ltd.

Sales & Distribution Rights only for India, Pakistan, Bangladesh, Sri Lanka, Nepal and Bhutan.This book is not to be sold outside these territories.

Contents

List of Abbreviations .... *vii*

List of Contributors.................................... ....*ix*

Preface.. ....*xiii*

Chapter 1   An Assessment of Air Pollution Exposure Information for
Health Studies.. 1

Chapter 2   Air Pollution in China: Mapping of Concentrations and Sources........ 25

Chapter 3   Air Pollution by Hydrothermal Volcanism and Human Pulmonary
Function... 41

Chapter 4   Evaluation of an Emission Inventory and Air Pollution in the
Metropolitan Area of Buenos Aires 59

Chapter 5   Spatial Analysis on the Concentrations of Air Pollutants in Basra
Province (Southern Iraq)... 93

Chapter 6   Estimation of Citywide Air Pollution in Beijing 109

Chapter 7   Meandering Dispersion Model Applied to Air Pollution.................... 125

Chapter 8   Air Pollution and Cultural Heritage: Searching for "The Relation
Between Cause and Effect" 143

Chapter 9   Effect of Air Pollution on Archaeological Buildings in Cairo............. 173

Chapter 10   Influence of Air Pollution on Degradation of Historic Buildings
at the Urban Tropical Atmosphere of San Francisco De
Campeche City, México 199

Chapter 11   Plasma-Based Depollution of Exhausts: Principles, State of the
Art and Future Prospects 229

Chapter 12   Generation and Dispersion of Total Suspended Particulate Matter
Due to Mining Activities in an Indian Opencast Coal Project 263

Citations 279

Index.. 281

List of Abbreviations

ATS	American Thoracic Society
BLUE	best linear unbiased estimates
CO	Carbon monoxide
CBA	City of Buenos Aires
DDS	Diffuse degassing structure
ETS	Environmental tobacco smoke
GIS	Geographic Information System
GBA	Greater Buenos Aires
INEL	Idaho Engineering Laboratory
LTO	Landing-take-off
MABA	Metropolitan Area of Buenos Aires
NHANES	National Health and Nutrition Examination Survey
NOAA	National Oceanic and Atmos-pheric Administration
NO	Nitric oxide
OR	Odds ratio
PM	Particulate matter
SPA	Single Point Areal Estimation
WHO	World Health Organization
CIMFR	Central Institute of Mining and Fuel Research
IMD	Indian Meteorological Department
MPS	Microwave plasma source
PM	Polarizing Microscope
PCD	Pulsed corona discharges
SEM	Scanning electron microscope
SEM	Scanning Electron Microscope
TSP	Total suspended particles
USEPA	United States Environment protection Agency
WCL	Western Coalfields Limited
XRD	X-ray diffraction

List of Contributors

Frederick W. Lipfert
Environmental Consultant, Greenport, New York, NY 11944, USA

Robert A. Rohde
Berkeley Earth, Berkeley, California, United States of America

Richard A. Muller
Berkeley Earth, Berkeley, California, United States of America
Department of Physics, University of California, Berkeley, California, United States of America

Diana Linhares
Department of Biology, University of the Azores, Ponta Delgada, 9501-801 Azores, Portugal
CVARG, Center for Volcanology and Geological Risks Assessment (CVARG), University of the Azores, Ponta Delgada, 9501-801 Azores, Portugal

Armindo dos Santos Rodrigues
Department of Biology, University of the Azores, Ponta Delgada, 9501-801 Azores, Portugal
CVARG, Center for Volcanology and Geological Risks Assessment (CVARG), University of the Azores, Ponta Delgada, 9501-801 Azores, Portugal

Patrícia Ventura Garcia
Department of Biology, University of the Azores, Ponta Delgada, 9501-801 Azores, Portugal
[3]CE3C, Centre for Ecology, Evolution and Environmental Changes (CE3C) and Azorean Biodiversity Group, University of the Azores, 9501-801 Ponta Delgada, Portugal

Fátima Viveiros
CVARG, Center for Volcanology and Geological Risks Assessment (CVARG), University of the Azores, Ponta Delgada, 9501-801 Azores, Portugal
Department of Geosciences, University of the Azores, Ponta Delgada, 9501-801 Azores, Portugal

Teresa Ferreira
CVARG, Center for Volcanology and Geological Risks Assessment (CVARG), University of the Azores, Ponta Delgada, 9501-801 Azores, Portugal
Department of Geosciences, University of the Azores, Ponta Delgada, 9501-801 Azores, Portugal

Laura E. Venegas
National Scientific and Technological Research Council (CONICET) National Technological University, Argentina

Nicolás A. Mazzeo
National Scientific and Technological Research Council (CONICET) National Technological University, Argentina

Andrea L. Pineda Rojas
National Scientific and Technological Research Council (CONICET) National Technological University, Argentina

Shukri I. Al-Hassen
Department of Geography, University of Basra, Basra, Iraq
Technical College, Southern Technical University, Basra, Iraq

Abdul Wahab A. Sultan
Technical College, Southern Technical University, Basra, Iraq

Abdul
Technical College, Southern Technical University, Basra, Iraq

Adnan A. Ateek
Technical College, Southern Technical University, Basra, Iraq

Hamid T. Al-Saad
Department of Environmental Chemistry, University of Basra, Basra, Iraq

Salah Mahdi
Department of Environmental Chemistry, University of Basra, Basra, Iraq

Abdulzahra A. Alhello
Department of Environmental Chemistry, University of Basra, Basra, Iraq

Jin-Feng Wang
State Key Laboratory of Resources and Environmental Information System, Institute of Geographic Sciences and Natural Resources Research, Chinese Academy of Sciences, Beijing, China

Cheng-Dong Xu
State Key Laboratory of Resources and Environmental Information System, Institute of Geographic Sciences and Natural Resources Research, Chinese Academy of Sciences, Beijing, China

Mao-Gui Hu
State Key Laboratory of Resources and Environmental Information System, Institute of Geographic Sciences and Natural Resources Research, Chinese Academy of Sciences, Beijing, China

George Christakos
Department of Geography, San Diego State University, San Diego, California, United States of America

Yu Zhao
State Key Laboratory of Resources and Environmental Information System, Institute of Geographic Sciences and Natural Resources Research, Chinese Academy of Sciences, Beijing, China
School of Geosciences and Info-Physics, Central South University, Changsha, China

Gervásio A. Degrazia
Universidade Federal de Santa Maria/UFSM Brazil

Andréa U. Timm
Universidade Federal de Santa Maria/UFSM Brazil

Virnei S. Moreira
Universidade Federal de Santa Maria/UFSM Brazil

Débora R. Roberti
Universidade Federal de Santa Maria/UFSM Brazil

Eleni Metaxa
School of Chemical Engineering, National Technical University of Athens, Greece

Mohamed Kamal Khallaf
Restoration Department, Faculty of Archaeology, Fayoum Universiy, Egypt

Javier Reyes
Autonomous University of Campeche,mexico

Ronny Brandenburg
Leibniz Institute for Plasma Science and Technology, Germany

Ratnesh Trivedi
Scientists, Central Institute of Mining and Fuel Research, India

M. K. Chakraborty
Scientists, Central Institute of Mining and Fuel Research, India

B. K. Tewary
Scientists, Central Institute of Mining and Fuel Research, India

Preface

Air pollution is the pollution of air by smoke and harmful gasses, mainly oxides of carbon, sulfur, and nitrogen. The text *Principles and Practices of Air Pollution Control and Analysis* deals with current air pollution control technologies, and presents environmental issues related to air pollution, such as acid rain, global climate change, and CFCs and ozone layer. An assessment of air pollution exposure information for health studies has been described in first chapter. Second chapter focuses on air pollution in China. The aim of third chapter is to assess whether chronic exposure to volcanogenic air pollution by hydrothermal soil diffuse degassing is associated with respiratory defects in humans. The evaluation of an emission inventory and air pollution in the metropolitan area of Buenos Aires has been proposed in fourth chapter. Fifth chapter aims to analyze the geographic distribution of air pollutant concentrations in Basra Province, Southern Iraq, and to cartographically determine the spatial variation of air pollution levels as well as to recognize the hottest spots of air pollution. Sixth chapter deals with the estimation of citywide air pollution in Beijing. Meandering dispersion model applied to air pollution has been discussed in seventh chapter. Air pollution and cultural heritage have been described in eighth chapter. The effect of air pollution on archaeological buildings in Cairo has been investigated in ninth chapter. The influence of air pollution on degradation of historic buildings at the urban tropical atmosphere of San Francisco De Campeche city in México has been discussed in tenth chapter. Eleventh chapter aims a comprehensive description of plasma-based air remediation technologies. Last chapter focuses on generation and dispersion of total suspended particulate matter due to mining activities in an Indian opencast coal project

Chapter 1

AN ASSESSMENT OF AIR POLLUTION EXPOSURE INFORMATION FOR HEALTH STUDIES

Frederick W. Lipfert

Environmental Consultant, Greenport, New York, NY 11944, USA; Tel.: +631-261-5735 Academic Editor: Pasquale Avino

ABSTRACT

Most studies of air pollution health effects are based on outdoor ambient exposures, mainly because of the availability of population-based data and the need to support emission control programs. However, there is also a large body of literature on indoor air quality that is more relevant to personal exposures. This assessment attempts to merge these two aspects of pollution-related health effects, emphasizing fine particles. However, the basic concepts are applicable to any pollutant. The objectives are to examine sensitivities of epidemiological studies to the inclusion of personal exposure information and to assess the resulting data requirements. Indoor air pollution results from penetration of polluted outdoor air and from various indoor sources, among which environmental tobacco smoke (ETS) is probably the most toxic and pervasive. Adequate data exist on infiltration of outdoor air but less so for indoor sources and effects, all of which have been based on surveys of small samples of individual buildings. Since epidemiology is based on populations, these data must be aggregated using probabilistic methods. Estimates of spatial variation and precision of ambient air quality are also needed. Hypothetical personal exposures in this paper are based on ranges in outdoor air quality, variable infiltration rates, and ranges of indoor source strength. These uncertainties are examined with respect to two types of mortality studies: time series analysis of daily deaths in a given location, and cross-sectional analysis of annual mortality rates among locations. Regressions of simulated mortality on personal exposures, as affected by all of these uncertainties, are used to examine effects on dose-response functions using quasi-Monte Carlo methods. The working hypothesis is that indoor sources are reasonably steady over time and thus applicable only to long-term cross-sectional studies. Uncertainties in exposure attenuate the simulated mortality regression coefficients; correlations between "true" and hypothesized exposures are used to compare their effects. For a given exposure uncertainty level, attenuation of regression coefficients is similar for both types of simulated mortality studies, but since cross-sectional studies involve indoor sources they are more sensitive, to the point where regression coefficients may be driven to zero. The most pressing need for confirming data is the distribution of indoor sources among cities, especially for ETS.

INTRODUCTION

There are several rationales for monitoring ambient air quality:

- To characterize the environment of a community.
- To determine compliance with regulatory standards.
- To investigate effects of new or modified pollution sources.
- To provide exposure data for studies of adverse effects, especially health effects.

In the first case, considering a wide range of pollutants may be more important than their precision or accuracy, and historical contexts may be of interest. However, accuracy is important with respect to developing effective pollution control strategies. Regulatory compliance issues are limited to specified pollutants and measurement methods.

Developing appropriate exposure information for health effects studies is more complex and depends on the type of study. It is important to note that outdoor ambient air quality data may suffice to characterize the environment of a community but are unlikely to adequately characterize exposures of individuals within the community [1]. Further, exposures of only a few of those individuals are relevant to epidemiology. This conclusion has seldom been considered by epidemiologists in recent years. Other important considerations include:

Regulatory standards may be based on studies that used incomplete exposure information [2].

A complete description of community environment should include both indoor and outdoor conditions [3].

The paper is organized as follows: This section describes the rationale for the study and lays out the requirements for different types of exposure data. Section 2 describes the assessment methodology. InSection 3, I describe the exposure data I found in the literature and their characteristics. Section

4then uses these data to estimate how different types of epidemiological studies would be affected by uncertainties inherent in the available exposure data and by the use of estimated personal exposures rather than the usual ambient air quality data. I discuss the implications of these findings in Section 5and the resulting conclusions in Section 6.

Data Requirements by Type of Study

Studies involving controlled exposures of defined individuals require accurate and precise measurements of the pollutants involved; surrogate measures are unlikely to suffice [4]. By contrast, epidemiological methods are required to study populations large enough to detect the subtle health effects found under current conditions [5]. The often quoted aphorism, *the dose makes the poison*, applies here and requires considering each of element connecting ambient air quality to dose. The main elements are:

- Selection of pollutants to be considered
- Accuracy of measurement methods
- Spatial and temporal variability of outdoor air quality
- Penetration of outdoor air into occupied spaces
- Strengths of indoor pollution sources.

Other elements of uncertainty include indoor ventilation rates, rates of human uptake, and doses to target organs, for all of which adequate data are lacking.

Each of these elements involve uncertainties and requires averaging over populations; both regulated ambient (*i.e.*, "criteria") and toxic pollutants should be considered. Assessment requires comparing the relative contributions of each element of uncertainty with respect to the uncertainty in total exposure. For example, it would not be cost-effective to require a measurement method to have an accuracy of say, ±1%, if the uncertainty of the community average is say, ±10%. Similarly, if most exposures occur indoors, the accuracy of outdoor air measurements becomes less important for health studies. As outdoor air becomes cleaner, indoor pollution sources become more important. The processes linking exposure to target organ dose are perhaps the most problematic but have seldom been considered.

Reasons for neglecting personal exposures in air pollution epidemiology include:

- Regulatory mandates are limited to outdoor air [6].
- Indoor exposure information is not available for populations.

- Physiological processes governing doses to target organs are not well understood [7].
- Current epidemiological methods based on outdoor air quality have produced consistent and highly statistically significant findings that have been interpreted as individual risks [8].

Conventional epidemiological studies require parallel data on each parameter for all subjects, typically numbering in the thousands. With these deterministic methods, significance levels are largely determined by model fit. Typically, individual exposures are inferred from air quality data from a limited number of fixed ambient monitoring stations; the resulting uncertainties have been assessed in a number of studies. However, personal exposure data must be obtained on an individual basis and probabilistic methods are thus required to estimate population exposures. Survey sample sizes and the properties of exposure distributions will also affect significance levels of population-based risk estimates. The difficulties of this task do not diminish its relevance.

Exposure Data Requirements for Epidemiological Studies

Studies of population health fall into two categories: variations over time at selected locations (time-series studies) and variations among locations during selected time periods (cross-sectional studies). Although there may be some overlap, their respective data requirements differ substantially.

Time-Series Studies

Short-term (daily) variations are mainly driven by changes in weather such as stagnations, storms, or frontal passages. In such cases, ambient concentrations of various pollutants tend to show similar patterns. Shorter term (hourly) variations are largely due to occupational patterns or emission cycles such as vehicular traffic. Longer term cycles are usually seasonal, driven by both emissions and weather. A valid time-series study must control all of these patterns that relate to short-term health effects like mortality or hospitalization. Nevertheless, time-series studies have important advantages. They have been validated by major episodes of the past century during which excess daily death rates were high enough to allow identification of individual victims. They are not confounded by indoor pollution sources that remain largely unaffected by daily weather changes. However, only a fraction of outdoor air (typically ~ 50% [9]) penetrates indoors, thus attenuating actual exposures. Indoor/outdoor relationships must be averaged over the communities under study, which reduces the variance of these perturbations but not biases resulting from partial penetration of outdoor air.

The appropriate duration of exposure has not been established for epidemiological studies; statistical significance is often seen for multi-day periods [10]. Daily means are generally more relevant than hourly means for population-based studies because penetration of outdoor air is not instantaneous and the timing of peak hourly concentrations will vary within a population.

Cross-Sectional Studies

Longer term, usually annual, studies involve differences among communities and cohorts and must control for all other spatial differences that may relate to air quality. Since disadvantaged areas often have the worst air quality within a city, confounding may result from intercommunity differences in smoking, poverty, or education, all of which are known to exert larger effects on public health than air pollution [11]. Such differences can be difficult to control because of the multitude of such factors and lack of adequate data. In contrast with time-series studies, cross-sectional studies have not been validated by identifying specific putative victims Other issues involve timing and duration of exposures, including contributions of short-term exposures and allowance for the lag and latency for development of new diseases, for which cumulative exposures may be appropriate [12]. Persistent exposures from indoor pollution sources must also be considered and averaged across each community in the study. Exfiltration (venting of indoor air) is also possible and relates to the degree of building tightness, but adequate data are not available. However, forced ventilation may result in additional intake of outside air in order to maintain equilibrium.

Longitudinal studies [13] of gradual changes in exposures, of which few have been published, must control for other long-term changes including improved medical care, better residential construction, and reduced rates of smoking.

Summary

Sources of exposure uncertainty in epidemiology include:

For time-series studies, sources include instrument accuracy and spatial variation of daily outdoor air quality averaged over the number of monitoring locations and annual rate of infiltration of outdoor air averaged over the number of residences in the community under study. Seasonal variations in infiltration rates might also be considered.

For cross-sectional studies, sources include instrument accuracy and annual spatial variation of outdoor air quality averaged over the number of monitoring locations in each city, infiltration of outdoor air averaged over the number of residences in each city, and annual average contributions of indoor sources averaged over the number of residences in each city.

Each of these parameters will vary by pollutant, and it is likely that outdoor and indoor air quality may involve different species. Although this paper emphasizes particles, pollutants of interest should not be limited to criteria pollutants but should include toxic species of known health effects, especially those found indoors.

METHODS AND DATA

The basic method of assessment involves postulating baseline datasets and hypothetical linear regressions of mortality on pollution, as if the mortality rates were adjusted for confounding variables. These regressions are then repeated using exposure data modified to reflect outdoor air quality variability, penetration into residences, and indoor air pollution sources, sequentially. The data required to estimate these exposure variations were obtained from the literature and expressed as exposure increments and their standard errors. The outcome of the assessment is the degree to which regression coefficients and their standard errors differ from baseline values according to definitions of exposure.

RESULTS

Data are available for each of these exposure parameters for several pollutants; fine particulate matter ($PM_{2.5}$) was selected as an illustrative example. Personal exposures are often assumed to be tantamount to indoor concentrations, which comprise infiltrated outdoor air and emissions from indoor sources [14]. However, indoor air quality relates to the average personal exposures of all the individuals in the household during the study period and is thus less difficult to model.

Indoor sources comprise key constituents of personal exposure for all types of health studies including toxicological studies of sick building syndrome and the like [15]. An epidemiology study based only on outdoor air quality, as has been the case, tacitly assumes that indoor sources may be neglected, ostensibly because they are not regulated under the Clean Air Act. As shown below, this assumption has important implications for epidemiology and estimates of health effects

Outdoor Air Concentrations

Many studies of daily variations in health parameters have been based on a few or single air quality monitoring stations; spatial and/or temporal air quality variability could thus be an important contributor to uncertainties in health effect estimates. Uncertainties in outdoor air concentrations, including spatial variability and instrumental or analytical precision, reflect directly on personal exposures to outdoor air.

The daily $PM_{2.5}$ data of Pinto et al. [16] for 27 urban areas that had multiple monitors for the year 2000 appear to be the best dataset for studying intracity spatial variations. Pinto et al. list the ranges in inter site means and site-pair correlations as indices of spatial variability. The correlations between pairs indicate their consistency over time irrespective of mean values, which is important for time-series studies. The ranges of mean values indicate variability for use in long-term studies. Neither of these statistics provides direct estimates of spatial uncertainties in outdoor air quality for comparison with those for indoor air quality, which are estimated as follows.

The definition of the correlation coefficient R is useful in this regard:

$$R^2 = 1 - \sigma_{xy}^2/\sigma_y^2 \tag{1}$$

where σ_{xy}^2 is the unexplained variance from a linear regression of y on x, which in this case are parallel records of daily $PM_{2.5}$ in a given city, and σ_y is the corresponding standard deviation. Unfortunately, σ_y values are not tabulated by Pinto et al. so that estimation is required from other sources. For this purpose, PM_{10} data was drawn upon for the 20 largest USA cities [17] and estimated a standard deviation for each city using the tabulated means and 10th and 90th percentiles. Note that PM_{10} and $PM_{2.5}$ have the same coefficients of variation (0.29) across cities [18] and thus similar frequency distributions. The ratios of estimated deviation to mean PM_{10} ranged from 0.20 to 0.43 with a mean of 0.30 and a standard deviation of 0.56. These estimates are considered adequate for the purposes of comparative analyses in this paper and used a mean value of 0.3 to estimate σ_y. Equation (1) was then used to estimate σ_{xy} for each city as shown in Figure 1.

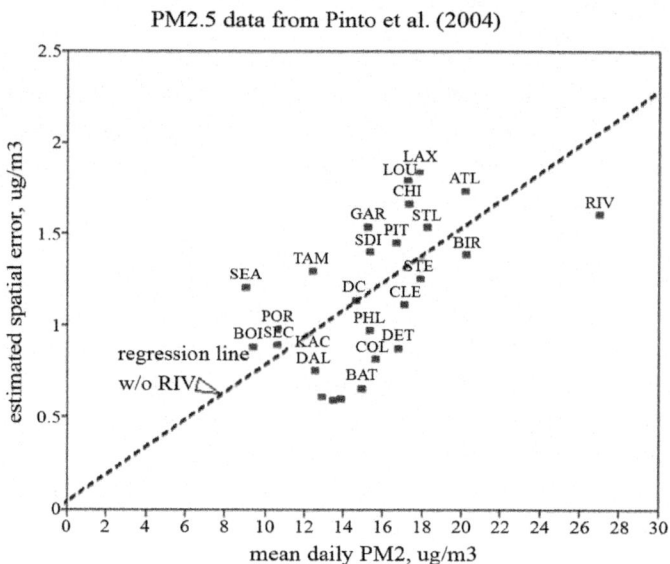

Figure 1: Estimated spatial PM$_{2.5}$ uncertainties for 27 USA cities [16].

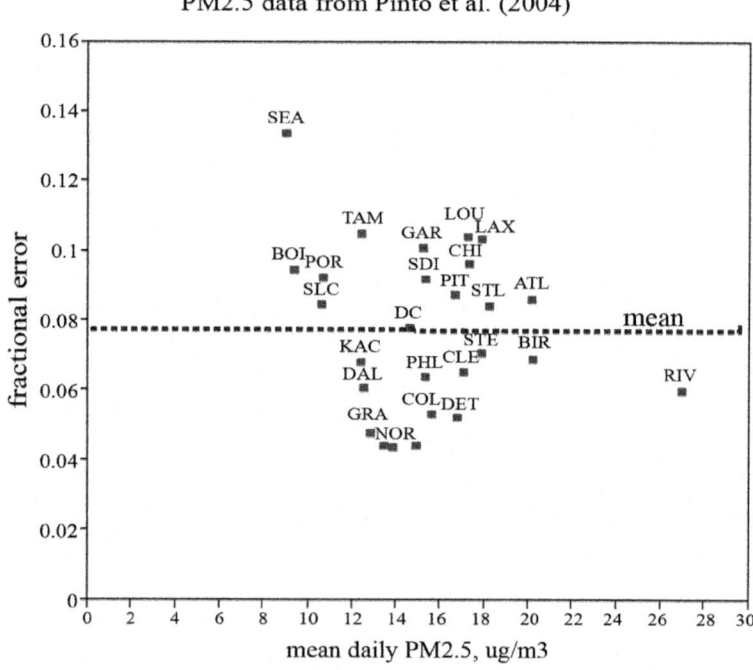

Figure 2: Fractional spatial uncertainties for 27 USA cities [16].

These estimated errors are highly correlated with mean $PM_{2.5}$ with an average ratio of error to mean of 0.077 (Figure 2). At 15 µg/m³, the 95% confidence intervals for a single monitoring station would thus be (12.7, 17.3). For the average over a network of 10 stations, the CIs would be (14.6, 15.4). As discussed below, corresponding indoor values for infiltrated outside air would be approximately half of these estimates. Note that the averaging process for outdoor air pertains to the number of monitoring stations within a city, while averaging for indoor air pertains to the number of households in that city.

Infiltration of Outdoor Air

Infiltration has been studied extensively in USA and abroad. A good approximation of the mean rate is about 0.50 and a typical distribution is shown in Figure 3; the standard deviation is about 0.07 [9]. However, epidemiology studies involve large populations and community averages over thousands of residences. The variability among individual buildings is thus of little importance, assuming minimal seasonal variations or correlations with other community attributes like income, education, proximity to pollution sources like traffic. Nevertheless, accounting for infiltration has the effect of doubling risk estimates because only half of the outdoor concentration could be responsible for health effects observed indoors.

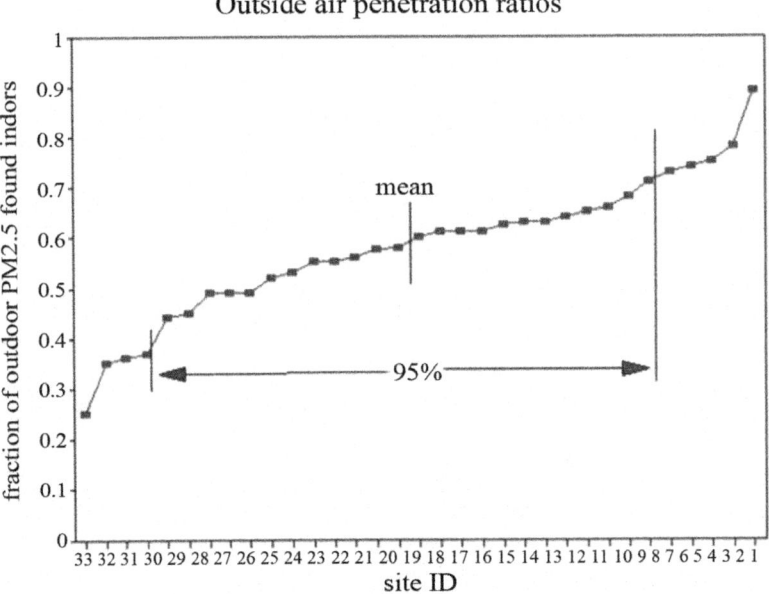

Figure 3: Distribution of outdoor air penetration ratios for individual buildings [9].

Indoor Air Pollution Sources

Many studies have characterized specific types of indoor particulate air pollution such as environmental tobacco smoke (ETS), pet dander, indoor combustion source including gas stoves, candles, incense, household dust [15]. Note that current regulatory practice considers the mass but not the chemistry of PM. In the developed world, the most common indoor pollutants are NO_2, CO, and particulate matter, especially $PM_{2.5}$. Aside from photocopy machines, there are no indoor sources of O_3, and SO_2 tends to be adsorbed onto interior surfaces. Important indoor non-criteria pollutants include NH_3, benzene, and formaldehyde, all of which are known to cause adverse health effects [19]. Acid aerosols tend to be neutralized indoors [20]. It is quite possible that important indoor pollutants may differ from outdoor species, requiring consideration of mixtures. ETS may be the most important source of indoor $PM_{2.5}$. Jenkins et $al.$ [21] measured personal exposures to respirable particulates (RSP, $PM_{3.5}$)* for 100 nonsmokers in each of 16 metropolitan areas and contrasted the results according to passive smoking in their work and home environments. The results are shown in Figure 4, with an RSP range of 8–15 µg/m³ due to ETS. There is also a negative relationship with outdoor levels. Spengler et $al.$ [22] reported an indoor RSP increment of 20 µg/m³ when two smokers were present, which is consistent with Figure 4.

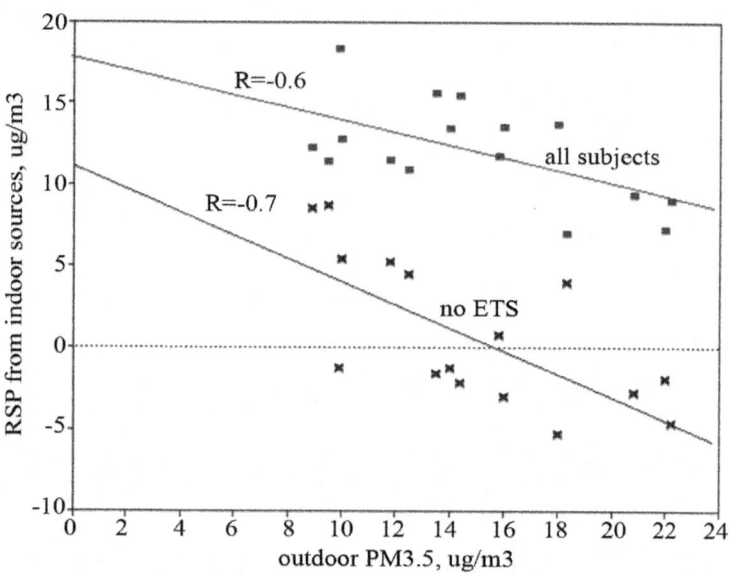

Figure 4: Effects of environmental tobacco smoke on indoor $PM_{3.5}$ concentrations in 16 USA cities [21]. In the 1980s, fine particles were designated "respirable particles" (RSP), defined as $PM_{3.5}$.

The contributions of indoor sources cannot be measured directly but may be inferred by regressing indoor concentrations on outdoor levels. The time-series data of Zeger et al. [23] for Riverside, CA shown in Figure 5 are useful for this purpose. The slope (0.58) represents the infiltration rate of outdoor air while the intercept (58 μg/m³) represents contributions of indoor sources. Both of these estimates are highly statistically significant. Figure 6 implies that the effects of indoor sources do not vary over time at this location.

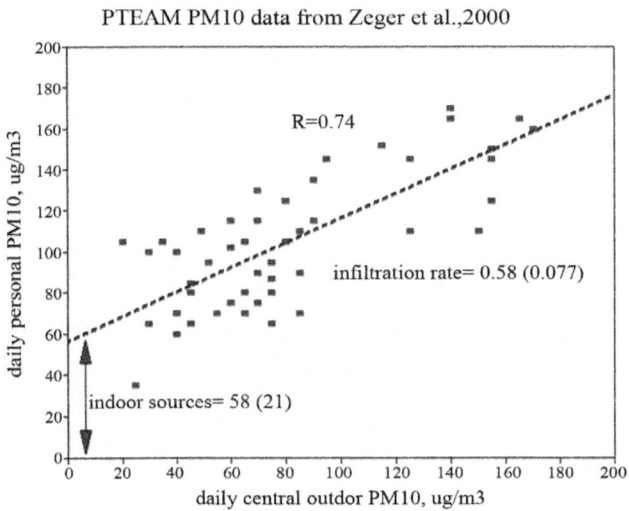

Figure 5: Indoor/outdoor PM$_{10}$ relationships from Riverside, CA [23].

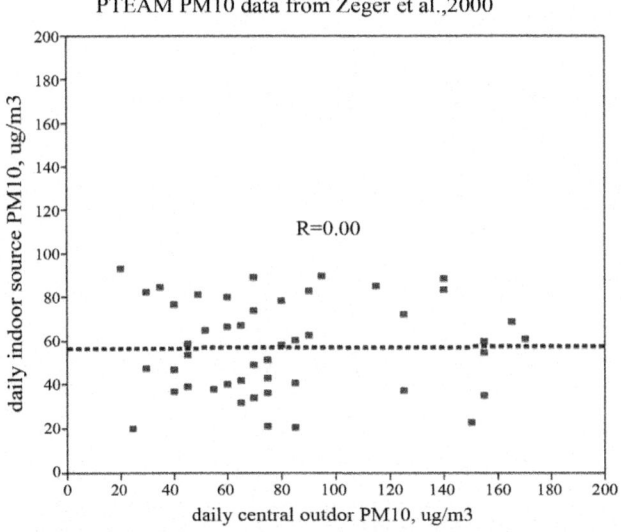

Figure 6: Estimated effects of indoor PM$_{10}$ sources in Riverside, CA [23].

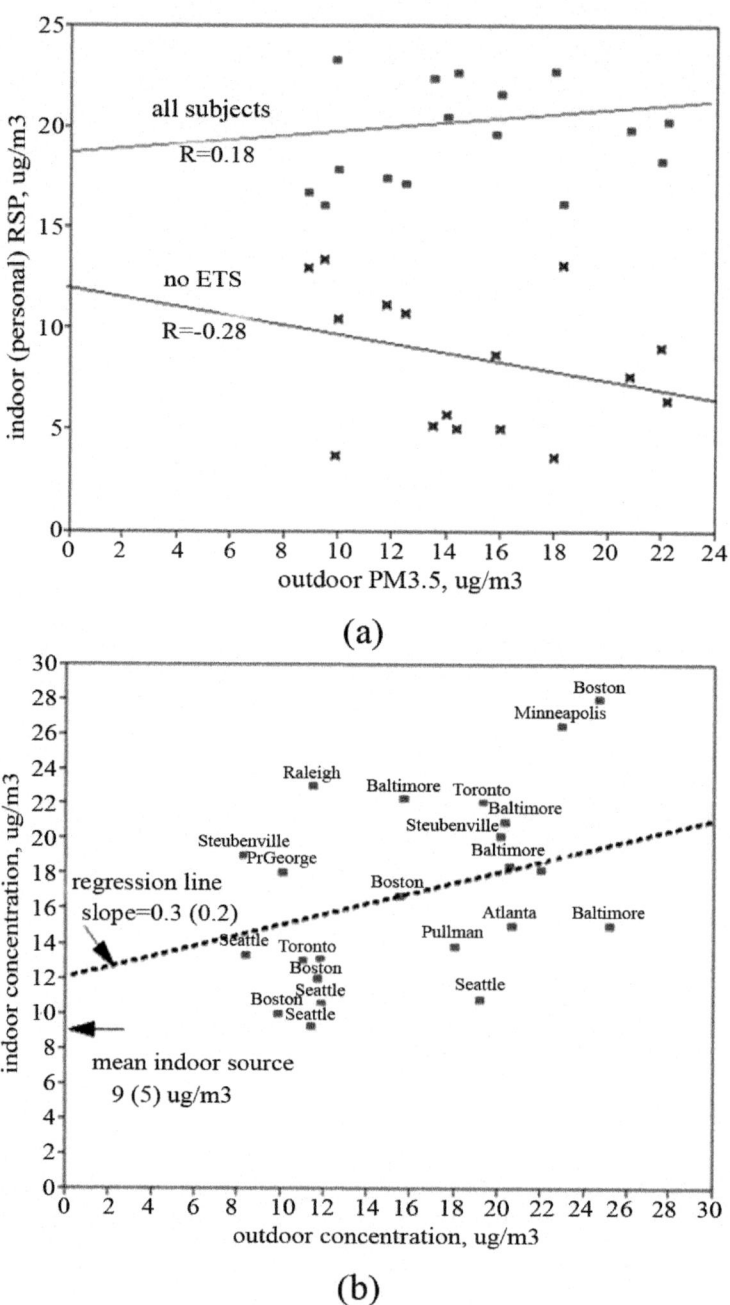

Figure 7: Indoor-outdoor PM$_{2.5}$ relationships among USA: **(a)** Cities [21]; **(b)** Cities [24] Note: ( ) denotes standard error.

Mean particulate levels tend to vary more between cities than within a city, especially for fine particles. Jenkins *et al.* [20] sampled outdoor and indoor respirable particulates in 16 US cities; Avery*et al.* [23] present detailed PM$_{2.5}$ sampling data from 34 USA cities. The contributions of indoor sources were estimated by subtracting 50% of the outdoor concentration from the indoor concentration. Caution is in order here because the small samples in each city may not adequately represent city-wide averages. These results are shown in Figure 7a,b which show that personal exposures cannot be predicted from outdoor concentration levels. The mean indoor source contribution from Avery *et al.* [24] is 9 µg/m³; from Jenkins *et al.* [21], 12 µg/m³ for all subjects and 0.9 µg/m³ in the absence of ETS. Thus, with respect to cross-sectional analysis, mean ambient concentration is a poor surrogate for actual (personal) exposure, even in nonsmoking households, even though they may be correlated over time in each city.

Indoor source frequency distributions

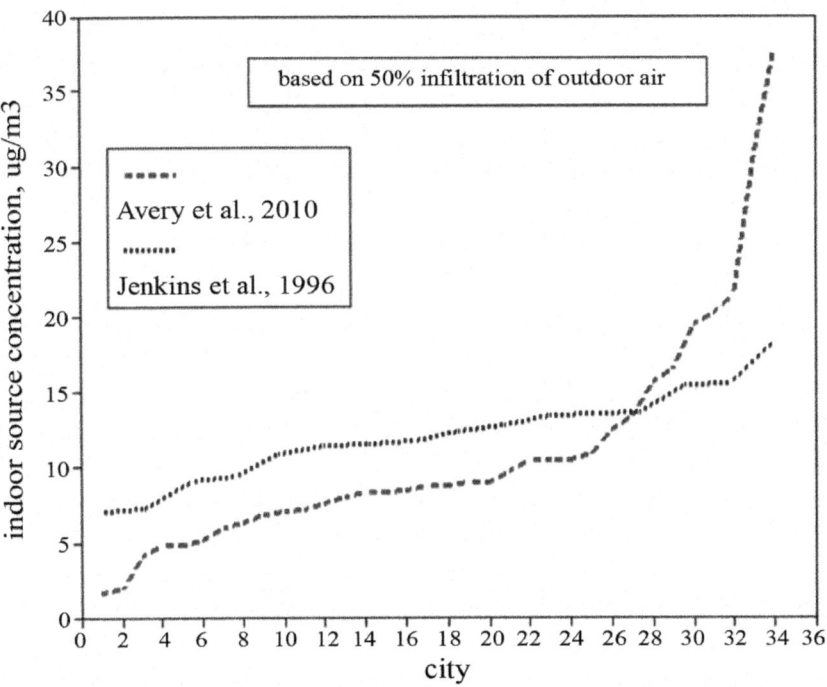

Figure 8: Estimated frequency distributions of city-average PM$_{2.5}$concentrations from indoor sources [21,24].

Adverse effects of ETS have been reported in epidemiological studies [25,26]. However, I found no air pollution studies that accounted for ETS increments to PM exposures. "Control" for passive smoking in regression analysis does not meet this requirement. In this regard, "% smokers" is probably a more appropriate confounding variable than "yes/no" for individual primary smokers.

Frequency distributions of these two sets of multi-city data are shown in Figure 8. The distribution of data from Jenkins et al. [21] is well behaved, in contrast to the data from Avery et al. [24], for which about 7% of cities had lower concentrations than expected and about 25% were higher. The remaining two-thirds of the cities had indoor source contributions in the expected range of 5–10 µg/m³. Specific reasons for the different distributions in these two studies are unknown and may comprise further sources of uncertainty.

SIMULATED EPIDEMIOLOGICAL ANALYSES

To illustrate potential effects of exposure uncertainties on epidemiological exposure-response relationships, a time-series of 1000 days and regressed simulated daily mortality variables against alternative exposure variables was simulated. Similar cross-sectional regressions for a cross-sectional dataset for 100 cities were ran. In the time-series analysis, the "true" $PM_{2.5}$ concentration was set at 15 µg/m³ for all days, as if seasonal and temperature effects had been removed. Random effects of spatial variability of outdoor air and the penetration to the indoors were simulated; indoor sources were assumed to be invariant on a daily basis and were not considered. Since these simulated series have no serial correlation, lag effects could not be considered.

The cross-sectional analysis included spatial variability of outdoor air and indoor penetration and parametric effects of arbitrary indoor sources. Mean PM_{10} values for 20 large cities [17] were used to establish the baseline distributions; $PM_{2.5}$ was then assumed to be about half of PM_{10} on average [9] and the dataset was replicated five times in order to provide an arbitrary set of 100 cities. The sources of variability included the spatial data of Pinto et al. [16], a mean infiltration rate of 0.5, and the indoor source distributions of either Avery et al. [24] or Jenkins et al. [21]. The analysis assumed that indoor source strengths are independent of outdoor air quality, as shown in the figures above.

The mean mortality rate was set at 35 deaths/day and the baseline $PM_{2.5}$ regression coefficient at 0.25. Random errors at various levels using a spread-sheet random function were introduced and found that 10 regression replications provided adequate numerical stability as judged by the similarity of the standard deviation across replications with respect to the average of the standard errors of each of the 10 regressions. Both analyses assume that the mortality variables have been adjusted for all confounders, such that pollution exposure is the only independent variable. These simulations are intended to serve as examples of what personal exposures might be expected, rather than precise estimates of actual situations. Table 1 provides summary statistics for the variables used in these simulations.

Table 1: Statistics of variables used in simulations

		Mean	Std. dev.	CV
	base $PM_{2.5}$ with instrument error	15	0.89	0.059
	with infiltration error 1	7.5	0.48	0.064
A. Simulated	with additional error 2	7.5	0.53	0.071
time-series	with additional error 2.5	7.5	0.59	0.079
analyses	with additional error 3	7.5	0.61	0.081
	with additional error 4	7.5	0.8	0.106
	mortality	38.75	1.74	0.045
	base $PM_{2.5}$	16.5	3.25	0.2
	w/spatial error	19	3.65	0.19
B. Simulated	w/infiltration	9.5	1.82	0.19
cross-sectional	w/100% indoor sources	33.9	12.2	0.36
analyses	w/50% indoor sources	22.1	6.87	0.31
	w/25% indoor sources	16.2	4.62	0.29
	w/15% indoor sources	13	2.68	0.21
	mortality	39.1	1.67	0.043

Time-Series Simulations

Figure 9 shows that, as spatial errors are added, the exposure sequences track well but regression coefficients are attenuated and tend to lose statistical significance. When infiltration to the indoors is considered, exposures are halved and thus regression coefficients are essentially doubled.

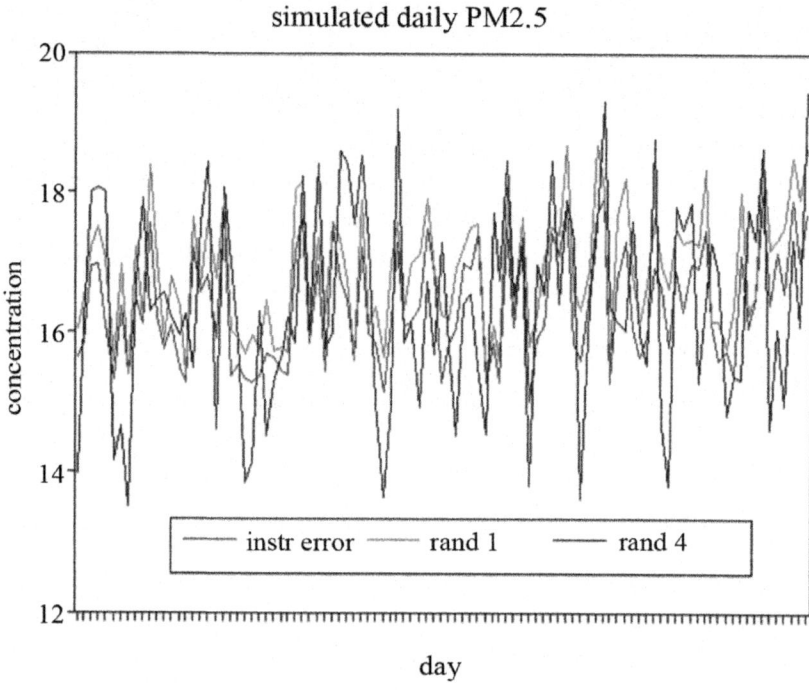

Figure 9: Sample of simulated daily variations in PM$_{2.5}$ with error added.

Cross-Sectional Simulations

I used the data of Pinto *et al.* [16] to estimate the spatial variability of outdoor PM$_{2.5}$ within each city. As seen in Figure 3, the mean fractional error in a given city is 7.7% with a standard deviation of 2.3%. This baseline distribution was then perturbed randomly among cities. The outdoor air infiltration rate was held constant at 0.5 for these simulations With respect to indoor pollution sources from Avery *et al.* [24], a quasi-normal distribution was able to be fit to Figure 8 by discarding the two lowest values and exponentiating the data to successively lower powers. The best fit was obtained using the 0.05 power of concentration, which produced a correlation coefficient of 0.95 and 95% confidence intervals of 1.5 and 46 µg/m^3. For the data of Jenkins *et al.* [21], the distribution of indoor sources was assumed to be normal with a mean of 12.1 µg/m^3 and standard deviation of 2.9.

Simulation Results

Figure 10 compares the results of these time-series and cross-sectional simulations in response to reduced correlation between exposure variables. In the absence of error, the 50% infiltration rate has the effect of doubling either regression coefficient, as shown at the right of the figure. For convenience, the regression coefficient scale is also converted to percent mortality change per 10 µg/m³ (right-hand scale). Attenuation of the outdoor time-series coefficients results from increasing spatial errors; for example, correlations between daily data from monitors in the same city are frequently around 0.8, which results in a coefficient reduction of about 30%, without considering indoor air quality. However, these effects of outdoor variability are more than compensated for by the biasing effect of partial penetration to the indoors. Intra-city variations in infiltration are of little concern because they would be averaged over the number of residences in each city under study.

Simulated mortality regressions

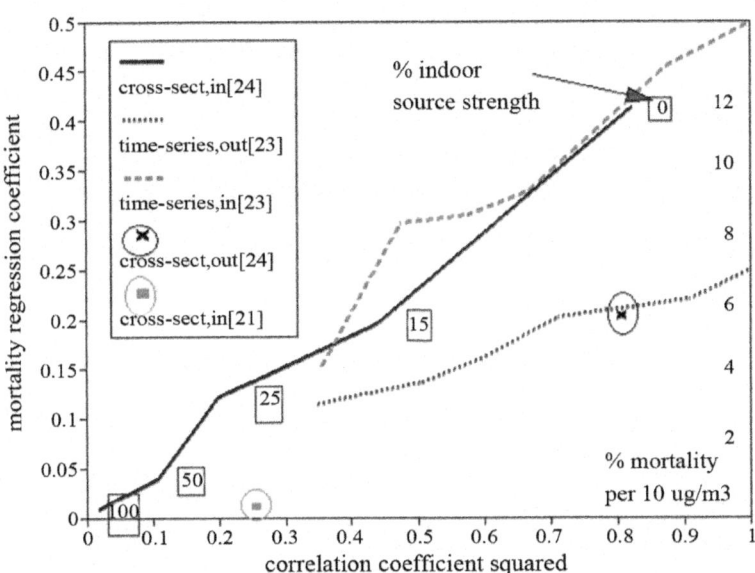

Figure 10: Simulated attenuation of regression coefficients by uncertain outdoor air quality, indoor infiltration, and indoor pollution sources.

Only one data point is shown for the cross-sectional model with outdoor data, for which the attenuation is due to spatial variability. The baseline cross-sectional analysis involves a much larger range in exposures based on the original set of cities selected. As with the time-series simulations, regression coefficients are doubled because of the 50% indoor penetration rate [9]. However, they can be greatly attenuated by effects of indoor sources. The initial simulation is based on the indoor source distribution of Figure 8; with this level of pollution from indoor sources, the relationship between personal exposure and mortality becomes nil. As a result, fractions of this indoor source level were used in order to generate a range of coefficient attenuation for comparison. The mean concentration of infiltrated outdoor air is about 9.6 µg/m³ and adding 15% of the Avery *et al.* [24] indoor source contribution (3.4 µg/m³) attenuates the cross-sectional mortality regression by more than 50%, placing it well within the range of attenuated time-series regression coefficients.

Using the indoor source distribution based on the data of Jenkins *et al.* [21] reduced the simulated mortality regression coefficient to essentially zero, as is the case with the full-strength data from Avery*et al.* [24].

Attenuation with Actual Epidemiological Data

A further example of regression attenuation may be seen with data from the Harvard Six Cities Study. Spengler *et al.* [22] reported personal RSP exposures for each city, and Dockery *et al.* [27] reported relative mortality risks. These data are combined in Figure 11, based on either outdoor data or personal exposures. The two exposure-response lines are roughly parallel, but using the higher personal exposures shifts the dose-response function to the right, implying a statistically significant threshold of about 20 µg/m³.

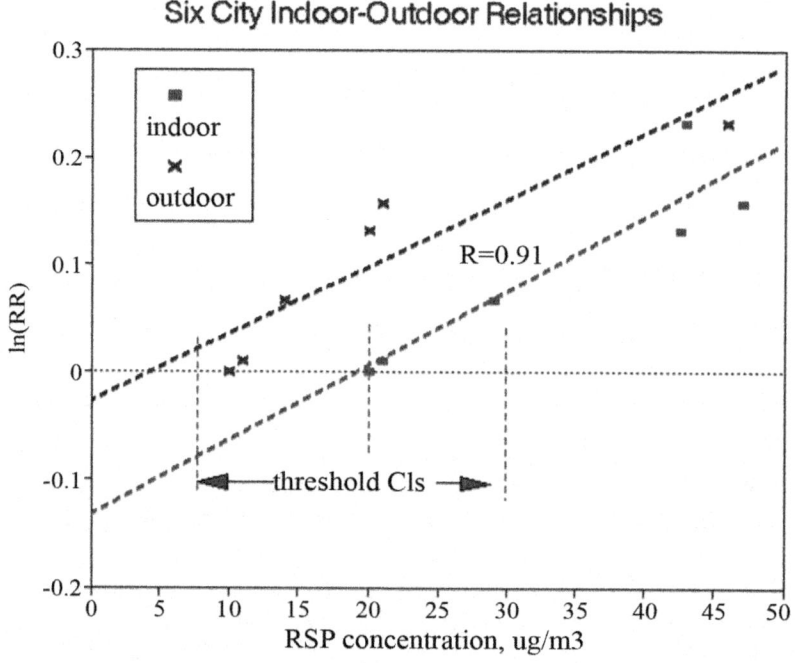

Figure 11: Exposure-response relationships from the Harvard Six Cities Study.

IMPLICATIONS

Uncertainties in exposure to air pollution have long been known to epidemiologists but seldom explored in detail, with the possible exception of spatial variability in outdoor concentrations. Most epidemiology studies use averages over the relevant ambient monitoring stations in a given city, but such spatial variability can be important when only a few monitoring sites are available in a given city. In time-series studies of daily events, data from nearby monitoring sites may be highly correlated, depending on the pollutant, but there are also issues of bias due to variability in mean values. These issues are relevant to most epidemiology studies.

However, variability in outdoor air quality comprises only a minor portion of the combined uncertainties in total (personal) exposures, which include variable rates of infiltration of outdoor air into occupied spaces and effects of many different indoor sources of air pollution. Infiltration is a factor for all types of epidemiology studies and these rates are well defined. Effects of indoor sources generally relate only to long-term studies, including cohort studies. This distinction creates a hierarchy in epidemiology studies, with daily time-series being the least affected and cross-sectional studies the most affected.

Figure 10 shows that time-series studies of 1000 days and cross-sectional studies of 100 cities will have equivalent regression attenuation for a given exposure uncertainty, but that cross-sectional studies are inherently less reliable since they are subject to additional uncertainties from indoor pollution sources.

In terms of optimizing exposure information for epidemiology, the most critical data needs relate to distributions of indoor pollution sources within and between cities. Because of averaging within cities, the accuracy for any one building may be less important than the number of cities with at least some rudimentary data; *i.e.*, in this sense, quantity may be more important than quality.

Figure 7 shows that indoor air quality is essentially unrelated to outdoor air quality across cities. Interpretation of all the extant cross-sectional studies, including cohort studies, is thus in question. Since such studies have played a large part in setting ambient air quality standards, this finding could have important regulatory implications.

Studying the health effects of indoor sources could be quite data intensive, requiring linkage between individual decedents and their residential characteristics. Privacy concerns may well preclude such an investigation. Nevertheless, the public should be aware of the risks of indoor air about which they have some control as well as the community levels of outdoor air quality subject to regulation.

CONCLUSIONS

Measured outdoor air quality is subject to uncertainties due to monitoring site location and analytical errors. Intersite variability in mean $PM_{2.5}$ showed a mean standard error of 7.7% among 27 USA cities, so that 95% confidence intervals would be (12.7–17.3) μg/m³ for a single monitoring station at 15 μg/m³. Multi-site data are thus strongly preferred for epidemiological studies and regulatory decisions.

Personal exposure is tantamount to indoor concentration for epidemiological purposes. Indoor air pollution emanates from infiltrated outdoor air and emissions from indoor sources. Rates of infiltration average about 50%, are reasonably well-defined, and contribute little uncertainty to exposure averaged across a city because of strong correlations between ambient and infiltrated outdoor concentrations. However, the decrease in mean effective exposure biases regression coefficients based on outdoor air such that the true effects of time-dependent exposures may be up to twice those of current estimates. Slowly varying indoor sources have essentially no effect on time-series analyses although base concentration levels would be affected.

By contrast, intercity variations in the mean contributions of indoor sources can strongly affect cross-sectional studies. Two independent studies of groups of USA cities show no correlation between mean indoor and outdoor concentrations. As a result, effect estimates from long-term cross-sectional and cohort studies may have been over-estimated. The largest contributions to indoor air pollution are from environmental tobacco smoke; contributions from other indoor pollution sources also vary widely among cities but have little effect when averaged across a group of cities.

All of these findings are based on small samples of indoor air and rudimentary Monte-Carlo analyses. Larger samples of indoor air quality are needed across many more cities for various criteria and non-criteria pollutants including PM size and composition, after which full-fledged probabilistic analyses would be appropriate for various health endpoints. Pending such verification, comparisons between time-series and cross-sectional studies remain problematic as does use of the latter for regulation.

ACKNOWLEDGEMENTS

I thank Ronald E. Wyzga for advice on this paper. However, I am fully responsible for its content and conclusions.

REFERENCES

1. Ferris, B.G., Jr.; Ware, J.H.; Spengler, J.D. Exposure measurement for air pollution epidemiology. In *Epidemiology and Health Risk Assessment*; Oxford University Press: Oxford, UK, 1988; pp. 120–128.

2. Cox, L.H. Statistical issues in the study of air pollution involving airborne particulate matter.*Environmetrics* 2000, *11*, 611–626.

3. Mage, D.; Wilson, W.; Gasselblad, V.; Grant, L. Assessment of human exposure to ambient particulate matter. *J. Air Waste Manag. Assoc.* 1999, *49*, 1280–1291.

4. Brown, K.W.; Sarnat, J.A.; Suh, H.H.; Coull, B.A.; Spengler, J.D.; Koutrakis, P. Ambient site, home outdoor and home indoor particulate concentrations as proxies of personal exposures. *J. Environ. Monit.* 2008, *10*, 1041–1051.

5. Poole, C. Ecologic analysis as outlook and method. *Am. J. Public. Health* 1994, *84*, 715–716.

6. McClellan, R.O. Setting ambient air quality standards for particulate matter. *Toxicology* 2002,*181–182*, 329–347.

7. Gong, J.; Zhu, T.; Kipen, H.; Wang, G.; Hu, M.; Guo, Q.; Ohman-Strickland, P.; Lu, S.E.; Wang, Y.; Zhu, P.; *et al.* Comparisons of ultrafine and fine particles in their associations with biomarkers reflecting physiological pathways. *Environ. Sci. Technol.* 2014, *48*, 5264–5273.

8. Expert Panel on Air Quality Standards.*Airborne Particles*; HM Stationery Office: London, UK, 2001.

9. Wallace, L.; Williams, R. Use of personal-indoor-outdoor sulfur concentrations to estimate the infiltration factor and outdoor exposure factor for individual homes and persons. *Environ. Sci. Technol.* 2005, *39*, 1707–1714.

10. Murray, C.J.; Lipfert, F.W. Inferring frail life expectancies in Chicago from daily fluctuations in elderly mortality. *Inhal. Toxicol.* 2013, *25*, 461–479.

11. Krewski, D.; Jerrett, M.; Burnett, R.T.; Ma, R.; Hughes, E.; Shi, Y.; Turner, M.C.; Pope, C.A.; Thurston, G.; Calle, E.E.; *et al.* Extended follow-up and spatial analysis of the American Cancer Society study linking particulate air pollution and mortality. *Res. Rep. Health Eff. Inst.* 2009,*140*, 5–114.

12. Koton, S.; Molshatzki, N.; Myers, V.; Broday, D.M.; Drory, Y.; Steinberg, D.M.; Gerber, Y. Cumulative exposure to particulate matter air pollution and long-term post-myocardial infarction outcomes. *Prev. Med.* 2013, *57*,

139–144.

13.   Goldberg, M.S.; Burnett, R.T. A new longitudinal design for identifying subgroups of the population who are susceptible to the short-term effects of ambient air pollution. *J. Toxicol. Environ. Health* 2005, *68*, 1111–1125.

14.   de Hartog, J.J.; Lanki, T.; Timonen, K.L.; Hoek, G.; Janssen, N.A.; Ibald-Mulli, A.; Peters, A.; Heinrich, J.; Tarkiainen, T.H.; van Grieken, R.; *et al.* Associations between $PM_{2.5}$ and heart rate variability are modified by particle composition and beta-blocker use in patients with coronary heart disease. *Environ. Health Perspect.* 2009, *117*, 105–111.

15.   *The Inside Story, A Guide to Indoor Air Quality*; USA Environmental Protection Agency: Washington, DC, USA, 1993.

16.   Pinto, J.P.; Lefohn, A.S.; Shadwick, D.S. Spatial variability of $PM_{2.5}$ in urban areas in the United States. *J. Air Waste Manag. Assoc.* 2004, *54*, 440–449.

17.   Daniels, M.J.; Dominici, F.; Samet, J.M.; Zeger, S.L. Estimating particulate matter-mortality dose-response curves and threshold levels: An analysis of daily time-series for the 20 largest USA cities. *Am. J. Epidemiol.* 2000, *152*, 397–406.

18.   Lipfert, F.W.; Morris, S.C. Temporal and spatial relations between age specific mortality and ambient air quality in the United States: Regression results for counties, 1960–97. *Occup. Environ. Med.* 2002, *59*, 156–174.

19.   Lipfert, F.W.; Wyzga, R.E.; Baty, J.D.; Miller, J.P. Air pollution and survival within the Washington University-EPRI veterans cohort: Risks based on modeled estimates of ambient levels of hazardous and criteria air pollutants. *J. Air Waste Manag. Assoc.* 2009, *59*, 473–489.

20.   Leaderer, B.P.; Naeher, L.; Jankun, T.; Balenger, K.; Holford, T.R.; Toth, C.; Sullivan, J.; Wolfson, J.M.; Koutrakis, P. Indoor, outdoor, and regional summer and winter concentrations of PM_{10}, $PM_{2.5}$, SO4(2)-, H+, NH4+, NO3-, NH3, and nitrous acid in homes with and without kerosene space heaters. *Environ. Health Perspect.* 1999, *107*, 223–231.

21.   Jenkins, R.A.; Palausky, A.; Counts, R.W.; Bayne, C.K.; Dindal, A.B.; Guerin, M.R. Exposure to environmental tobacco smoke in sixteen cities in the United States as determined by personal breathing zone air sampling. *J. Expo. Anal. Environ. Epidemiol.* 1996, *6*, 473–502.

22.   Spengler, J.D.; Dockery, D.W.; Turner, W.A.; Wolfson, J.M.; Ferris, B.G., Jr. Long-term measurements of respirable sulfates and particles inside and outside homes. *Atmos. Environ.*1981, *15*, 23–30.

23.   Zeger, S.L.; Thomas, D.; Dominici, F.; Samet, J.M.; Schwartz, J.;

Dockery, D.; Cohen, A. Exposure measurement error in time-series studies of air pollution: Concepts and consequences.*Environ. Health Perspect.* 2000, *108*, 419–426.

24. Avery, C.L.; Mills, K.T.; Williams, R.; McGraw, K.A.; Poole, C.; Smith, R.L.; Whitsel, E.A. Estimating error in using residential outdoor $PM_{2.5}$ concentrations as proxies for personal exposures: A meta-analysis. *Environ. Health Perspect.* 2010, *118*, 673–678.

25. Enstrom, J.E.; Kabat, G.C. Environmental tobacco smoke and coronary heart disease mortality in the United States: A meta-analysis and critique. *Inhal. Toxicol.* 2006, *18*, 199–210.

26. Steenland, K.; Thun, M.; Lally, C.; Heath, C., Jr. Environmental tobacco smoke and coronary heart disease in the American Cancer Society CPS-II cohort. *Circulation* 1996, *94*, 622–628.

27. Dockery, D.W.; Pope, C.A.; Xu, X.; Spengler, J.D.; Ware, J.H.; Fay, M.E.; Ferris, B.G., Jr.; Speizer, F.E. An association between air pollution and mortality in six USA cities. *N. Eng. J. Med.*1993, *329*, 1753–1759.

Chapter 2

AIR POLLUTION IN CHINA: MAPPING OF CONCENTRATIONS AND SOURCES

Robert A. Rohde[1], Richard A. Muller[1,2]

[1] Berkeley Earth, Berkeley, California, United States of America

[2] Department of Physics, University of California, Berkeley, California, United States of America

ABSTRACT

China has recently made available hourly air pollution data from over 1500 sites, including airborne particulate matter (PM), SO_2, NO_2, and O_3. We apply Kriging interpolation to four months of data to derive pollution maps for eastern China. Consistent with prior findings, the greatest pollution occurs in the east, but significant levels are widespread across northern and central China and are not limited to major cities or geologic basins. Sources of pollution are widespread, but are particularly intense in a northeast corridor that extends from near Shanghai to north of Beijing. During our analysis period, 92% of the population of China experienced >120 hours of unhealthy air (US EPA standard), and 38% experienced average concentrations that were unhealthy. China's population-weighted average exposure to $PM_{2.5}$ was 52 $\mu g/m^3$. The observed air pollution is calculated to contribute to 1.6 million deaths/year in China [0.7–2.2 million deaths/year at 95% confidence], roughly 17% of all deaths in China.

INTRODUCTION

Air pollution is a problem for much of the developing world and is believed to kill more people worldwide than AIDS, malaria, breast cancer, or tuberculosis [1–4]. Airborne particulate matter (PM) is especially detrimental to health [5–8], and has previously been estimated to cause between 3 and 7 million deaths every year, primarily by creating or worsening cardiorespiratory disease [2–4,6,7]. Particulate sources include electric power plants, industrial facilities, automobiles, biomass burning, and fossil fuels used in homes and factories for heating. In China, air pollution was previously estimated to contribute to 1.2 to 2 million deaths annually [2–4].

In 2012, China adopted the Ambient Air Quality Standard [9], and began development of a national Air Reporting System that now includes 945 sites in 190 cities. These automated stations report hourly via the internet, and focus on six pollutants: particulate matter < 2.5 microns ($PM_{2.5}$), particulate matter < 10 microns (PM_{10}), sulfur dioxide (SO_2), nitrogen dioxide (NO_2), ozone (O_3), and carbon monoxide (CO). Provincial governments perform air quality monitoring at 600 additional locations that are not yet integrated into the national system. Previous studies of regional scale air pollution have generally relied on satellite data [10,11] or modeling [12,13], but the high density of hourly data in China now allows regional patterns to be constructed directly from ground observations.

MATERIALS AND METHODS

Though China deserves praise for its monitoring system and transparent communication, most archived observations are not publicly available. To compensate, real-time data was downloaded every hour during a four month interval from April 5, 2014 to August 5, 2014. Due to download restrictions on the official Chinese air quality reporting system, two different third-party sources were used: PM25.in and AQICN.org. PM25.in is a direct mirror of data from the 945 stations in China's national network, while AQICN.org is the world's largest aggregator of real-time air quality data and included many additional sites in China and surrounding areas. Nearly all of the additional data from within China originates from stations operated by provincial environmental agencies that have not yet been incorporated in China's national network.

Consistency, quality control, and validation checks were applied to the raw data prior to further analysis in order to reduce the impact of outliers, badly calibrated instruments, and other problems. The most common quality problem was associated with stuck instruments that implausibly reported the same concentration continuously for many hours. A regional consistency check was also applied to verify that each station was reporting data similar to its neighboring stations. Approximately 8% of the data was removed as a result of the quality control review. Further details are described in the supplemental material (S1 Text).

As little monitoring is conducted in western China (Fig 1); we will focus on China east of 95° E, which includes 97% of the population. After removing stations with a high percentage of missing values or with other quality control problems, this study used 880 national network sites, 640 other sites in China and Taiwan, and 236 sites in other countries within 500 km of China (mostly South Korea). The air quality network is skewed towards urban areas, often with several sites per city and fewer, if any, in rural areas. For the n-th site, we use $p_n(t)$ to denote the pollutant concentration time series and $\bar{p}_n$ to denote the mean pollutant concentration.

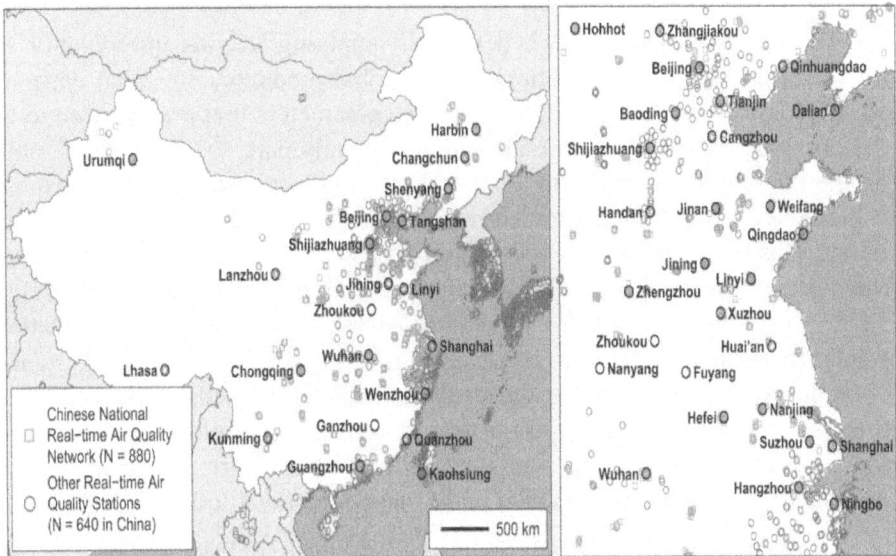

Figure 1: Map of real-time air pollution monitoring stations.

Map shows the locations of air quality monitoring sites in China and surrounding areas with sufficient hourly data to be included in this study. Selection criteria and data sources are described in S1 Text. The map was prepared in MATLAB using political boundaries from the Global Database of Administrative Areas (version 2; http://gadm.org/).

For each pollutant, a correlation vs. distance function was estimated by computing all possible pairwise correlations between different stations and fitting the resulting correlations to a two part exponential decay as a function of distance. The resulting functional forms are stated in S2 Table and shown in S2 and S3 Figs. The correlation functions are used to construct correlation matrices that in turn are used to compute Kriging coefficients [14, 15], $K_n(\vec{x})$.

The interpolated pollutant field, $P(\vec{x}, t)$, is then estimated in two parts.

$$P(\vec{x}, t) = S(\vec{x}) + A(\vec{x}, t)$$

$$S(\vec{x}) = \left(\sum_n K_n(\vec{x}) \left(\bar{p}_n - G(\vec{x}_n) \right) \right) + G(\vec{x})$$

$$A(\vec{x}, t) = \sum_n K_n^*(\vec{x}, t)(p_n(t) - \bar{p}_n)$$

The stationary part, $S(\vec{x})$, is derived by applying Kriging interpolation to the mean pollutant concentrations and a global predictor, $G(\vec{x}_n)$, that depends on latitude and longitude and contains free parameters that are adjusted to fit the observed means. The time-dependent anomaly part, $A(\vec{x}, t)$, depends only on the fluctuations at each station relative to the local mean, and its Kriging coefficients, $K_n^*(\vec{x}_n, t)$, are computed with restriction to stations that are active at time t. This two-step process reduces errors associated with stations that have intermittently missing data. This method is similar to that used by Berkeley Earth for its historical earth temperature analysis [16]. Since the correlation vs. distance function has been constructed with the correlation at zero distance obtaining a value less than one, the resulting interpolated fields will be smoother than the original data. This design was chosen for its ability to compensate for noise in the underlying station measurements. Additional details of the interpolation process are provided in the supplement methods (S1 Text).

For mapping and computation, this continuous field was sampled with an approximately 6 km resolution, though in practice, the characteristic size of resolvable features is often larger (e.g. 30 km) and varies with station density and noise.

A simple estimate of pollutant fluxes, $F(\vec{x}, t)$, was computed by comparing observed changes in the hourly pollutant concentration to the concentrations expected due to short-term wind transport $\vec{v}(\vec{x}, t)$ and an exponential decay with lifetime τ. Differences from the simple transport and decay model are assumed to represent source fluxes.

$$F(\vec{x}, t) = \frac{P(\vec{x} + \vec{v}(\vec{x}, t) \Delta t, t + \Delta t) - e^{-\Delta t / \tau} P(\vec{x}, t)}{2\Delta t} + \frac{P(\vec{x}, t) - e^{-\Delta t / \tau} P(\vec{x} - \vec{v}(\vec{x}, t) \Delta t, t - \Delta t)}{2\Delta t}$$

The near-surface (80 m) wind field from the Global Forecast System [17] was used for this calculation, and the effective pollutant lifetime was estimated as described in S1 Text and reported in S2 Table. Flux averages were computed by time-averaging the resulting field after excluding outlying values and cells affected by rain events as determined from Tropical Rainfall Measuring Mission data [18, 19].

The change in mortality due to $PM_{2.5}$ air pollution was calculated by adopting the integrated exposure response function approach [20] which considers relative risk of death for five disease classes (stroke, ischemic heart disease, lung cancer, chronic obstructive pulmonary disease, and lower respiratory infection) and which was adopted by World Health Organization (WHO) for the Global Burden of Disease study [21]. The model incorporates non-linear response versus concentration and provides an estimate of uncertainty. Relative risk was calculated at the prefecture level using local average $PM_{2.5}$ concentration. The data for different diseases and prefectures was then combined to construct national average mortality estimates.

Analysis and figure rendering was performed using original software written for this project on the MATLAB platform (version 2014a; http://www.mathworks.com/). Additional details of these calculations and associated background information is provided in the supplemental methods document (S1 Text).

Results

Fig 2 shows a time series of $PM_{2.5}$ concentration at Beijing and interpolated maps at three time points separated by 6 hours each. This shows the volatile nature of air pollution and the role of weather patterns in redistributing pollution on short timescales. Our approach creates a smooth field that approximates the data at each station, but allows a degree of difference attributable to noise. The pollution is extensive and rapidly evolves in response to winds and other atmospheric conditions. In the figure, fresh air from the North displaces a period of heavy pollution. Hourly data allows us to capture this evolution and ultimately estimate source fluxes.S1 Movie shows the time evolution of $PM_{2.5}$ across the entire country.

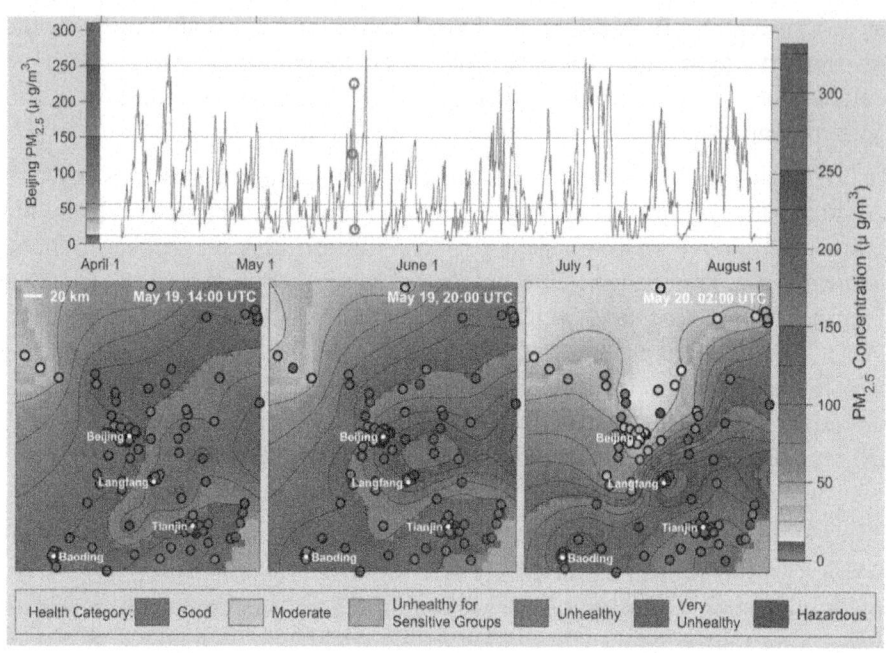

Figure 2: Time evolution of $PM_{2.5}$ pollution in the vicinity of Beijing.

(Top) Time series of $PM_{2.5}$ concentration at Beijing extracted from the interpolated field. Red circles indicate times shown in bottom row. (Bottom) Maps of interpolated $PM_{2.5}$ concentration during a period of high pollution. Pollution concentrations were computed as described in the text from hourly data and maps were rendered in MATLAB. Concentrations are shown using color gradients and contour lines, where color tones (green, yellow, etc.) correspond to health impact categories defined by the US EPA. Bold circles show station locations with the observed value at each station indicated by the color within the circle.

Fig 3 presents averages of the interpolated data for $PM_{2.5}$, PM_{10}, and O_3 across the study duration. The maps are color-coded based on US EPA health categories for 24-hour exposure [22]. Maps for SO_2 and NO_2 are included in the supplemental materials and show "good" levels nearly everywhere (S11 and S13 Figs).

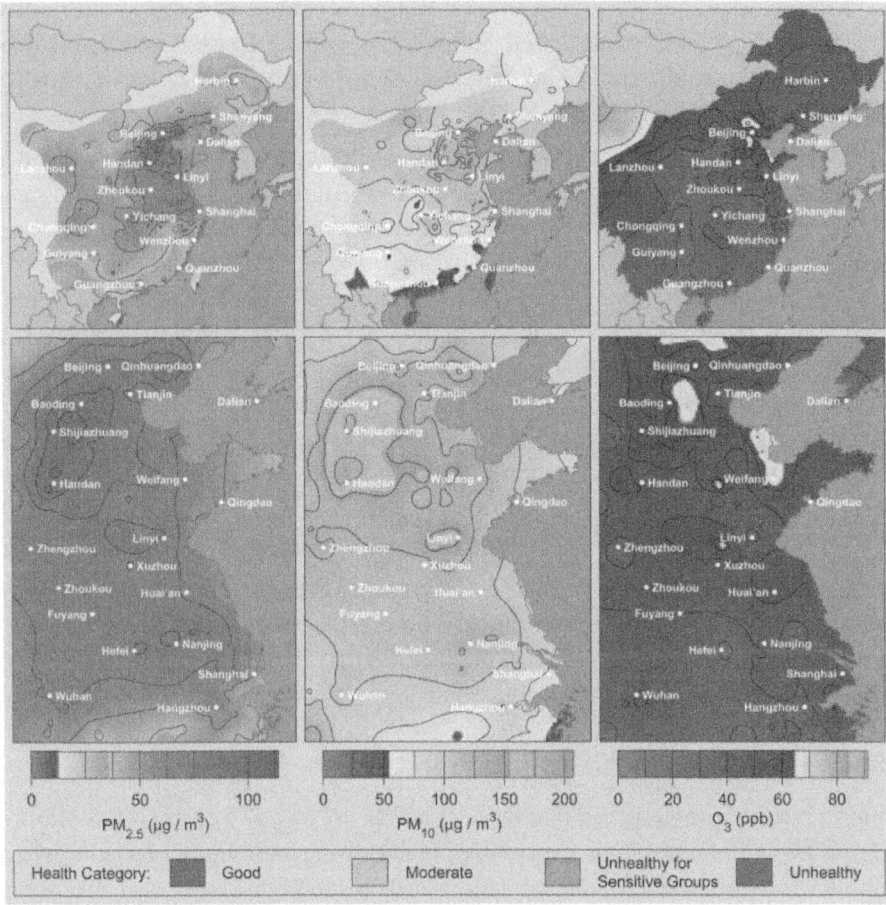

Figure 3: Average air pollution maps.

Maps of average pollutant concentration for $PM_{2.5}$, PM_{10}, and O_3 for eastern China (top row) and the Beijing to Shanghai corridor (bottom row). Concentrations are shown using color gradients and contour lines; the colors (green, yellow, etc.) represent US EPA qualitative health impacts. Pollution concentrations were computed as described in the text using hourly data and then the hourly concentration fields were averaged over the four month study duration. The resulting maps were rendered using MATLAB.

Air pollution is extensive in China, with the highest particulate concentrations observed south of Beijing (e.g. Xingtai / Handan), but significant levels extend throughout the interior, which is consistent with previous satellite and modeling estimates [11–13]. Extensive pollution is not surprising since particulate matter can remain airborne for days to weeks and travel thousands of kilometers. The corridor south of Beijing contains the highest pollution concentrations and, as discussed below, many of the largest sources. During this study, the southern coastal area experienced somewhat better air quality, possibly linked to greater rainfall (S1 Fig).

For $PM_{2.5}$, portions of China encompassing roughly 38% of the population are classified as "unhealthy" on average (>55 $\mu g/m^3$, red) with an additional 45% of the population averaging "unhealthy for sensitive groups" (>35 $\mu g/m^3$, orange). Almost none of the study area averaged below the US EPA's 12 $\mu g/m^3$ standard for annual average $PM_{2.5}$ exposure (green). The area-weighted average was 46 $\mu g/m^3$ and the population-weighted average exposure to $PM_{2.5}$ was 52 $\mu g/m^3$. 92% of China's population experienced unhealthy $PM_{2.5}$ for at least 120 hours during the study period. 46% of China's population experienced $PM_{2.5}$ above the highest EPA threshold ("hazardous", >250 $\mu g/m^3$), during at least one hour in the observation period.

Patterns for PM_{10} are similar but less severe, with average PM_{10} levels "moderate" for most of China. Ozone concentrations are modest across most of China, though higher levels occur in the Northwest desert area, and in a small number of Northeastern cities. Though the average levels of PM_{10} and O_3 are "moderate" or "good" for much of China, intermittently high levels of these pollutants can occur in some areas.

Source Regions for Air Pollution in China

Fig 4 shows estimated pollutant fluxes for $PM_{2.5}$, PM_{10}, SO_2, and NO_2. Pollution emission is often localized, especially in the Beijing to Shanghai corridor where many of the highest PM concentrations also occur (Fig 3). Most of the largest emissions appear in or near urban areas (e.g. Handan, Shijazhuang, Zibo, Tangshan, Linyi, Hangshou), though not all major cities have high pollution fluxes (e.g. Chongqing, Chengdu, Wuhan). The source map presumably reflects patterns of industrial activity, though detailed differences will not be explored here.

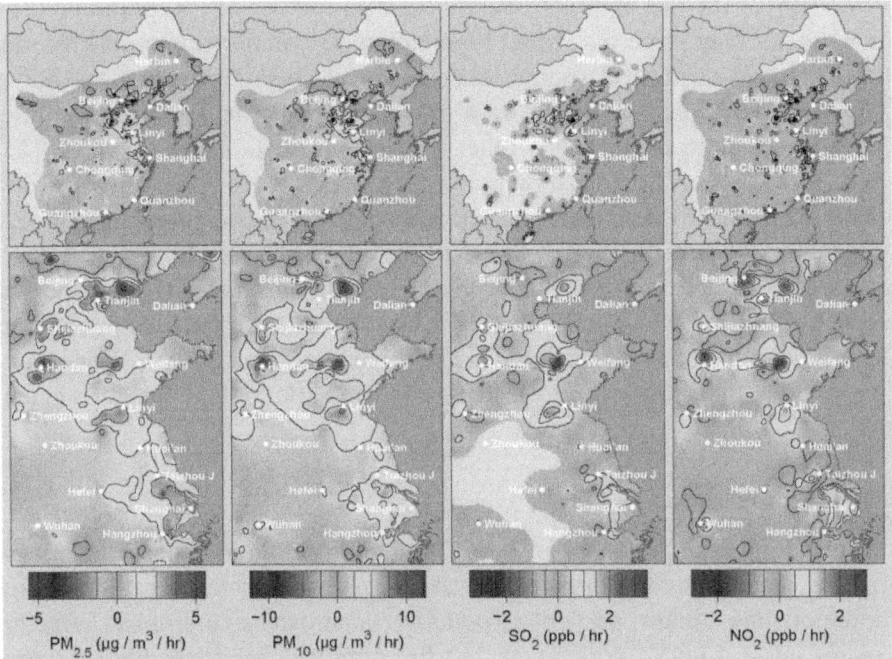

Figure 4: Air pollution source maps.

Maps of average pollutant flux for $PM_{2.5}$, PM_{10}, SO_2, and NO_2 for eastern China (top row) and the Beijing to Shanghai corridor (bottom row). Pollutant fluxes were computed as described in the text from changes in the interpolated hourly pollution fields along with contemporaneous wind and weather data. Due to sparse sampling and secondary transformations of pollutants in the atmosphere, apparent source fluxes are likely to appear more diffuse than the true emissions source.

NO_2 and SO_2 emissions help suggest the pollution source. Nitrogen oxides, including nitric oxide (NO) and NO_2 are created when air is heated, and on average have been attributed to transportation fuels (15–25%), fossil fuel burning in power plants (30–50%), and to industrial facilities (25–35%) [23, 24]. It is expected that NO dominates at the combustion source, but in the presence of sunlight NO and NO_2 will equilibrate within a few minutes (as well as reacting with O_3), implying that NO_2 measurements reflect a combination of NO and NO_2 emissions. SO_2 emissions have been previously associated with coal (~90%) in power plants and industrial facilities [25]. Beijing has negligible SO_2 flux, despite a large NO_2 signal, possibly a result of policies that limit coal burning in the immediate vicinity of Beijing and more extensively apply mitigation technologies.

Many of the SO_2 and NO_2 sources are also sources of PM pollution. This is not surprising since fossil fuel burning is also a major source of $PM_{2.5}$ and PM_{10}. However, the PM sources appear more diffuse than either the SO_2 or NO_2 sources. In part, this is caused by secondary particulate matter formed within the atmosphere from other pollutants, such as SO_2 or NO_2[26]. Secondary particulate formation may cause PM fluxes to appear more widely distributed than the underlying emitters. Nonetheless, many strong PM sources are identified through this analysis. Within the study region, 10% of the area is responsible for 34% of the $PM_{2.5}$ emissions, and 5% of the area is responsible for 22% of emissions. However, small and moderate sources are also important. Approximately 37% of the study region had $PM_{2.5}$ fluxes >0.5 $\mu g/m^3/hr$, sufficient to exceed US EPA standards after only 3 days of stagnant air.

Discussion

We have presented a technique for mapping air pollution concentrations and sources using data from monitoring stations. As has been known from satellite and modeling studies, particulate pollution is an extensive problem affecting nearly all of China's population, but the observed heterogeneity of source locations could help develop strategies to reduce pollution.

We examined a four month interval as long-term station data were not available for most of China. Previous studies of both in situ and satellite data have indicated that winter and early spring months in China have somewhat higher PM concentrations due to increased use of fossil fuels for seasonal heating, weather patterns that concentrate pollution at low altitudes, and increased desert dust fluxes [12, 27]. In contrast, the air in China is typically cleanest from late summer to early fall. The April 5 to August 5 study period is somewhat intermediate. A review of hourly $PM_{2.5}$ station data from Beijing (2009–2014), Shanghai (2012–2014), Guangzhou (2012–2014), Chengdu (2013–2014), and Shenyang (2014) indicates that the months studied in this paper averaged 91%, 84%, 89%, 72%, and 73% respectively of the annual averages (U.S. Air Quality Monitoring Program, http://www.stateair.net/web/mission/1/). Monthly-resolved $PM_{2.5}$ satellite data for the whole study region was not immediately available, but a monthly satellite history for Beijing reported that April-July averaged 99% of the annual mean during 2000 to 2012 [28]. Hence, particulate pollution estimates drawn from the current short study period will likely be similar to or somewhat lower than long-term averages. Future work could explore seasonal variations and long-term trends.

During the four months studied, the population-weighted and area-weighted $PM_{2.5}$ averages were 52 and 46 $\mu g/m^3$ respectively. Satellite pollution datasets generally focus on annual or multi-year averages, which limits the ability to make direct comparisons. However, the available satellite estimates tend to be similar to or somewhat lower than the ground observations. An analysis of the larger East Asia region estimated a population-weighted $PM_{2.5}$ exposure of 50 µg/m³ for 2001 to 2010 [28].

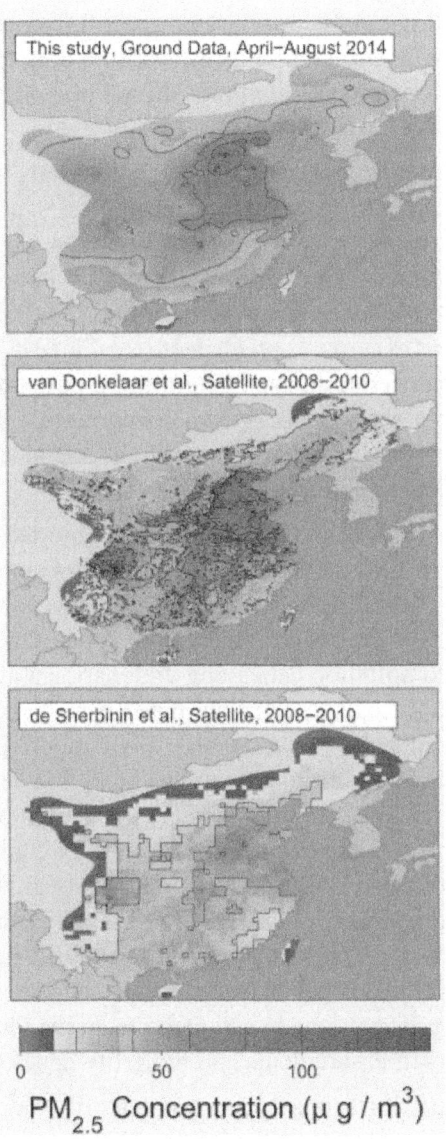

Figure 5: Comparison of $PM_{2.5}$ observations to satellite data.

A version of the same dataset masked to the current study region had an area-weighted average of 40 µg/m^3 from 2010 to 2012 [28, 29]. A different satellite estimate using similar observations but different calibrations and modeling gave 25 µg/m^3 for the 2008 to 2010 average over the present study region [30, 31]. Both of these datasets show similar spatial patterns to what we observe (Fig 5), though the magnitudes in van Donkelaar et al.'s work [28] are clearly more consistent with our ground data estimates. As noted in previous satellite to in situ comparisons, satellite data may be more likely to underestimate pollution concentrations during the most extreme pollution events [28, 32].

Maps of average PM$_{2.5}$ concentration from this study (top) and two satellite-derived datasets restricted to the same region. The average over the 2008 to 2010 time interval was chosen for the satellite data due to the limitations of the available satellite data. Both the concentrations reported by van Donkelaar [28] (middle) and those reported by de Sherbinin [30] (bottom) rely on similar satellite observations of aerosol optical depth (obtained by NASA), but interpret those observations differently when determining pollutant concentration. The satellite-derived data was imported from concentration data files provided by their respected sources and rendered via MATLAB to use the same US EPA health category color scheme applied in Figs 2 and 3.

The conversion of pollution concentrations to mortality is complicated. We adopt the framework [20] developed for the WHO Global Burden of Disease study [21] that considers PM$_{2.5}$ mortality due to impacts on five distinct diseases and accounts for nonlinearities as a function of concentration. Using prefecture level population and pollution data along with national average death rates for the five modeled diseases, we calculate that 1.6 million deaths / year can be attributed to PM$_{2.5}$ air pollution under the WHO model [95% confidence: 0.7 to 2.2 million deaths/year]. This is equivalent to 4 thousand deaths / day or 17% of all deaths in China. Additional details appear in S1 Text and S1 Table. For perspective, the categories of mortality events considered by the WHO model, e.g. cardiorespiratory deaths, account for roughly 55% of all Chinese deaths [21]. This compares to only 42% of mortality in the United States in the same cardiorespiratory categories, despite much higher incidence of obesity in the United States [21]. The calculated mortality is somewhat higher than the 1.2 million deaths/year previously estimated from the Huai River study using Chinese air pollution measurements and mortality data [3,4,33].

Though most of China is subject to potentially harmful levels of PM$_{2.5}$, some large population centers (Chongqing, Wuhan, Chengdu) emit less than half the PM$_{2.5}$ of others. Among northeastern cities, Beijing has relatively low emissions except for NO$_2$. Low SO$_2$ fluxes may indicate cities that benefit from lower coal usage or better smokestack pollution controls. Compared to natural gas, coal produces 150 to 400 times more PM for the same energy delivered [34,35]. China has plans for new coal plants in the next decade that could effectively double their coal consumption [36], potentially exacerbating the problem of air pollution. A table of pollution concentrations and fluxes by province and prefecture is included in S1 Table.

The methods of this study should be applicable to air quality monitoring in other regions of the world. However, these techniques require an extensive air-quality monitoring network with frequent updates (e.g. hourly), and such networks presently exist in only a few places. We hope that other countries will follow China's lead and provide both extensive and transparent real-time air quality monitoring.

ACKNOWLEDGMENTS

We thank Elizabeth Muller for suggesting this study, Steve Mosher and Zeke Hausfather for discussion and feedback, and John Li and Xinyu Zhang for help with Chinese materials.

REFERENCES

1. World Health Organization. WHO methods and data sources for global causes of death 2000–2012. Global Health Estimates Technical Paper WHO/HIS/HSI/GHE/2014.7. 2014. Available:http://www.who.int/entity/healthinfo/global_burden_disease/GlobalCOD_method_2000_2012.pdf.

2. World Health Organization. Burden of disease from the joint effects of Household and Ambient Air Pollution for 2012. WHO Technical Report. 2012. Available:http://www.who.int/phe/health_topics/outdoorair/databases/AP_jointeffect_BoD_results_March2014.pdf.

3. O'Keefe B. Recent Trends in Air Quality Standards in Europe and Asia: What's next? HEI Annual Conference 2012. 2013. Available:http://www.healtheffects.org/Slides/AnnConf2013/OKeefe-Sun.pdf.

4. Yang G, Wang Y, Zeng Y, Gao GF, Liang X, Zhou M, et al. Rapid health transition in China, 1990–2010: findings from the Global Burden of Disease Study 2010. Lancet 2013; 381: 1987–2015. doi: 10.1016/S0140-6736(13)61097-1. pmid:23746901

5. Dockery DW, Pope CA, Xu X, Spengler JD, Ware JH, Fay ME, et al. An association between air pollution and mortality in six U.S. cities. N Engl J Med 1993; 329(24), 1753–1759. pmid:8179653 doi: 10.1056/nejm199312093292401

6. Pope CA, Burnett RT, Thun MJ, Calle EE, Krewski D, Ito K, et al. Lung cancer, cardiopulmonary mortality, and long-term exposure to fine particulate air pollution. JAMA 2002; 287(9): 1132–1141. doi: 10.1001/jama.287.9.1132 pmid:11879110.

7. Hoek G, Krishnan RM, Beelen R, Peters A, Ostro B, Brunekreef B, et al. Long-term air pollution exposure and cardiorespiratory mortality: a review. Env Health. 2012; 12:43. doi: 10.1186/1476-069x-12-43

8. Beelen R, Raaschou-Nielsen O, Stafoggia M, Andersen ZJ, Weinmayr G. Effects of long-term exposure to air pollution on natural-cause mortality: an analysis of 22 European cohorts within the multicentre ESCAPE project. Lancet 2013; 383(9919): 785–795. doi: 10.1016/S0140-6736(13)62158-3. pmid:24332274

9. Government of China. Ambient Air Quality Standards (in Chinese). GB 3095–2012. 2012. Available:http://kjs.mep.gov.cn/hjbhbz/bzwb/dqhjbh/dqhjzlbz/201203/W020120410330232398521.pdf.

10. Zhang Q, Geng GN, Wang SW, Richter A, He KB. Satellite remote sensing of changes in NO$_x$ emissions over China during 1996–2010. Chinese Sci. Bull. 2012; 57(22): 2857–2864. doi: 10.1007/s11434-012-5015-4.

11. Wang J, Xua X, Spurr R, Wang Y, Drury E. Improved algorithm for MODIS satellite retrievals of aerosol optical thickness over land in dusty atmosphere: Implications for air quality monitoring in China. Remote Sensing Env. 2010; 114: 2575–2583. doi: 10.1016/j.rse.2010.05.034

12. Liu X-H, Zhang Y, Cheng S-H, Xing J, Zhang Q, Streets DG, et al. Understanding of regional air pollution over China using CMAQ, part I: performance evaluation and seasonal variation. Atmos Env. 2010; 44(20): 2415–2426. doi: 10.1016/j.atmosenv.2010.03.035

13. Lei Y, Zhang Q, He K, Streets D. Primary anthropogenic aerosol emission trends for China, 1990–2005. Atmos Chem Phys. 2011; 11, 931–954. doi: 10.5194/acp-11-931-2011

14. Schabenberger O, Gotway CA. Statistical Methods for Spatial Data Analysis. CRC Press. 2004.

15. Krige DG. A statistical approach to some basic mine valuation problems on the Witwatersrand. J Chem Metall Min Soc S Afr. 1951; December: 119–159.

16. Rohde RA, Muller RA, Jacobsen R, Perlmutter S, Rosenfeld A, Wurtele J., et al. Berkeley Earth Temperature Averaging Process. Geoinfor Geostat: An Overview 2013; 1(2) doi: 10.4172/gigs.1000103.

17. NOAA/NCEP. Global Forecast System (GFS) Atmospheric Model. 2012. Available:www.ncdc.noaa.gov/data-access/model-data/model-datasets/global-forcast-system-gfs

18. Huffman GJ, Adler RF, Bolvin DT, Gu G, Nelkin EJ, Bowman KP, et al. The TRMM Multi-satellite Precipitation Analysis: Quasi-global, multi-year, combined-sensor precipitation estimates at fine scale. J. Hydrometeor. 2007; 8(1): 38–55. doi: 10.1175/jhm560.1

19. Huffman GJ, Bolvin DT. Real-Time TRMM Multi-Satellite Precipitation Analysis Data Set Documentation. NASA/GSFC Laboratory for Atmospheres, 43 pp. 2011. Available:ftp://meso.gsfc.nasa.gov/pub/trmmdocs/rt/3B4XRT_doc.pdf.

20. Burnett RT, Pope CA III, Ezzati M, Olives C, Lim SS, Mehta S, et al. An integrated risk function for estimating the global burden of disease attributable to ambient fine particulate matter exposure. Env Health Perspect 2014; 122:397–403. doi: 10.1289/ehp.1307049.

21. Naghavi M, Wang H, Lozano R, Davis A, Liang X, Zhou M, et al. Global, regional, and national age–sex specific all-cause and cause-specific mortality for 240 causes of death, 1990–2013: a systematic analysis for the Global Burden of Disease Study 2013. Lancet 2015; 385 (9963): 117–171. doi: 10.1016/S0140-6736(14)61682-2. pmid:25530442

22. US EPA. Air Quality Index: A Guide to Air Quality and Your Health. U.S. EPA Report EPA-456/F-14-002. 2014. Available:http://www.epa.gov/airnow/aqi_brochure_02_14.pdf.

23. Emission Database for Global Atmospheric Research (EDGAR), release 4.2. Available:http://edgar.jrc.ec.europa.eu.

24. Shi Y, Xia Y-F, Lu B, Liu N, Zhang L, Li S-J, et al. Emission inventory and trends of NO_x for China, 2000–2020. J Zhejiang Univ-SCI A 2014; 15(6):454–464. doi: 10.1631/jzus.a1300379

25. Lu Z, Zhang Q, Streets D. Sulfur dioxide and primary carbonaceous aerosol emissions in China and India, 1996–2010. Atmos Chem Phys. 2011; 11: 9839–9864. doi: 10.5194/acp-11-9839-2011

26. Davidson C, Phalen R, Solomon P. Airborne particular matter and human health: a review. Aerosol Sci Technol. 2005; 39: 737–749. doi: 10.1080/02786820500191348

27. Zhang W-J, Sun Y-L, Zhuang G-S, Xu D-Q. Characteristics and Seasonal Variations of PM2.5, PM10, and TSP Aerosol in Beijing. Biomed Env Sci 2006; 19: 461–468.

28. van Donkelaar A, Martin RV, Brauer M, Boys BL. Global fine particulate matter concentrations from satellite for long-term exposure assessment. Env Health Perspectives 2015. doi: 10.1289/ehp.1408646.

29. Boys BL, Martin RV, van Donkelaar A, MacDonell R., Hsu NC, Cooper MJ, Yantosca RM, Lu Z, Streets DG, Zhang Q, Wang S. Fifteen-year global time series of satellite-derived fine particulate matter, Env Sci Technol. 2014. doi: 10.1021/es502113p.

30. de Sherbinin A, Levy M, Zell E, Weber S, Jaiteh M. Using Satellite Data to Develop Environmental Indicators. Environmental Research Letters 2014; 9(8): 084013. doi: 10.1088/1748-9326/9/8/084013.

31. Battelle Memorial Institute, Center for International Earth Science Information Network—CIESIN—Columbia University. Global Annual Average PM2.5 Grids from MODIS and MISR Aerosol Optical Depth (AOD). Palisades, NY: NASA Socioeconomic Data and Applications Center (SEDAC). 2013. doi: 10.7927/H4H41PB4.

32. Song W, Jia H, Huang J, Zhang Y. A satellite-based geographically weighted regression model for regional PM2.5 estimation over the Pearl River Delta region in China. Remote Sensing Env. 2014; 154: 1–7. doi: 10.1016/j.rse.2014.08.008

33. Chen Y, Ebenstein A, Greenstone M, Lie H. Evidence on the impact of sustained exposure to air pollution on life expectancy from China's Huai River policy. PNAS 2013; 110(32): 12936–12941. doi: 10.1073/pnas.1300018110. pmid:23836630

34. Cai H, Wang M, Elgowainy A, Han J. Updated Greenhouse Gas and Criteria Air Pollutant Emission Factors and Their Probability Distribution Functions for Electric Generating Units. Argonne National Laboratory ANL/ESD/12-2. 2012.

35. US Energy Information Agency. Natural Gas 1998, Issues and Trends. US EIA Report DOE/EIA-0560(98). 1998. Available:http://www.eia.doe.gov/oil_gas/natural_gas/analysis_publications/natural_gas_1998_issues_and_trends/it98.html.

36. Yang A, Cui Y. Global Coal Risk Assessment: Data Analysis and Market Research. World Resources Institute Working Paper. 2012. Available:

Chapter 3

AIR POLLUTION BY HYDROTHERMAL VOLCANISM AND HUMAN PULMONARY FUNCTION

Diana Linhares,[1,2] Patrícia Ventura Garcia,[1,3] Fátima Viveiros,[2,4] Teresa Ferreira,[2,4] and Armindo dos Santos Rodrigues[1,2]

[1]Department of Biology, University of the Azores, Ponta Delgada, 9501-801 Azores, Portugal

[2]CVARG, Center for Volcanology and Geological Risks Assessment (CVARG), University of the Azores, Ponta Delgada, 9501-801 Azores, Portugal

[3]CE3C, Centre for Ecology, Evolution and Environmental Changes (CE3C) and Azorean Biodiversity Group, University of the Azores, 9501-801 Ponta Delgada, Portugal

[4]Department of Geosciences, University of the Azores, Ponta Delgada, 9501-801 Azores, Portugal

ABSTRACT

The aim of this study was to assess whether chronic exposure to volcanogenic air pollution by hydrothermal soil diffuse degassing is associated with respiratory defects in humans. This study was carried in the archipelago of the Azores, an area with active volcanism located in the Atlantic Ocean where Eurasian, African, and American lithospheric plates meet. A cross-sectional study was performed on a study group of 146 individuals inhabiting an area where volcanic activity is marked by active fumarolic fields and soil degassing (hydrothermal area) and a reference group of 359 individuals inhabiting an area without these secondary manifestations of volcanism (nonhydrothermal area). Odds ratio (OR) and 95% confidence intervals (CIs) were adjusted for age, gender, fatigue, asthma, and smoking. The OR for restrictive defects and for exacerbation of obstructive defects (COPD) in the hydrothermal area was 4.4 (95% CI 1.78–10.69) and 3.2 (95% CI 1.82–5.58), respectively. Increased prevalence of restrictions and all COPD severity ranks (mild, moderate, and severe) was observed in the population from the hydrothermal area. These findings may assist health officials in advising and keeping up with these populations to prevent and minimize the risk of respiratory diseases.

INTRODUCTION

About 10% of the worldwide population inhabits or lives in the vicinity of some active or historically active volcano [1]. Despite the hazards associated with volcanic activities, the richness of soils in nutrients attracts people to live in these areas. Several studies have established an association between acute [2–4] and long-term [5, 6] exposure to anthropogenic air pollutants and lung function, while only few have analyzed the respiratory effects from volcanogenic air pollution [7–9].

The Azores archipelago (Portugal) comprises nine volcanic inhabited islands, located between 36°45′–39°45′N and 24°45′–31°17′W (Figure 1(a)), where the Eurasian, African, and American lithospheric plates meet [10]. On account of this complex tectonic setting, seismic and volcanic activities are frequent in the archipelago [11]. São Miguel Island, the largest of the archipelago, is formed by three major active central volcanoes (Sete Cidades, Fogo, and Furnas), linked by rift zones [12] (Figure 1(b)). Furnas Volcano is located in the eastern part of the island, where present-day volcanic activity is marked by several hydrothermal manifestations consisting of active fumarolic fields, thermal and cold CO_2-rich springs, and soil diffuse degassing areas [11,13]. Gases released in these diffuse degassing areas are essentially carbon dioxide (CO_2) and radon (^{222}Rn), this last one a radioactive gas. Carbon dioxide is one of the most abundant volcanic gases and is amongst the most important diffused gases released by soil degassing in Furnas Volcano (hydrothermal soil CO_2 emissions in Furnas Volcano are estimated to be approximately 968 t/d) [14]; this gas, if present at high concentrations, can become particularly dangerous for public health, since it works as asphyxiant preventing oxygen respiration [15]. Previous studies [13] showed that CO_2 is released permanently to the atmosphere from soils in volcanic areas not only during eruptive periods, but also during quiescent periods of activity. Considering that CO_2 released by soils may enter the buildings through pipes, cracks in the floor, and/or the contact between floor and walls, it is considered important to assess the CO_2 flux in buildings. Carbon dioxide level is usually greater inside a building than outside, and it can act as an indicator of ventilation efficiency, showing whether the supply of outside air is sufficient to dilute indoor air contaminants [16]. According to WHO [17], indoor air pollution is responsible for 2.7% of the diseases worldwide; such effects of indoor air pollution are particularly highlighted in studies regarding the occupational exposure, as it was shown in the review made by Balmes et al. [18] that estimated that 15% of COPD was attributable to the air quality at the workplace.

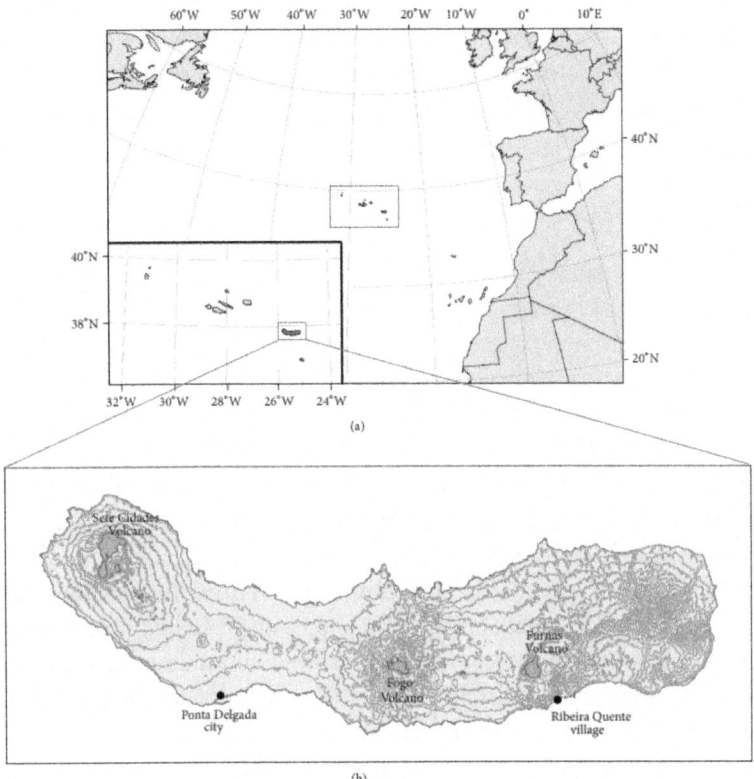

Figure 1: (a) Location map of the Azores archipelago and (b) São Miguel Island. The places represented on the map correspond to the two studied areas (Ponta Delgada and Ribeira Quente).

Furnas and Ribeira Quente are two villages located, respectively, inside the caldera and in the south flank of Furnas active volcano, where the ground gas emissions that characterize the diffuse degassing areas occur permanently and, thus, inhabitants of such areas are often exposed to elevated concentrations of CO_2 from volcanic origin [14]. Previous studies evidenced that Furnas inhabitants have a high incidence of chronic bronchitis and of some cancer types (e.g., lip, oral cavity, and pharynx) [19, 20] and a higher risk of DNA damage in human buccal epithelial cells [21]. Moreover, a very recent study by Camarinho et al. [22] showed that chronic exposure to volcanogenic air pollutants causes lung injury in wild mice. However, to our knowledge, up to date no study was carried out to assess the association between volcanogenic air pollution by soil diffuse degassing (DDS) and the risk of development of respiratory defects. Therefore, the present study was carried out to evaluate whether chronic exposure to permanent volcanogenic air pollution is a risk factor for human restrictive and obstructive (COPD) respiratory defects.

METHODS

Study Population

To perform this study, two areas were selected: an area with no secondary manifestations of volcanism, therefore a nonhydrothermal area (Ponta Delgada), and an area where volcanic activity is marked by active fumarolic fields and degassing soils (hydrothermal area) (Ribeira Quente). A diagnosis campaign was established for evaluation of pulmonary function by spirometry tests for the inhabitants of both areas. Spirometry tests were carried out in either the participants' workplaces or their homes. A standard questionnaire was applied to each individual that volunteered to participate. Medical history data for respiratory symptoms were taken using a standard questionnaire modified from a standardized respiratory symptom questionnaire from American Thoracic Society (ATS) [23] and British Medical Research Council's Committee [24]. Each person was interviewed about their age, height, weight, education, occupation, smoking habits (smoking of cigarettes and/or use of smokeless tobacco), amount of cigarettes smoked/day, fatigue, and general respiratory health status, where asthma was defined by a positive response to "Did any doctor diagnose you with asthma?".

The study group consisted of 146 participants (94 women and 52 men), residents in Ribeira Quente village (Table 1). Ribeira Quente village has 767 inhabitants [25]: 148 children and 619 adults (298 women and 321 men); therefore, the participation rate was 26.3%. This village is located on the south flank of Furnas Volcano (São Miguel Island) (Figure 1(b)), in an important diffuse degassing structure (DDS); about 98% of buildings in this village are placed above anomalous soil CO_2 degassing with volcanic-hydrothermal origin [14, 26]. In addition, several fumaroles may be found along Ribeira Quente stream and dispersed in the village [13]. Thus, Ribeira Quente inhabitants are chronically exposed to gases of volcanic origin, particularly from soil diffuse degassing. Even if no indoor measurements were performed concomitantly to the diagnosis tests, previous works showed that in the diffuse degassing structures indoor CO_2 is anomalously high and can reach lethal values during extreme weather events [13]. Permanent release of gases from the soils increases the indoor CO_2 concentrations independently of the anthropogenic contribution.

Table 1: Description of the study populations (study and reference groups) (mean ± SE for continuous variables or n (%) for categorical variables)

		Reference group (Ponta Delgada, $n = 359$)	Study group (Ribeira Quente, $n = 146$)	P value[a]
General characteristics				
Age (years)	41.3 ± 12.7	39.8 ± 11.2	45.1 ± 15.3	0.001
Age, >41[b]	234 (46.3)	152 (42.3)	82 (56.1)	0.005
Gender, male	207 (40.9)	155 (43.1)	52 (35.6)	0.117
BMI (kg/m^2)	26.1 ± 4.4	25.7 ± 3.9	27.1 ± 5.4	0.006
BMI, >26[c]	213 (42.2)	141 (39.3)	72 (49.3)	0.041
Smoking status				
Smoker	163 (32.2)	135 (37.6)	28 (19.1)	<0.001
Previous smoker	91 (18)	69 (19.2)	22 (15)	0.271
Easy fatigue, yes	116 (23)	40 (11.1)	76 (52)	<0.001
Asthma, yes	43 (8.5)	29 (8)	14 (9.6)	0.581
Occupation				
White collar	135 (26.7)	122 (34)	13 (8.9)	<0.001
Blue collar	256 (50.7)	233 (65)	23 (15.8)	<0.001
Other	114 (22.6)	4 (1)	110 (75.3)	<0.001
Study				
FEV$_1$ predicted[d]	88.5 ± 20.4	92.5 ± 16.6	78.6 ± 25.2	0.001
FVC predicted	96.6 ± 19.2	98.8 ± 15	91.1 ± 26.2	<0.001
FEV$_1$/FVC predicted	91.8 ± 14.2	93.8 ± 11.7	87 ± 18.1	<0.001
Lung function				
Restriction, yes	26 (5.1)	11 (3)	15 (10.2)	0.001
COPD, yes	92 (18.2)	43 (11.9)	49 (33.6)	<0.001
CO$_2$ flux, g/m^2/d[e]		15.36	508	<0.001

[a] P value comparing reference and study groups, by Mann-Whitney for continuous variables and by χ^2 for categorical variables.
[b] Cut-off defined according to the mean value (i.e., 41.3 years) of the observed age distribution in the whole population.
[c] Cut-off defined according to the mean value (i.e., 26.1 BMI) of the observed BMI distribution in the whole population.
[d] Predicted values from the National Health and Nutrition Examination Survey Cohort.
[e] CO$_2$ flux is expressed in mean.

The reference group comprised 359 individuals (204 women and 155 men) working in downtown of Ponta Delgada city (Table 1). Ponta Delgada's downtown has 7818 adult workers [25]; therefore, the participation rate was 4.5%. Ponta Delgada is located in the south side of São Miguel Island (Azores, Portugal; Figure 1(b)), in the Região dos Picos Volcanic Complex, a basaltic rift zone located between Sete Cidades and Fogo Volcanoes, where no manifestations of volcanism have been identified [11].

The Ethics Board of Divino Espirito Santo Hospital (Ponta Delgada, Azores, Portugal) approved the study. All individuals signed a written informed consent, in compliance with the Helsinki Declaration and Oviedo Convention, to participate in this study. Table 1 summarizes the demographic characteristics and main lifestyle habits of the studied populations. Only individuals resident for more than five years in each locality were considered in this study.

Spirometry Tests

The forced expiratory volume in one second (FEV_1) and the forced vital capacity (FVC) values were obtained by spirometry in all participants. Spirometry tests were conducted with the participants in an up position wearing a nose clip and a disposable mouth piece using the EasyOne automated portable spirometer (ndd, Zürich, Switzerland), which meets ATS/ERS spirometry standards [27] and is equipped with software that checks for unacceptable maneuvers and compares the measured values with reference tables. Standardized operating procedures were implemented and controlled. Participants performed three to five attempts to provide at least three technically acceptable maneuvers, following the criteria recommended by the ATS [28] and the guidelines of the European Respiratory Society [27]. Postbronchodilator tests were not applied.

Spirometry data were classified categorically as being consistent with either a normal pulmonary function, a restrictive defect, or an obstructive defect. Spirometric detection of restriction was considered in all subjects with normal FEV_1/FVC, with FVC < 80% predicted, and FEV_1/FVC < 70% was used as a fixed cut-point for obstruction (COPD), according to the third United States National Health and Nutrition Examination Survey (NHANES III) for adult Caucasians. According to the GOLD guidelines, COPD was further classified in the following ranks given by the spirometer output: mild ($FEV_1 \geq$ 80%), moderate (FEV_1 50–79%), and severe (FEV_1 30–49%).

Exposure Assessment to Volcanogenic Air Pollution by DDS

Carbon dioxide degassing maps may be useful for volcanic/seismic monitoring purposes as they represent a reference for future variations on the state of activity of the volcano [14].

Measurements of soil CO_2 released by soil degassing in Ribeira Quente village were carried out recently by Viveiros et al. [14, 26]. The surveys were made using portable instruments that perform measurements based on the accumulation chamber method [29]. Almost all buildings (about 98%) in Ribeira Quente village are placed in an anomalous high CO_2 degassing zone (DDS), with soil CO_2 fluxes that can reach values higher than 25000 g/m²/d [14]. Considering that CO_2 released by soils may enter the buildings (e.g., through pipes, cracks in the floor, and/or the contact between floor and walls), CO_2 degassing maps produced for Ribeira Quente village [14, 26] were used to attribute a CO_2 flux value to each building. Even if no indoor measurements were performed, the CO_2 released from soils is positively correlated with the indoor CO_2 concentrations [13], so measurements applied outside the buildings are indirectly representative of the CO_2 concentrations that can be found indoors [13]. Sporadic soil CO_2 flux measurements were performed at Ponta Delgada in the areas surrounding the buildings occupied by individuals from the reference group. The 68 measurements performed showed that CO_2 flux values were lower than $25\,g/m^2/d$ and thus representative of biogenic origin and without hydrothermal contribution.

Statistical Analysis

Pearson Chi-Square test was used to compare restrictions and COPD prevalence between individuals inhabiting the environment with volcanic degassing and individuals from the reference group. To estimate the association between chronic exposure to an environment with volcanic air pollution and restrictive and obstructive defects (no versus yes), odds ratio (OR) and 95% confidence intervals (95% CIs) were calculated using a binary logistic regression model, adjusting for age, gender (male versus female), fatigue (yes versus no), and smoking status (yes versus no).

To estimate the association between chronic exposure to an environment with volcanic air pollution and the increase in the severity COPD, odds ratio (OR) and 95% confidence intervals (95% CIs) were calculated using an ordinal logistic regression model, adjusting for age, fatigue (yes versus no), asthma (yes versus no), and smoking status (yes versus no). The occurrence of obstructions was graded on scales, according to their occurrence and severity: 1: without obstruction, and 2 to 4: with obstruction, from the least severe (2) to the most severe (4).

Mann-Whitney U test was used to compare soil CO_2 fluxes released by diffuse degassing between the reference and the study group.

All statistical analysis was performed using IBM SPSS Statistics 20.0 for Windows [30], and the level of statistical significance was set at $P \le 0.05$.

RESULTS

The general characteristics of the study populations are presented in Table 1. The study group has an older population and a higher BMI than the reference one (56.1% versus 42.3% of individuals with more than 41 years and 27.1 kg/m^2 versus $25.7\,kg/m^2$, resp.). On the other hand, the reference group has a higher percentage of smokers than the study group (37.6% versus 19.1%).

Soil CO_2 flux was significantly different $(P < 0.001)$ between the studied areas (Table 1). According to criteria for diffuse degassing susceptibility areas defined by Viveiros et al. [14], all the analyzed buildings in Ponta Delgada are located in low susceptibility areas (CO_2 flux $< 25\,g/m^2/d$), while in Ribeira Quente 1.9% of the buildings were located in moderate susceptibility areas (soil CO_2 flux between $25\,g/m^2/d$ and $50\,g/m^2/d$) and the remaining 98.1% were in a high susceptibility area (soil CO_2 flux $\geq 50\,g/m^2/d$) (Supplementary Material 1 in Supplementary Material available online at http://dx.doi. org/10.1155/2015/326794).

Prevalence of Restrictive and Obstructive Respiratory Defects

The prevalence of restrictions in the study group was significantly higher than in the reference group (10.2% versus 3.0%, resp. $P = 0.001$;). Similarly, the prevalence of COPD was significantly higher in the study group than in the reference one (33.6% versus 11.9%, resp. $P < 0.001$;) (Figure 2). The prevalence of more severe obstructions was also higher in the study group compared to the reference one (mild, 15.7 versus 4.4, moderate, 6.8 versus 2.2, and severe, 4.7 versus 0, resp.).

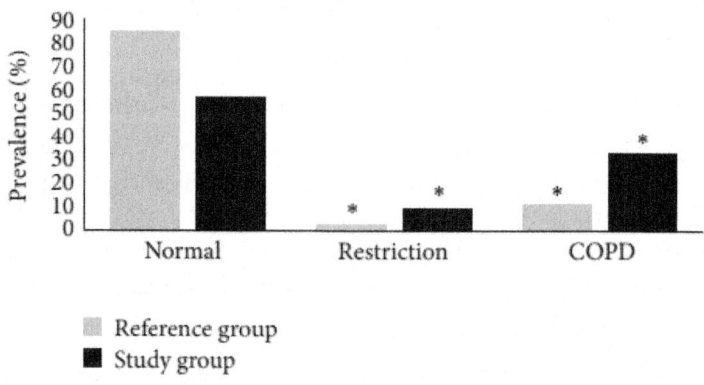

Figure 2: Proportion (%) of individuals with obstructive and restrictive airway diseases in study (Ribeira Quente) and reference (Ponta Delgada) groups; significantly different at $P < 0.05$.

Restrictive or Obstructive Respiratory Defects and Exposure to Volcanogenic Air Pollution by DDS

Exposure to volcanogenic air pollution by DDS was a significant predictor of the prevalence of respiratory restrictions and of COPD exacerbation in the multivariate analysis. After adjustment for age, gender, fatigue, and smoking status, a higher prevalence of respiratory restrictions was found associated with exposure to volcanogenic air pollution by DDS (OR = 4.4; 95% CI, 1.78–10.69; $P = 0.001$) (Table 2). Also, after the adjustment for the same factors, a higher prevalence of respiratory obstructions was found associated with exposure to the volcanic environment (OR = 2.8; 95% CI, 1.60–4.99) (Table 3). Exacerbation in COPD severity was also found significantly associated with exposure to volcanogenic air pollution by DDS (OR = 3.2; 95% CI, 1.82–5.58 $P < 0.001$;) (Table 4).

Table 2: Adjusted association between characteristics of study participants, exposure to volcanogenic soil diffuse degassing (DDS), and restrictive lung defects

Binomial logistic regression			Number of obs. 505 LR chi2(5) = 10.97 Prob > chi2 = 0.05
Parameter	n (%)	OR (95% CI)ª	P value
Age		0.99 (0.96–1.02)	0.933
Gender			
Male	207 (41)	1.16 (0.50–2.69)	0.729
Female	298 (59)	1.00	
Easy fatigue			
Yes	116 (23)	0.62 (0.23–1.67)	0.349
No	389 (77)	1.00	
Smoking status			
Smoker	163 (32.3)	0.94 (0.36–2.42)	0.892
Nonsmoker	342 (67.7)	1.00	
Exposure to DDS			
Yes (study group)	146 (28.9)	4.37 (1.78–10.69)	0.001
No (reference group)	359 (71.1)	1.00	

ªOR, odds ratio; 95% CI, 95% confidence interval.

Table 3: Adjusted association between characteristics of study participants, exposure to volcanogenic soil diffuse degassing, and obstructive lung defects

Binomial logistic regression			Number of obs. 505 LR chi2(5) = 60.80 Prob > chi2 = <0.001
Parameter	n (%)	OR (95% CI)ª	P value
Age		1.04 (1.02–1.06)	<0.001
Easy fatigue			
Yes	116 (23)	1.80 (1.05–3.28)	0.033
No	389 (77)	1.00	
Gender			
Male	207 (41)	0.66 (0.39–1.11)	0.123
Female	298 (59)	1.00	
Smoking status			
Smoker	163 (32.3)	2.34 (1.35–4.06)	0.002
Nonsmoker	342 (67.7)	1.00	
Exposure to DDS			
Yes (study group)	146 (28.9)	2.83 (1.60–4.99)	<0.001
No (reference group)	359 (71.1)	1.00	

ªOR, odds ratio; 95% CI, 95% confidence interval.

Table 4: Adjusted association between characteristics of study participants, exposure to volcanogenic soil diffuse degassing, and COPD exacerbation

Ordinal logistic regression			Number of obs. 505 LR chi2(5) = 70.19 Prob > chi2 = <0.001
Parameter	n (%)	OR (95% CI)[a]	P value
Age		1.04 (1.02–1.06)	<0.001
Easy fatigue			
Yes	116 (23)	1.80 (1.00–3.24)	0.046
No	389 (77)	1.00	
Asthma			
Yes	43 (8.5)	2.25 (1.05–4.82)	0.037
No	462 (91.5)	1.00	
Smoking status			
Smoker	163 (32.3)	2.11 (1.25–3.56)	0.005
Nonsmoker	342 (67.7)	1.00	
Exposure to DDS			
Yes (study group)	146 (28.9)	3.19 (1.82–5.58)	<0.001
No (reference group)	359 (71.1)	1.00	

[a]OR, odds ratio; 95% CI, 95% confidence interval.

The analyzed confounding factors did not show any significant association with respiratory restrictions (Table2), but respiratory obstructions were significantly associated with age, fatigue, and smoking status (Table 3). The increase in COPD severity was significantly associated with asthma and the abovementioned factors (Table4).

DISCUSSION

The association between exposure to air pollution and adverse respiratory effects has been widely demonstrated [31, 32]. Such association is usually related to anthropogenic air pollution, while for volcanogenic pollution there are much fewer studies. Volcanoes and volcanic manifestations, such as fumaroles and hot and cold CO_2-rich springs as well as degassing soils, release into the environment metals, hazardous aerosols, and gases that daily affect the quality of the environment and the health of human populations [33].

The measurements of CO_2 flux revealed that the study group (from the hydrothermal area) is chronically exposed to elevated volcanogenic air pollution by soil diffuse degassing compared to the reference group (average values are 508 $g/m^2/d$ ($\approx$35.73 ppm/s) versus 15.36 $g/m^2/d$ ($\approx$1.08 ppm/s), resp.). Carbon dioxide levels are described to be usually greater inside a building than outside [34, 35]; therefore, these populations are much probably subjected to higher levels of indoor CO_2 than the presented values. Also, independently of the biological CO_2 contribution, the hydrothermal CO_2 emission is permanent and thus the level of CO_2 in the Ribeira Quente buildings must be always higher than the "normal" values and sporadically reach anomalous high values as recognized in previous works, only due to meteorological changes [13].

The threshold limit value for 8 h time weighted-average exposures to CO_2 is 5000 ppm (WHO), and possibly the average CO_2 concentrations encountered in the buildings are below this limit and are not expected to cause health symptoms such as headaches, fatigue, and eye and throat irritation. However, the consequences of the permanent presence of indoor CO_2 in concentrations higher than the outdoor environment are still far from being understood and may eventually lead to some stress effect in the organisms that needs to be better studied. In addition, the CO_2 emission may be also associated with other volcanic emissions, as it is the case of the radioactive gas radon. The measurements performed showed that the study group is chronically exposed to elevated volcanogenic air pollution, namely, high CO_2 soil diffuse degassing; high concentrations of CO_2 can cause headaches due to hypercapnia and the decrease of O_2 in the bloodstream [36]. When these two events combine, the body responds by increasing blood flow to the brain by dilating the blood vessels, resulting in pain [36]. A significant association was found between restrictive respiratory defects and the exposure to volcanogenic air pollution by DDS, measured as 4.4 times higher in the inhabitants of Ribeira Quente (study group), when compared to the reference one. A significant association was also found between obstructive respiratory defects and exposure to volcanogenic air pollution by DDS, measured as 2.8 times higher in the study group. None of the analyzed confounding factors (age, gender, fatigue, and smoking status) was significantly associated with respiratory restriction, but apart from gender all confounding factors were positively associated with respiratory obstructions.

Results also showed significant associations between the increase in the severity of COPD and the exposure to soil diffuse degassing in the study group when compared to the reference one, given as 3.2 times higher in Ribeira Quente inhabitants (study group). Age, asthma, fatigue, and smoking status were also significantly associated with the increase in COPD severity.

Age and smoking habits are well associated with COPD prevalence [37, 38]. According to our results, the classification of subjects who are at risk is relevant in middle-aged and elderly persons (i.e., > 41 years old), individuals with asthma, and smokers. Aging affects the structure, function, and control of the respiratory system; according to Stanojevic et al. [39], both FEV_1 and FVC decrease with age, and thus the observed increase in COPD associated with age was expected (since older individuals will have a reduced FVC and a reduced FEV_1/FVC ratio). Also recently results from Eurostat [40] demonstrated that respiratory diseases such as bronchitis and asthma affect mainly older people, as almost 90% of EU-28 deaths from these diseases occur among those aged 65 and above. Moreover, this report presents a standardized death rate from diseases of the respiratory system of at least 150 deaths per 100 000 inhabitants; the highest death rates were observed in two Portuguese regions, the volcanic islands of Madeira (294.6 deaths per 100 000 inhabitants) and Azores (195.8 deaths per 100 000 inhabitants); the rest of the Portuguese territory (mainland) presented a death rate below 150 deaths per 100 000 inhabitants; therefore, the observed results are in accordance with this general overview.

Regarding asthma, there might be a significant overlap with COPD, since both are characterized by an underlying airway inflammation, although the structural and pathophysiologic findings in both diseases can be usually differentiated [41]. According to the Global Initiative for Chronic Obstructive Lung Diseases [42], in individuals with chronic respiratory symptoms and fixed airflow limitation, it is difficult to differentiate asthma from COPD. However, there is epidemiological evidence that long-standing asthma can lead to fixed airflow limitation [43], which increases the frequency of asthma attacks [44].

Fatigue is considered the second most common symptom of COPD, after breathlessness [45]. The increased level of fatigue found in patients with COPD is associated with an increase in the severity of lung impairment and a reduction in exercise tolerance. Symptoms of fatigue can be physical and mental, such as lack of energy or poor concentration. Baghai-Ravary et al. [46] and Breslin et al. [47] found a relationship between fatigue and COPD, which persists at COPD exacerbations. Since COPD was observed mainly in elderly people that have diminished lung function and therefore a reduction in health status, fatigue was expected to be associated with COPD.

It is well established that tobacco is the most important causative factor for COPD, with individual susceptibility continuously interacting synergistically with other risk factors [48], such as age. According to Lundbäck et al. [49], lifelong smoking increases the chance of developing COPD and, thus, about 50% of the smokers eventually develop COPD during their lifetime. However, according to GOLD [42], other risk factors of COPD include exposure to air pollution, secondhand smoke and occupational dusts and chemicals, heredity, a history of childhood respiratory infections, and socioeconomic status. In the present study, air pollution by hydrothermal volcanism was proven to be a risk factor in COPD exacerbation.

It is well known that air pollution can perpetuate a chronic inflammatory process that potentially leads to lung diseases [31]. However, the majority of studies regard air pollutants of anthropogenic origin, such as PM_{10}, NO_2, SO_2, and O_3 [50, 51], while only few studies were developed in areas with volcanic activity and associate volcanogenic pollutants with human airway diseases [7, 52]. Longo and Yang [52] and Iwasawa et al. [8] reported a higher risk of acute bronchitis across the lifespan in humans exposed to sulfurous volcanic air pollution. Furthermore, in 2013, Camarinho et al. [22] showed that noneruptive active volcanism is associated with increased lung injury in mice. Thus, it is not unexpected to observe an increase in the exacerbation of COPD severity in the population living in Ribeira Quente, a volcanic area with elevated soil diffuse degassing. These findings also corroborate the study by Amaral and Rodrigues [19] that observed higher rates of chronic bronchitis in Furnas village inhabitants (in comparison to a reference group) and suggested that it could be partially associated with chronic exposure, in a very humid atmosphere, to environmental factors resulting from volcanic activity.

CONCLUSIONS

To our knowledge, this is the first study assessing long-term effects of CO_2 emissions in hydrothermal areas on the development of respiratory defects. The results of this study show that long-term exposure to air pollution by soil diffuse degassing increases the risk of developing restrictive defects as well as the exacerbation of COPD in the inhabitants of hydrothermal volcanic areas. Therefore, mitigation measures should be implemented in populations inhabiting elevated diffuse degassing areas, such as the construction of natural/forced ventilation systems, as well as follow-up health programs in order to provide medical counseling when necessary. The results of the study also support the need to particularly follow up the asthmatic and the older individuals, since these factors are associated with exacerbation in COPD severity in individuals chronically exposed to volcanogenic soil diffuse degassing.

In addition, considering the positive trend of the CO_2 in the atmosphere, areas as Ribeira Quente village can be used as natural analogous to studying the possible effects of these increases on the population.

ACKNOWLEDGMENTS

The authors thank Catarina Silva, Ana Ferreira, Carolina Parelho, Carla Raposo, and Ricardo Camarinho for their support in field and laboratory work. Diana Linhares and Fátima Viveiros were supported, respectively, by Ph.D. and postdoc fellowships from the Fundo Regional da Ciência (Regional Government of the Azores) (PROEMPREGO Programme) (M3.1.2/F/019/2011 and M3.1.7/F/018/2011). The authors also thank the financial support of BioAir—Biomonitoring Air Pollution: Development of an Integrated System (M2.1.2/F/008/2011) from Fundo Regional da Ciência (Regional Government of the Azores).

REFERENCES

1. National Research Council, Active Tectonics: Impact on Society, The National Academies Press, Washington, DC, USA, 1986.

2. K. B. Brunekreef, D. W. Dockery, and M. Krzyzanowski, "Epidemiologic studies on short-term effects of low levels of major ambient air pollution components," Environmental Health Perspectives, vol. 103, no. 2, pp. 3–13, 1995.

3. J. McCreanor, P. Cullinan, M. J. Nieuwenhuijsen et al., "Respiratory effects of exposure to diesel traffic in persons with asthma," The New England Journal of Medicine, vol. 357, no. 23, pp. 2348–2358, 2007.

4. S. Weichenthal, R. Kulka, A. Dubeau, C. Martin, D. Wang, and R. Dales, "Traffic-related air pollution and acute changes in heart rate variability and respiratory function in urban cyclists," Environmental Health Perspectives, vol. 119, no. 10, pp. 1373–1378, 2011.

5. R. Dales, L. M. Kauri, S. Cakmak et al., "Acute changes in lung function associated with proximity to a steel plant: a randomized study," Environment International, vol. 55, pp. 15–19, 2013.·

6. B. Jacquemin, J. Lepeule, A. Boudier et al., "Impact of geocoding methods on associations between long-term exposure to urban air pollution and lung function," Environmental Health Perspectives, vol. 121, no. 9, pp. 1054–1060, 2013.

7. G. Gudmundsson, "Respiratory health effects of volcanic ash with special reference to Iceland. A review,"Clinical Respiratory Journal, vol. 5, no. 1, pp. 2–9, 2011.

8. S. Iwasawa, Y. Kikuchi, Y. Nishiwaki et al., "Effects of SO_2 on respiratory system of adult miyakejima resident 2 years after returning to the Island," Journal of Occupational Health, vol. 51, no. 1, pp. 38–47, 2009.

9. B. M. Longo, W. Yang, J. B. Green, F. L. Crosby, and V. L. Crosby, "Acute health effects associated with exposure to volcanic air pollution (VOG) from increased activity at kilauea volcano in 2008," Journal of Toxicology and Environmental Health, Part A: Current Issues, vol. 73, no. 20, pp. 1370–1381, 2010.

10. R. Searle, "Tectonic pattern of the Azores spreading centre and triple junction," Earth and Planetary Science Letters, vol. 51, no. 2, pp. 415–434, 1980.

11. T. Ferreira, J. L. Gaspar, F. Viveiros, M. Marcos, C. Faria, and F. Sousa, "Monitoring of fumarole discharge and CO_2 soil degassing in

the Azores: contribution to volcanic surveillance and public health risk assessment," Annals of Geophysics, vol. 48, no. 4-5, pp. 787–796, 2005.

12. J. E. Guest, J. L. Gaspar, P. D. Cole et al., "Volcanic geology of Furnas Volcano, Sao Miguel, Azores,"Journal of Volcanology and Geothermal Research, vol. 92, no. 1-2, pp. 1–29, 1999.

13. F. Viveiros, T. Ferreira, C. Silva, and J. L. Gaspar, "Meteorological factors controlling soil gases and indoor CO_2 concentration: a permanent risk in degassing areas," Science of the Total Environment, vol. 407, no. 4, pp. 1362–1372, 2009.

14. F. Viveiros, C. Cardellini, T. Ferreira, S. Caliro, G. Chiodini, and C. Silva, "Soil CO_2 emissions at Furnas volcano, São Miguel Island, Azores archipelago: volcano monitoring perspectives, geomorphologic studies, and land use planning application," Journal of Geophysical Research B: Solid Earth, vol. 115, no. 12, Article ID B12208, 2010.

15. M. Durand, "Indoor air pollution caused by geothermal gases," Building and Environment, vol. 41, no. 11, pp. 1607–1610, 2006.

16. Y. Y. You, C. Niu, J. Zhou et al., "Measurement of air exchange rates in different indoor environments using continuous CO_2 sensors," Journal of Environmental Sciences, vol. 24, no. 4, pp. 657–664, 2012.

17. World Health Organization, Programmes and Projects: Indoor Air Pollution, WHO, 2008.

18. J. Balmes, M. Becklake, P. Blanc, et al., "American Thoracic Society Statement: occupational contribution to the burden of airway disease," American Journal of Respiratory and Critical Care Medicine, vol. 167, no. 5, pp. 787–797, 2003.

19. F. S. Amaral and A. S. Rodrigues, "Chronic exposure to volcanic environments and chronic bronchitis incidence in the Azores, Portugal," Environmental Research, vol. 103, no. 3, pp. 419–423, 2007.

20. Amaral, V. Rodrigues, J. Oliveira et al., "Chronic exposure to volcanic environments and cancer incidence in the Azores, Portugal," Science of the Total Environment, vol. 367, no. 1, pp. 123–128, 2006. ·

21. S. Rodrigues, M. S. C. Arruda, and P. V. Garcia, "Evidence of DNA damage in humans inhabiting a volcanically active environment: a useful tool for biomonitoring," Environment International, vol. 49, pp. 51–56, 2012.

22. R. Camarinho, P. V. Garcia, and A. S. Rodrigues, "Chronic exposure to volcanogenic air pollution as cause of lung injury," Environmental Pollution C, vol. 181, pp. 24–30, 2013.

23. G. Ferris, "Epidemiology standardization project II," The American Review of Respiratory Disease, vol. 118, no. 6, pp. 7–53, 1978.

24. British Medical Research Council Committee on the Aetiology of Chronic Bronchitis, "Standardized questionnaire on respiratory symptoms," British Medical Journal, no. 2, p. 1965, 1960.

25. Instituto Nacional de Estatística, Censos, INE, 2011, http://censos.ine.pt.

26. F. Viveiros, C. Cardellini, T. Ferreira, and C. Silva, "Contribution of CO_2 emitted to the atmosphere by diffuse degassing from volcanoes: the furnas volcano case study," International Journal of Global Warming, vol. 4, no. 3-4, pp. 287–304, 2012.

27. M. R. Miller, J. Hankinson, V. Brusasco et al., "Standardisation of spirometry," European Respiratory Journal, vol. 26, no. 2, pp. 319–338, 2005.

28. American Thoracic Society, "Standardization of spirometry, 1994 update," American Journal of Respiratory and Critical Care Medicine, vol. 152, no. 3, pp. 1107–1136, 1995.

29. G. Chiodini, R. Cioni, M. Guidi, B. Raco, and L. Marini, "Soil CO_2 flux measurements in volcanic and geothermal areas," Applied Geochemistry, vol. 13, no. 5, pp. 543–552, 1998.

30. IBM SPSS 20.0, IBM SPSS Statistics 20 Core System User›s Guide, USA, 2011.

31. T. Götschi, J. Heinrich, J. Sunyer, and N. Künzli, "Long-term effects of ambient air pollution on lung function: a review," Epidemiology, vol. 19, no. 5, pp. 690–701, 2008.

32. G. Weinmayr, E. Romeo, M. de Sario, S. K. Weiland, and F. Forastiere, "Short-Term effects of PM_{10} and NO_2 on respiratory health among children with asthma or asthma-like symptoms: a systematic review and Meta-Analysis," Environmental Health Perspectives, vol. 118, no. 4, pp. 449–457, 2010.

33. Amaral and A. Rodrigues, "Volcanogenic contaminants: chronic exposure," in Encyclopedia of Environmental Health, J. Nriagu, S. Kacwe, T. Kawamoto, J. A. Patz, and D. M. Rennie, Eds., vol. 5, pp. 645–653, Elsevier, 2011.

34. H. H. Denli, D. Z. Seker, and S. Kaya, "GIS based carbon dioxide concentration research in ITU campus, Istanbul-Turkey," in Proceedings of the FIG Congress Engaging the Challenges-Enhancing the Relevance, Kuala Lumpur, Malaysia, June 2014.

35. H. Mudarri, "Potential correlation factors for interpreting CO_2 measurements in buildings," ASHRAE Transactions, vol. 103, no. 2, pp. 244–255, 1997.

36. S. A. Rice, "Human health risk assessment of CO_2: survivors of acute high-level exposure and populations sensitive to prolonged low-level exposure," in Proceedings of the 3rd Annual Conference on Carbon Sequestration Rice, Alexandria, Va, USA, May 2004.

37. P. Lange, "Chronic care for COPD patients in Denmark," Pneumonologia i Alergologia Polska, vol. 80, no. 4, pp. 292–295, 2012.

38. Lindberg, B. Eriksson, L.-G. Larsson, E. Rönmark, T. Sandström, and B. Lundbäck, "Seven-year cumulative incidence of COPD in an age-stratified general population sample," Chest, vol. 129, no. 4, pp. 879–885, 2006.

39. S. Stanojevic, A. Wade, J. Stocks et al., "Reference ranges for spirometry across all ages: a new approach,"The American Journal of Respiratory and Critical Care Medicine, vol. 177, no. 3, pp. 253–260, 2008.

40. Eurostat, Regional Yearbook. General and Regional Statistics: Statistical Books, Publications Office of the European Union, Luxembourg, Luxembourg, 2014.

41. T. Welte and D. A. Groneberg, "Asthma and COPD," Experimental and Toxicologic Pathology, vol. 57, no. 2, pp. 35–40, 2006.

42. Global Initiative for Chronic Obstructive Lung Diseases (GOLD), Global Strategy for Diagnosis, Management, and Prevention of Chronic Obstructive Pulmonary Disease, GOLD, 2010,

43. P. Lange, J. Parner, J. Vestbo, P. Schnohr, and G. Jensen, "A 15-year follow-up study of ventilatory function in adults with asthma," The New England Journal of Medicine, vol. 339, no. 17, pp. 1194–200, 1998.

44. K.-H. Kim, S.A. Jahan, and E. Kabir, "A review on human health perspective of air pollution with respect to allergies and asthma," Environment International, vol. 59, pp. 41–52, 2013.

45. J. S. Paddison, T. W. Effing, S. Quinn, and P. A. Frith, "Fatigue in COPD: association with functional status and hospitalisations," European Respiratory Journal, vol. 41, no. 3, pp. 565–570, 2013.

46. R. Baghai-Ravary, J. K. Quint, J. J. P. Goldring, J. R. Hurst, G. C. Donaldson, and J. A. Wedzicha, "Determinants and impact of fatigue in patients with chronic obstructive pulmonary disease,"Respiratory Medicine, vol. 103, no. 2, pp. 216–223, 2009. ·

47. Breslin, C. P. van der Schans, S. Breukink et al., "Perception of fatigue and quality of life in patients with COPD," Chest, vol. 114, no. 4, pp. 958–964, 1998.

48. S. Marsh, S. Aldington, P. Shirtchiffe, M. Weatherall, and R. Beasley, "Smoking and COPD: what really are the risks?" European Respiratory Journal, vol. 28, no. 4, pp. 883–884, 2006.

49. Lundbäck, A. Lindberg, M. Lindström et al., "Not 15 but 50% of smokers develop COPD?—report from the obstructive lung disease in Northern Sweden studies," Respiratory Medicine, vol. 97, no. 2, pp. 115–122, 2003.

50. M. Kampa and E. Castanas, "Human health effects of air pollution," Environmental Pollution, vol. 151, no. 2, pp. 362–367, 2008.

51. W. MacNee and K. Donaldson, "Mechanism of lung injury caused by PM10 and ultrafine particles with special reference to COPD," European Respiratory Journal, vol. 21, supplement 40, pp. 47s–51s, 2003.

52. M. Longo and W. Yang, "Acute bronchitis and volcanic air pollution: a community-based cohort study at Kilauea Volcano, Hawai›i, USA," Journal of Toxicology and Environmental Health, Part A: Current Issues, vol. 71, no. 24, pp. 1565–1571, 2008.

Chapter 4

EVALUATION OF AN EMISSION INVENTORY AND AIR POLLUTION IN THE METROPOLITAN AREA OF BUENOS AIRES

Laura E. Venegas[1], Nicolás A. Mazzeo and Andrea L. Pineda Rojas

[1]National Scientific and Technological Research Council (CONICET) National Technological University, Argentina

INTRODUCTION

Air is a vital resource, so its quality must fall within a tightly bound range. This quality is the level needed to protect public health. In addition, the quality must be able to support other life, notably diverse and sustainable ecosystems. The atmosphere is an extremely complex system in which numerous physical and chemical processes occur simultaneously. Ambient measurements give us only a snapshot of atmospheric conditions at a particular time and location. Such measurements are often difficult to interpret without a clear conceptual model of atmospheric processes. Moreover, measurements alone cannot be used directly by policymakers to establish an effective strategy for solving air quality problems. An understanding of individual atmospheric processes (chemistry, transport, removal, etc.) does not imply an understanding of the system as a whole. Mathematical models provide the necessary framework for integration of our understanding of individual atmospheric processes and study of their interactions. A combination of state-of-the-science measurements with state-of-the-science models is the best approach for making real progress toward understanding the atmospheric environment. Over the past four decades, there has been a significant increase in the number of locations where air quality data have been obtained. Also, there has been a substantial improvement in the technique for modelling the different physical and chemical processes occurring in the atmosphere. Despite this progress, currently available observations are still spatially and temporally sparse and the predictions of current generation of air quality models are still uncertain. Consequently, observations and model outputs should be combined to create high-resolution spatial-temporal maps of air quality. However, at present air quality observations and model results are generally used separately.

Urban air pollution is still on rise at many cities worldwide, or has experienced only small improvements. Some causes of urban air pollution problems are the amount and density of air pollutant sources, particularly vehicles, residences and industries. Because of the complexity of urban systems, air quality management in these areas is still a serious problem.

Emission inventories are important tools to describe the emission situation and eventually to manage air quality. An emission inventory is a list of the amount of pollutants from different sources entering the air in a given time period and a particular geographical area. It usually includes information on the amount of the pollutants released from major industrial sources, and averages figures for the emissions from smaller sources throughout the area. The information included in an emission inventory helps to identify the sources and in the development of abatement strategies. Ball & Radcliffe (1979) have identified several applications for urban air pollution emission inventories. Their information is useful to elaborate a map showing the geographic distribution of emissions. This map can be an important aid in land use planning by identifying parts of the region that are likely to be subject to high levels of pollution, and the location of pollution sources in relation to sensitive areas. Emission inventories can point out the major sources whose control can lead to a considerable reduction of pollution in the area. They can be used in conjunction with an atmospheric dispersion model, to estimate air pollutant concentrations at ground level and/or assess trends in air quality. They can also help in the design of air quality monitoring networks, by indicating, for example, where the highest concentrations of pollution are likely to be found, or which areas are the most representative.

The method used to develop an emission inventory does have some elements of error, but other two alternatives are expensive and subject to their own errors. The first alternative would be to monitor continually every major source in the area. The second alternative would be to monitor continually the pollutants in ambient air at many points and apply appropriate dispersion equations to calculate the emissions. In practice, the most informative system would be a combination of all three, knowledgeably applied. Air pollution emission inventories have been developed for several urban areas (Andrade et al, 2010; Ariztegui et al., 2004; Beaton et al., 1992; Borrego et al., 2003; Butler et al., 2008; D'Avignon et al, 2010; Gurjar et al., 2008; Kim, 1996; Miller et al. 2006; Mohan et al., 2007; Nishikawa & Kannari, 2010; Saija & Romano, 2002; Sallés et al., 1996; Seika et al., 1996; Sturm et al., 1999; Tsilingiridis et al., 2002; Wang et al., 2010; Zárate et al., 2007). Particularly, a few years ago, the first versions of an urban emission inventory (year 2000) for the city of Buenos Aires (Mazzeo & Venegas, 2003) and for area sources located in the Metropolitan Area of Buenos Aires (year 2005) (Pineda Rojas et al., 2007) have been prepared.

An emission inventory is an essential tool in the management of local air quality, particularly when its information is used in conjunction with atmospheric dispersion models.

A model is a simplified representation of real conditions. It contains assumptions and sometimes also some experimentally derived constants. Operational decisions based on predictions of a model should be made therefore when the underlying assumptions are met and when the model is being applied within the range of values for which the model has been tested. Atmospheric dispersion models provide a link between the source emissions and ambient concentrations. The heart of the matter is to estimate the concentration of a pollutant at a particular receptor point by calculating from basic information about the source of the pollutant, the meteorological conditions and the surface characteristics. Atmospheric dispersion models help us to understand the way air pollutants behave in the environment and are a useful tool in the urban air quality management system. There are many reasons for using atmospheric dispersion models, such as working out which sources are responsible for what proportion of concentration at any receptor; estimating population exposure on a higher spatial or temporal resolution than is practicable by measurement; targeting emission reductions on the highest contributors; and predicting concentration changes over time. Urban atmospheric dispersion models range from simple empirical models to complex three-dimensional urban air-shed models. Sometimes, available input data make application of complex numerical tools not possible, and simple urban background pollution models become an acceptable alternative giving as good results as computations with more sophisticated models (Berkowicz, 2000; Hanna et al., 2002). Some examples of urban scale dispersion modelling systems developed during last decades are the UAM model (Morris & Myers, 1990), the DAUMOD model (Mazzeo & Venegas, 1991; 2010); the Danish OML model (Olesen, 1995), the UK-ADMS Urban model (Carruthers et al., 1994; CERC, 2003; McHugh et al., 1997) and the UDM-FMI (Karppinen et al., 2000) model. Dispersion models for the urban scale estimate urban background concentrations.

This chapter presents a summary of the development and results of a high spatial and temporal resolution version of the emission inventory of carbon monoxide (CO) and nitrogen oxides (NO_x) for the Metropolitan Area of Buenos Aires (MABA), including area source emissions (motor vehicles, aircrafts, residential heating systems, commercial combustion and small industries). The spatial distributions of CO and NO_x annual emission rates from area sources within the Metropolitan Area of Buenos Aires are shown with a spatial resolution of 1 x 1 km. The urban atmospheric dispersion model DAUMOD is applied to evaluate the air quality in the MABA due to the contribution of area source emissions in the urban area. Estimations of horizontal distributions of CO and nitrogen dioxide (NO_2) background concentrations in the MABA are presented.

DESCRIPTION OF THE METROPOLITAN AREA OF BUENOS AIRES

The Metropolitan Area of Buenos Aires (MABA) is considered the third megacity in Latin America, following Mexico City (Mexico) and Sao Paulo (Brazil). It is integrated by the city of Buenos Aires (CBA) and the Greater Buenos Aires (GBA). The city of Buenos Aires (Lat. 34°35'S – Long. 58°26'W), capital of Argentina, is located on the west coast of de la Plata River. The city has an extension of $203km^2$ and 2891082 inhabitants (Instituto Nacional de Estadística y Censos [INDEC], 2010). The city of Buenos Aires is surrounded by the Greater Buenos Aires. This area is compound by 24 districts. It has an extension of $3627km^2$ and 9910282 inhabitants (INDEC, 2010). The area of the MABA is 0.14% of the territory of Argentina and its population is approximately 32% of the population of the country.Fig. 1 shows the different districts of the Metropolitan Area of Buenos Aires and the grid net considered in calculations.

The terrain is flat with height differences lesser than 30 m. The de la Plata River is a shallow estuary, which covers $35000 km^2$ approximately. The estuary is 320 km long, and its width varies between 38 km and 230 km in the upper and lower regions, respectively. In front of the CBA, the width of the river is about 42 km. The mean water temperature in the river varies from 12°C in winter to 24°C in summer.

The de la Plata River plain has a temperate climate. In summer (December to February), the city of Buenos Aires is warm and moist, with a mean temperature of 24°C. During autumn and spring atmospheric conditions are variable, with fluctuating temperatures. The winter months (June to August) are temperate and moist, with a mean temperature of 12°C. The annual mean temperature in the city is 18°C, and between 15-16°C in its surroundings. In the MABA, frosts occur between June and August, and snowfalls are very rare. The annual precipitation varies between 900 mm and 1600 mm, influenced by winds that advect humidity from the Atlantic Ocean. Rains are heavier on March. Winds are generally of low intensity. Strong winds are more frequent between September and March, when the greatest storm frequency is observed. The annual frequency of winds blowing clean air from the river towards the urban area is 58%, and that of calm conditions is 3%. Cases of wind direction persistence may last more than 6 hours at any wind direction sector (Mazzeo & Venegas, 2004).

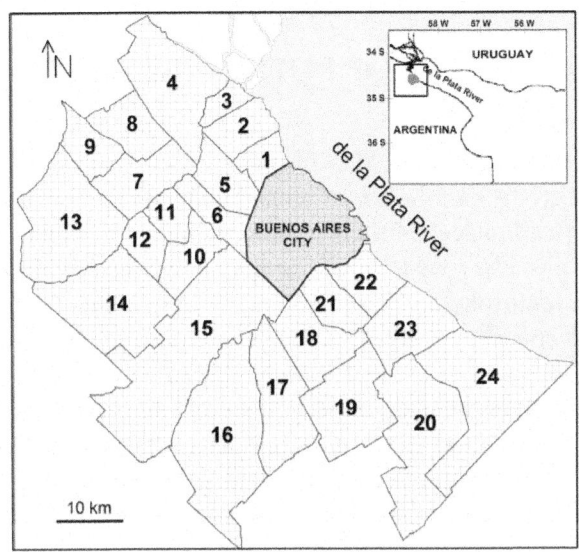

Figure 1: Map of the Metropolitan Area of Buenos Aires, including the city of Buenos Aires and the Greater Buenos Aires. Districts of the Greater Buenos Aires (*inhabitants*): 1: Vicente López (*270929*); 2: San Isidro (*291608*); 3: San Fernando (*163462*); 4: Tigre (*380709*); 5: Gral. San Martín (*422830*); 6: Tres de Febrero (*343774*); 7: San Miguel (*281120*); 8: Malvinas Argentinas (*321833*); 9: José C. Paz (*263094*), 10: Morón (*319934*); 11: Hurlingham (*176505*); 12: Ituzaingó (*168419*); 13: Moreno (*462242*); 14: Merlo (*524207*); 15: La Matanza (*1772130*); 16: Ezeiza (*160219*); 17: Esteban Echeverría (*298814*); 18: Lomas de Zamora (*613192*); 19: Almirante Brown (*555731*); 20: Florencio Varela (*423992*); 21: Lanús (*453500*); 22: Avellaneda (*340985*); 23: Quilmes (*580829*); 24: Berazategui (*320224*). Grid cell side: 1km.

Mazzeo & Venegas (2008, 2010) and Venegas & Mazzeo (2006a, 2010 a) proposed and applied methodologies to design different air quality monitoring networks for the city of Buenos Aires. At present, air pollutant concentrations are registered at the first three air quality monitoring stations of the network in the city (Venegas & Mazzeo, 2010b). The air quality in the city of Buenos Aires has been the subject of several studies carried out during the last years using different methodologies: analysis of data obtained from some measurement surveys of pollutants in urban air (Arkouli et al., 2010; Bocca et al., 2006; Bogo et al., 1999, 2001, 2003; Mazzeo & Venegas, 2002 , 2004; Mazzeo et al., 2005; Venegas & Mazzeo, 2000, 2003 ; Vogt et al., 2007) and application of atmospheric dispersion models (Mazzeo & Venegas, 2010 ; Mazzeo et al.,

2010 ; Venegas & Mazzeo, 2005, 2006b, 2010 a). In the Greater Buenos Aires, very few air quality measurements have been made (Fagundez et al., 2001; Japan International Cooperation Agency-Secretaría de Desarrollo Sustentable y Política Ambiental [JICA-SAyDS], 2002).

BRIEF DESCRIPTION OF POLLUTANTS CONSIDERED IN THIS CHAPTER

The pollutants considered in this chapter are carbon monoxide (CO) and nitrogen oxides (NO_x). Carbon monoxide is generated primarily by incomplete combustion of carbonaceous fuels in automobile engines and is a colourless and odourless gas. It is a very stable compound having a lifetime of two to four months in the atmosphere. There are studies (e.g. Harte et al., 1991), which show that high concentrations of CO can cause physiological and pathological changes and ultimately death of human. Carbon monoxide is a poisonous inhalant that deprives the body tissues of necessary oxygen. It is toxic because haemoglobin absorbs CO more readily than oxygen. With the bloodstream carrying less oxygen, brain functions is affected and heart rate increases in an attempt to offset the oxygen deficit. In very high doses it is fatal due to cerebral and cardiac hypoxia.

The stable gaseous oxides of nitrogen include nitrous oxide (N_2O), nitric oxide (NO), nitrogen trioxides (N_2O_3), nitrogen dioxide (NO_2) and nitrogen pentoxide (N_2O_5). An unstable NO_3, also exist. The nitrogen oxides present in the atmosphere in any significant amount are N_2O, NO and NO_2. N_2O is an inert gas with anaesthetic characteristics. Its atmospheric concentrations are considerably below the threshold concentration for biological effects, but it may be a significant contributor to global warming. NO is a colourless gas and at its air concentrations its biological toxicity in terms of human health is insignificant. However, NO is a precursor to the formation of NO_2 and is an active compound in photochemical smog formation as well. NO_2 is a reddish brown gas and is quite visible in sufficient amounts. The toxicological and epidemiological effects of NO_2 on human being are not completely known (WHO, 2006a). NO_2 may penetrate to the pulmonary region increasing susceptibility to respiratory pathogens.

THE EMISSION INVENTORY FOR THE MABA

Inventory Technique

To develop an emission inventory for an area, one must: a) determine the type of air pollutants of concern, such as CO and NO_x; b) list the types of sources for the area, such as motor vehicles, aircrafts, residential, commercial and industrial combustions; c) examine the literature to find valid emission factors for each pollutant of concern; d) through an actual count, or means of some estimating technique, determine the number and size of specific sources in the area; and e) multiply the corresponding numbers from c) and d) to obtain the total emissions for each activity and then sum the similar emissions to obtain the total for the area. Valid emission factors for each source of pollution are the key to an emission inventory. Emission factors are then applied to the activity data in order to estimate the likely emissions:

Emission = Activity level x Emission factor   (1)

This chapter focuses on area source emissions of CO and NO_x in the Metropolitan Area of Buenos Aires. The following source categories are considered:

- Mobile sources: road traffic and aircrafts
- Fixed sources: residential, commercial and small industries activities.

Point source emissions could not be included because, at present, there is not available sufficient data on the large industries located in the MABA. Information on the actual point source emissions is only available in a limited number of cases. There is also a lack of homogeneity among the amount and quality of basic information available for each district of the MABA. The existence of multiple local Administrations in the region is the main reason for such heterogeneity. Furthermore, it should be noted that the city of Buenos Aires is a city-state and the 24 districts of the Greater Buenos Aires are part of the Province of Buenos Aires. CBA and GBA have different Governments.

Emissions from Mobile Sources

Main mobile sources in the Metropolitan Area of Buenos Aires have been divided into the following groups:

Road traffic: passenger cars (including taxis), buses (including coaches) and heavy-duty vehicles.

Air traffic: aircraft's landing-take-off (LTO) cycles at the domestic airport located in the city of Buenos Aires and at the international airport located in the Greater Buenos Aires.

Road Traffic Emissions

There are usually about three million vehicles circulating in the MABA during working days. The city of Buenos Aires and its surroundings concentrate approximately 43% of private cars, 60% of taxis, 50% of urban and interurban buses and 29% of cargo transportation of Argentina. As mentioned above, the methodological approach to estimate the emission rates from road traffic is based on the multiplication of activity data by emission factors. The first step involves the determination of an estimate of vehicle activity. Five traffic parameters are considered: volume, composition, vehicle velocity, vehicle age and travel distance in each grid cell. All traffic data have been obtained from the National Secretary of Transportation, the Buenos Aires City Government and the Secretary of Transportation of the Province of Buenos Aires. Available information includes mean daily traffic flow at several locations as well as traffic flow and composition measured at different hours of the day on different streets, routes, avenues and highways in the MABA. Most private cars are petrol-driven and taxis burn natural gas. All the buses and heavy duty vehicles are considered to run on diesel. The approximately age of the vehicle fleet is illustrated in Fig.2.

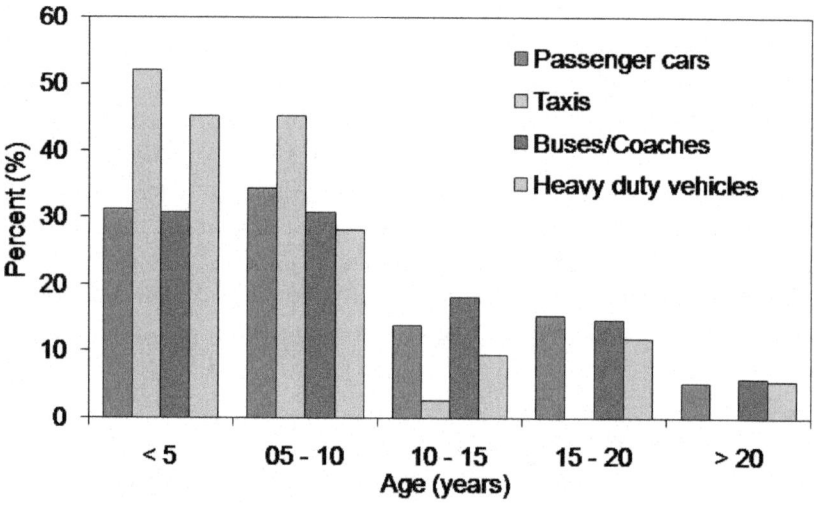

Figure 2: Distribution of the age of each vehicle category.

The road network of the Metropolitan Area of Buenos Aires is assumed to be integrated by: highways, routes, avenues, main streets and streets. These specifications are quite useful since they generally correspond to particular traffic density levels and fleet compositions. Statistical shapes for traffic rates and average speed on these road types are then applicable to the CBA and the GBA. Based on measurements at several sites, the following vehicle fleet compositions are considered for the entire domain: a) highways: 89.7% (passenger cars) and 10.3% (buses and heavy duty vehicles); b) routes, avenues and main streets: 95.0% (passenger cars) and 5.0% (buses and heavy duty vehicles); and c) streets: 99.0% (passenger cars) and 1.0% (buses and heavy duty vehicles). As it is difficult to define a law to estimate a spatial vehicle speed evolution along a road, an average vehicle speed has been set for each road type (highways, routes, avenues, main streets, streets). The second step involves selecting emission factors. The emission factors used for mobile sources are based on measurements of in-service emissions in Buenos Aires (Rideout et al. 2005) and on the European Environment Agency's Atmospheric Emission inventory Guidebook (COPERT method) (European Environment Agency, 2001). Results for the city of Buenos Aires and for the Greater Buenos Aires are described below.

In the city of Buenos Aires passenger cars employ the following fuels: 78.9% gasoline; 16.0% diesel and 5.1% compressed natural gas (CNG). Traffic flow data in the city are available at different sites located in highways, avenues and streets (Gobierno de la Ciudad de Buenos Aires, [GCBA], 2006). In the interest of completeness, where there are no traffic flow data available for a particular road street, a local mean flow is assigned according to the traffic map elaborated by the Secretary of Transport for the city of Buenos Aires (see detail in Fig. 3). Using this information, population density distribution and representative traffic flow measured at different points of the city, the vehicle kilometres travelled in each grid cell are estimated.

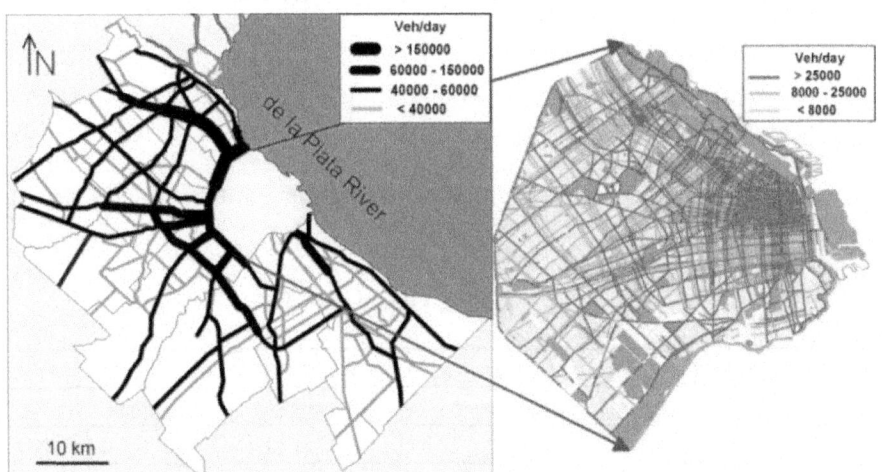

Figure 3: Traffic flow map in the MABA and the city of Buenos Aires (detail).

Examples of traffic profile registered at different locations within the CBA are shown in Fig. 4 (left). The average vehicle speed considered for the different roads within the city of Buenos Aires is: 80km/h (highways); 35km/h (avenues); and 15km/h (streets). The representative emission factors for CO and NO$_x$ considered for the vehicles in the CBA are included in Table 1.

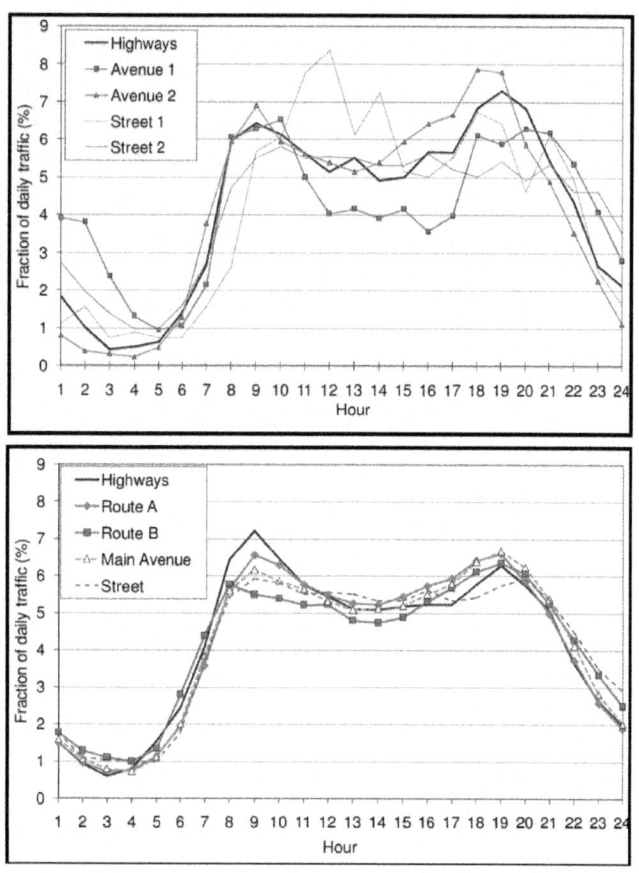

Figure 4: Hourly traffic profiles at different sites registered within the CBA (left) and the GBA (right)

Table 1: Emission factors (F) (g veh^{-1} km^{-1}) for CO and NO$_x$

City of Buenos Aires		Highways		Avenues		Streets	
Vehicle type	Fuel	F (CO)	F (NO$_x$)	F (CO)	F (NO$_x$)	F (CO)	F (NO$_x$)
Passenger cars	Gasoline	12.0	2.8	18.7	1.8	38.0	1.5
	Diesel	0.5	0.8	0.6	1.0	1.0	1.1
	CNG	8.0	2.6	2.0	2.1	7.0	2.0
Buses	Diesel	2.2	3.0	5.5	5.0	8.0	14.7
Heavy Duty Vehicles	Diesel	1.8	3.6	2.5	6.1	6.0	10.0
Greater Buenos Aires							

		Highways		Routes and Avenues		Streets	
Vehicle type	Fuel	F (CO)	F (NOx)	F (CO)	F (NOx)	F (CO)	F (NO$_x$)
Passenger cars	Gasoline	10.0	3.4	16.7	1.9	38.0	1.5
	Diesel	0.4	0.8	0.5	0.9	1.0	1.1
	CNG	8.0	2.8	2.0	2.2	7.0	2.0
Buses	Diesel	2.0	5.6	5.0	6.7	8.0	14.7
Heavy Duty Vehicles	Diesel	1.5	4.0	2.0	6.0	6.0	10.0

In the Greater Buenos Aires, passenger cars employ the following fuels: 66.1% gasoline; 15.6% diesel and 18.3% CNG. Traffic flow data in the GBA are available at different sites in highways, routes and avenues. Even when data are given for a specific road, they are usually measured within a particular duration of time and in discrete locations. Since information is not available everywhere along a given road, both spatial and temporal assumptions have been made in order to obtain the characteristics for the whole road and so finally to describe emissions over the entire region in a daily evolution. In order to elaborate a map of the traffic flow in the main roads of the GBA, vehicle rates (R) are extrapolated anywhere along every main road. These estimations are based mostly on empirical assumptions. In the GBA, highways and most major roads are roughly either radial or semi-circular. This feature facilitates the direct setting of roads on a polar reference frame. From available data, it may be assumed that radial road traffic rates decrease with distance (x) to the border of the city of Buenos Aires. An empirical exponential law is used to describe this typical star-form network behaviour (Sallés et al, 1996). The traffic rate (R(x)) at a given distance from the border of the city of Buenos Aires is estimated by:

$$R(x) = R(0) \exp [-\alpha x] \qquad (2)$$

where R(0) is a reference value (traffic rate at the border of the CBA) and α is an empirical coefficient. The values of α have been obtained by fitting to traffic flow measurements registered at several points in highways and routes. Fig. 5 shows the values of R(0) and α for each sector considered in calculations.

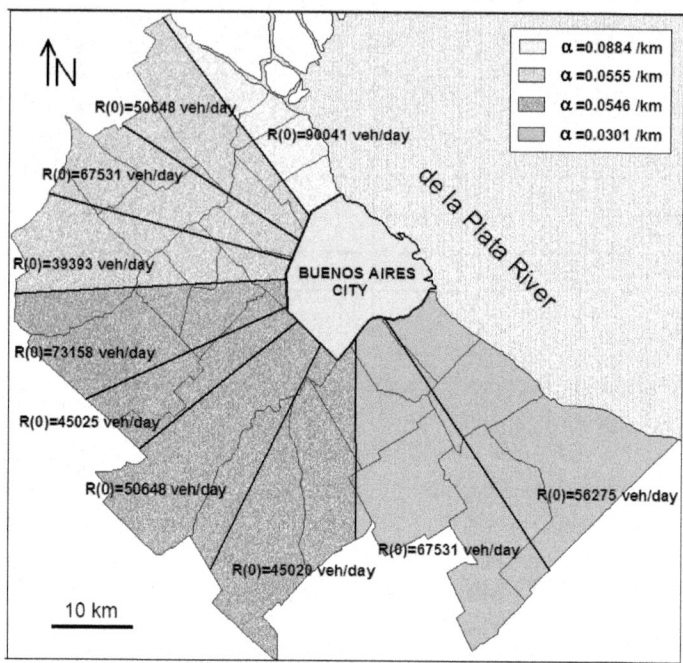

Figure 5: Values of R(0) and α for each sector, considered in Equation (2) to estimate the traffic flow along the highways and main roads in the GBA.

Available data suggest that semi-circular road traffic rates remain constant along the main roads, within each sector. The traffic map for the Greater Buenos Aires showing the obtained mean daily traffic flow in highways, routes, avenues and main streets is included in Fig. 3. These vehicle flux data are further used to account for vehicle distribution in the streets of the urban area in the GBA. The extrapolation assumption includes both the traffic in the main roadways and the spread traffic in streets and is based on a flux balance criterion between ingoing and outgoing vehicles in each grid cell. Hourly variation of traffic rate is obtained applying hourly typical traffic rate profiles. Representative traffic profiles at different roadways within the GBA are shown in Fig. 4. The average vehicle speed considered for the different roads within the Greater Buenos Aires is: 100km/h (highways); 40km/h (routes and avenues); and 15km/h (streets). The representative emission factors for CO and NO_x considered for the vehicles in the GBA are presented in Table 1.

CO and NO_x emissions from buses in the MABA are obtained from the emission factors, the total distance travelled by each bus within each grid cell, the bus service frequency and the mean speed of the vehicles in each grid cell. Finally, CO and NO_x emissions from road traffic are estimated for each grid cell over the entire region in a daily evolution.

Air Traffic Emissions

Aircraft emissions have been estimated using the "alternative simple methodology" proposed byRomano et al. (1999). The aircraft operations of interest that may affect ground level pollutant concentrations are defined as the landing and takeoff (LTO) cycle. The cycle begins when the aircraft approaches to the airport on its descent from the cruising altitude, lands and taxis to the gate. It continues as the aircraft taxis back out to the runway for subsequent takeoff and climb-out as it heads back to the cruising altitude. For all forms of commercial aircrafts, the time spent in each of the LTO modes is reckoned at 19 min for idling and taxiing out, 42 sec for take-off, 2.2 min for climb-out and; at the other end of the cycle, 4 min for approach to landing and 7 min for taxiing and idling (Romano et al. 1999). Fuel consumption and CO and NO_x emission factors for each operation mode depend on engine type (European Environment Agency, 2001; Romano et al., 1999; US.EPA, 1995). Emissions from aircrafts are calculated considering the modes related to the departure and arrival parts of the LTO cycle separately. For example, the total emission for an aircraft type during its departure is calculated multiplying the emission rates by the amount of time in each mode of the departure part of the LTO cycle, and then summing results from the considered modes. The aircraft type that operates at the domestic airport is mainly Boeing B-737, as it is used in domestic and regional flights. At the international airport the airlines operate the following aircraft types: Boeing B-737, B-747, B-757, B-767, B-777 and Airbus A319, A320, A321, A340. Fig. 6 shows the hourly distribution of the mean daily frequency of departures and arrivals at each airport, respectively. The daily evolution of aircraft emissions is added to the area source emissions estimated for the grid cells where each airport is located.

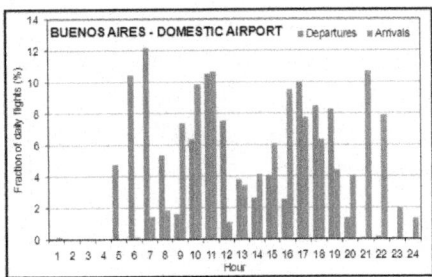 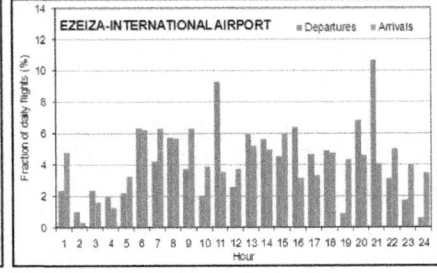

Figure 6: Hourly distribution of the fraction (%) of mean daily departures and arrivals at the domestic and the international airports.

Emissions from Fixed Sources

The small size fixed sources (residential, commercial and small industries combustion activities) are considered as area sources. These sources consume natural gas for heating, cooking and other activities. The monthly natural gas consumed by residential houses was spatially distributed considering population density. Then, using the CO and NO_x emission factors for natural gas combustion for domestic heating units given in US.EPA (1995) residential emission rates at each grid cell are estimated. Considering monthly natural gas consumed by commercial activity, its spatial distribution in the MABA and the emission factors (US.EPA, 1995), CO and NO_x emission rates for this activity are computed for each grid cell. Finally, considering the monthly natural gas consumed by small industries, their spatial distribution in the MABA and the emission factors (US.EPA, 1995), CO and NO_x emission rates of this activity were estimated for each grid cell. Natural gas consumption and a typical diurnal variation of the consumption for each activity have been provided by the National Gas Administration (ENARGAS).

Carbon Monoxide and Nitrogen Oxides Emissions in the Metropolitan Area of Buenos Aires.

Annual area source emission rates estimated for the city of Buenos Aires (CBA) are 324.7 Gg-CO yr^{-1} and 22.9 Gg-NO_x yr^{-1} and for the Greater Buenos Aires (GBA) are 294.6 Gg-CO yr^{-1} and 43.9 Gg-NO_xyr^{-1}. Therefore, for the Metropolitan Area of Buenos Aires (MABA) annual area source emissions result 619.3 Gg-CO yr^{-1} and 66.8 Gg-NO_x yr^{-1}. Fig. 7 shows the percentage distribution of the annual emission of carbon monoxide and nitrogen oxides by source category in the MABA. Road traffic accounts for 99.4% of CO and 80.6% of NO_x annual area source emissions in the MABA.

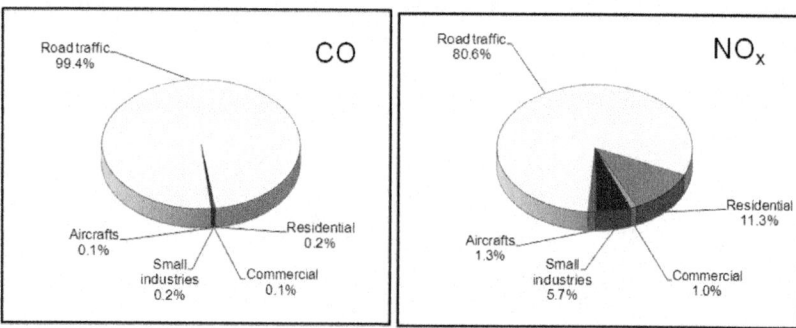

Figure 7: Estimated annual area source emission of CO and NO_x by source category in the MABA.

The spatial distributions of CO and NO_x annual emission rates (in ton $km^{-2} yr^{-1}$) from area sources within the MABA are shown in Fig. 8. The intensity of emissions varies considerably across the urban area. There is a wide range in CO and NO_x emissions between different grid cells depending on the density of road transportation sources in each grid cell. It is clear that high emission rates per unit area can be found in downtown of the city of Buenos Aires.

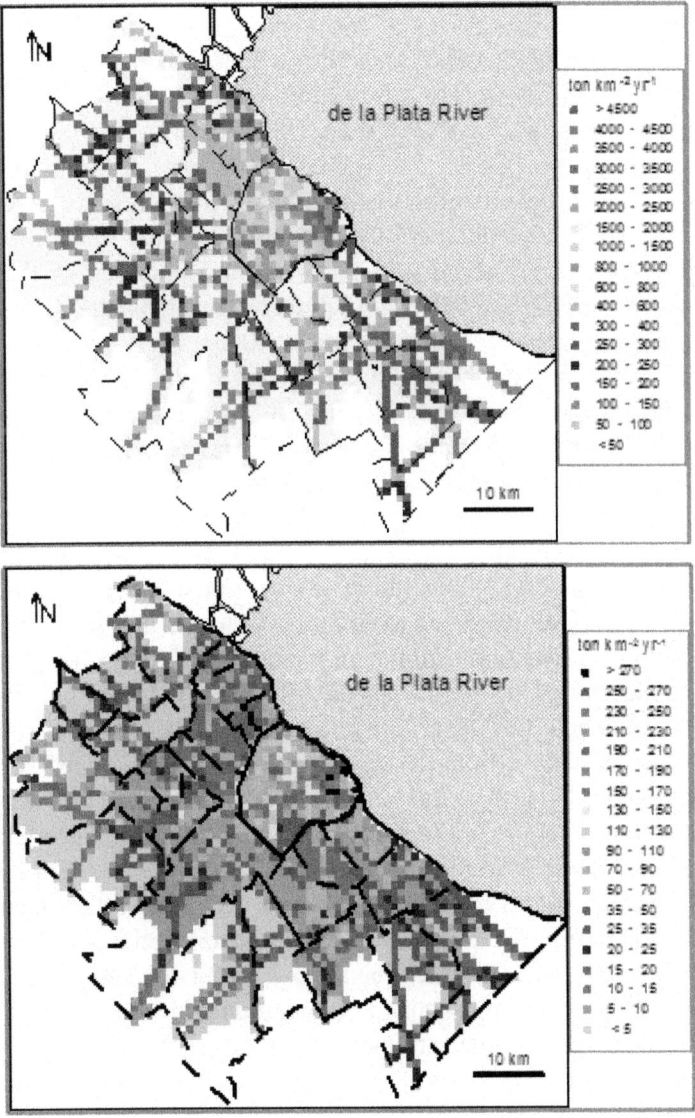

Figure 8: Annual emission rates (ton $km^{-2} yr^{-1}$) of CO (top) and NO_x (bottom) from area sources in the MABA. Grid resolution is 1 x 1 km.

Uncertainty Assessment

Uncertainty is a statistical term that is used to represent the degree of accuracy and precision of data. It often expresses the range of possible values of a parameter or a measurement around a preferred value. Various approaches for representing uncertainty in the context of different domains are widely described (Azondèkon & Martel, 1999; Draper, 1995).

Emission inventories are based on assumptions that are needed to be made and statistical data. Real measurements are available for a few emission sources and/or for certain time periods only. Therefore, uncertainty estimations are of importance and should always be foreseen (Sturm, 2003). However, it is not easy to assess uncertainty at the level of aggregated datasets. Information related to emissions and their uncertainties that originates from a few measurements has to be applied for a large number of sources. As the latter may differ strongly, even within the same category, the uncertainty increases as the emission inventory becomes more detailed. On a more aggregated level, averaging helps to improve the uncertainty situation (Sturm, 2003). There are different studies (Frey & Li, 2003; Frey & Zheng, 2002; Regan et al., 2003; Romano et al, 2004) done in differentiating and quantifying the contributions from uncertainty and natural variability to emission data. The reliability of the information provided by emission inventories is strongly biased by a wide range of causes. Particularly, when the emissions are estimated through emission factors the following points have to be taken into account: a) uncertainty related to the choice of the indicators, b) uncertainty related to the quantitative value of the indicators, c) uncertainty related to emission factors and d) uncertainty related to the structure of emission estimate models. In the MABA, the use of traffic flow values registered at several sites on different days to compute the average vehicle fleet in each road type, results in a mean error of approximately 20-30%. The uncertainty estimation of the average vehicle speed for different road types is found to be near 20%. These uncertainties may introduce an error in the selection of emission factors of about 20%. Other uncertainties may come from the spatial grid resolution. The estimation of travel distance along each road type in each grid cell has an error of 10-15%. The uncertainty of the spatial distribution of population density, commercial activity and small industries in each grid cell of the urban area is about 40%. In general, the error in the estimation of the emissions of carbon monoxide and nitrogen oxides in the MABA are expected to be around 40%.

AIR POLLUTANT CONCENTRATION ESTIMATIONS

Brief Description of the Urban Atmospheric Dispersion Model Used

Urban background concentrations of CO and NO_2 in the Metropolitan Area of Buenos Aires have been estimated applying the urban atmospheric dispersion model DAUMOD(v.2) to the area sources described above. This model has been developed and introduced in former papers (Mazzeo & Venegas, 1991, 2010 ; Venegas & Mazzeo, 2002; 2006b). However, a brief description of its main considerations and assumptions is included below.

The DAUMOD model (Mazzeo & Venegas, 1991) is an urban atmospheric dispersion model valid for steady-state conditions. It is assumed that effluents are emitted continuously from the surface. The x-axis is in the direction of the mean wind and the z-axis is vertical. At a given distance, the vertical extension of the plume of contaminants is given by h(x). Concentration at h(x) is negligible and there is no transport of mass through the upper limit of the plume. The variation of h(x) is parameterised in the model by potential functions given by (Mazzeo & Venegas, 1991),

$$\frac{h}{z_0} = a\left(\frac{x}{z_0}\right)^b$$

(3)

where z_0 is the surface roughness length and coefficients a and b depend on atmospheric stability (Mazzeo & Venegas, 2010). Other basic assumption included in the model is that background air pollutant concentration $[C(x,z)]$ can be expressed by the following polynomial form:

$$C(x,z) = C(x,0) \sum_{j=0}^{6} A_j\left(\frac{z}{h}\right)^j$$

(4)

Coefficients A_j (j=0,...6) depend on surface roughness and atmospheric stability (Mazzeo & Venegas, 2010) and have been computed by fitting Equation (4) to the results given by the following expression (Pasquill & Smith, 1983):

$$C(x,z) = C(x,0) \exp\left[-4.605\left(\frac{z}{z_m}\right)^s\right]$$

(5)

where s is a shape factor which depends on atmospheric stability and surface roughness (Gryning et al., 1987) and z_m is the height at which concentration is $0.01C(x,0)$. The height z_m is usually considered to be the upper limit of the plume, so it is assumed $h = z_m$. Considering different atmospheric stability conditions, the coefficients $(A_0, A_1, ...A_6)$ of the polynomial of grade 6 are obtained for each fitting. There are excellent fittings of polynomial forms (given by Equation (4)) to values obtained fromEquation (5), with coefficients of determination of ≈ 1.0 (the reader can find details of these results inMazzeo & Venegas, 1991).

In an urban area, a horizontal distribution of area sources with strength varying according to a typical square grid pattern may be assumed. Each grid square has a uniform source strength Q_i $(i = 0, 1, 2, ..., N)$ expressed as mass per unit area per unit time. According to the DAUMOD model $C(x,z)$ can be estimated by:

$$C(x,z) = \frac{a\left[Q_0 x^b + \sum_{i=1}^{N} (Q_i - Q_{i-1})(x - x_i)^b\right]}{\left(|A_1|k_v z_0^b u_*\right)} \sum_{j=0}^{6} A_j \left(\frac{z}{h}\right)^j$$

(6)

where k_v is the von Kármán's constant and u_* is the friction velocity.

A constant wind direction is required for application of Equations (6). It has been noted from the applications of Equation (6) that estimated concentration at any receptor is mainly originated from the emission in the grid square in which the receptor is located. This is because area source distributions in a city are generally quite smooth and, the contribution of upstream grid squares (from Equation (6)) rapidly reduces with distance to the receptor. The simplification of assuming that the uniform area source strength Q_i only varies with x (in the wind direction), suppose to consider a "narrow plume" hypothesis. This assumption has also been included in other simple urban dispersion models (Arya, 1999; Gifford, 1970; Gifford & Hanna, 1973). The spatial resolution of the model calculations is given by the resolution of the area source emission inventory.

The performance of DAUMOD model in estimating concentrations has been evaluated comparing estimated and observed concentration data from several cities. Results for Bremen (Germany), Frankfurt (Germany) and Nashville (USA) have been reported in Mazzeo & Venegas (1991) and for Copenhagen (Denmark) can be found in Venegas & Mazzeo (2002). The comparison of DAUMOD estimations of background air pollutant concentrations with observations in the city of Buenos Aires can be found in Venegas & Mazzeo (2006b). Results show that the performance of the model in estimating short-term concentrations (hourly and daily) is good and it improves when estimating long averaging time values (monthly and annual). Several applications of different versions of DAUMOD to Buenos Aires have been reported in former papers (Mazzeo & Venegas, 2004, 2008, 2010; Mazzeo et al., 2010 ;Pineda Rojas & Venegas, 2008, 2009, 2010; Venegas & Mazzeo, 2005, 2006a; 2006b).

At present, photochemical transformations involving NO, NO_2 and O_3 are not included in DAUMOD model. However, DAUMOD(v.2) estimates concentrations of NO_2 on the basis of an empirical relationship between NO_2 and NO_x (Derwent & Middleton, 1996; Dixon et al., 2001; Middleton at al., 2008). The concentration of NO_2 is calculated using the polynomial expression (CERC, 2003; Derwent & Middleton, 1996):

$$[NO_2] = 2.166 - [NO_x] (1.236 - 3.348 B + 1.933 B^2 - 0.326 B^3)$$

where $B = \log_{10}([NO_x])$ and $[NO_x]$ is hourly-averaged concentration in ppb.

An application of DAUMOD(v.2) to estimate the influence of NO_x emitted from area sources in the Metropolitan Area of Buenos Aires on the air quality of the city of Buenos Aires have been reported inVenegas & Mazzeo (2007).

Application of Daumod Model to Area Source Emissions In The Metropolitan Area Of Buenos Aires

The DAUMOD(v.2) model is applied to area source emissions in the MABA, to estimate hourly ground level background concentrations of CO and NO_2 in the area. Calculations are performed considering three years of hourly meteorological information registered at the weather stations of the Argentine Meteorological Office located at the domestic airport (in the city of Buenos Aires) and at the international airport (in the Greater Buenos Aires, 30 km southwest the city of Buenos Aires). The spatial resolution used in calculations is 1x1 km.

One consideration to take into account is that this modelling approach does not produce 3-dimensional wind fields, so land–sea breezes are not modelled. Breeze circulations over the wide estuary of the river could bring pollutants back to the receptor area. However, the frequency of atmospheric recirculation events over the city is small: 8% in summer, 7% in autumn, 5% in winter and 7% in spring (Venegas & Mazzeo 1999). In this way, it is expected that this modelling limitation will not significantly affect the results.

Concentrations of Co in the MABA

Three years of hourly and running 8-h average ground level CO concentrations are estimated for the entire MABA. As expected, estimated CO concentration values are higher in the city of Buenos Aires than in the Greater Buenos Aires.

Hourly CO concentrations are all below the air quality standard value of 35ppm (Res. 198/06 city of Buenos Aires and Res. 242/97 Province of Buenos Aires). The highest hourly concentration value resulted 25.7ppm and appeared at downtown of the city of Buenos Aires. In order to illustrate the spatial distribution of CO concentration in the MABA, Fig. 9 shows the computed hourly CO concentrations at rush hour in the evening (20:00), averaged over the three years.

Figure 9: Mean hourly ground level CO concentrations for rush hour in the evening (20:00)

Also, as an example, the spatial distribution of mean (three years average) running 8-h average ground level CO concentrations in the MABA for the period 08:00-16:00 is shown in Fig. 10. The highest running 8-h average CO concentration (C_{8h}) estimated for the three years is 16.1ppm. This value is greater than the air quality standard value (9ppm) established for the MABA. However, as can be seen in Fig. 11, the highest mean annual frequency of $C_{8h} >$ 9ppm at one grid cell reaches 118 cases and appears in the downtown area of the CBA. This value represents the 1.3% of the annual cases of running 8-h average concentrations. Therefore, the air quality regulation for the CBA (Res. 198/06) is accomplished as it requires that 98th percentile of annual cases (considering three years) should be below 9ppm. The analysis of the situations with $C_{8h} >$ 9ppm reveals that 41% of these cases affect areas of 1km² (Fig.12). Only in 10% of the cases the extension of the affected areas is between 16-35 km².

Fig. 13 shows the frequency distribution of the running 8-h average CO concentrations greater than 9ppm obtained during the three years according to the end hour of the 8-h period. Most exceedances are associated to high emission values during the evening (when most people returns home) and nocturnal atmospheric conditions (low wind speed, neutral or stable atmospheric stability). Monthly distribution of the estimated running 8-h average CO concentrations greater than 9ppm is included in Fig.14. These situations are more frequent between May and August, during late autumn and winter.

Figure 10: Estimated mean running 8-h average ground level CO concentrations (period: 08:00-16:00)

Figure 11: Annual mean number of cases with running 8-h average CO concentration greater than 9ppm.

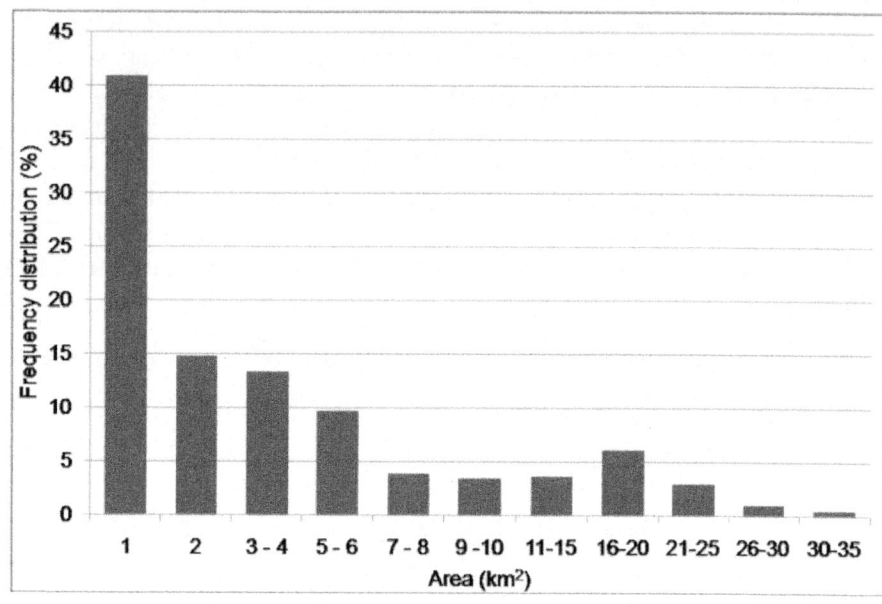

Figure 12: Frequency distribution of the affected area (km²) of the situations with running 8-h average CO concentration greater than 9ppm.

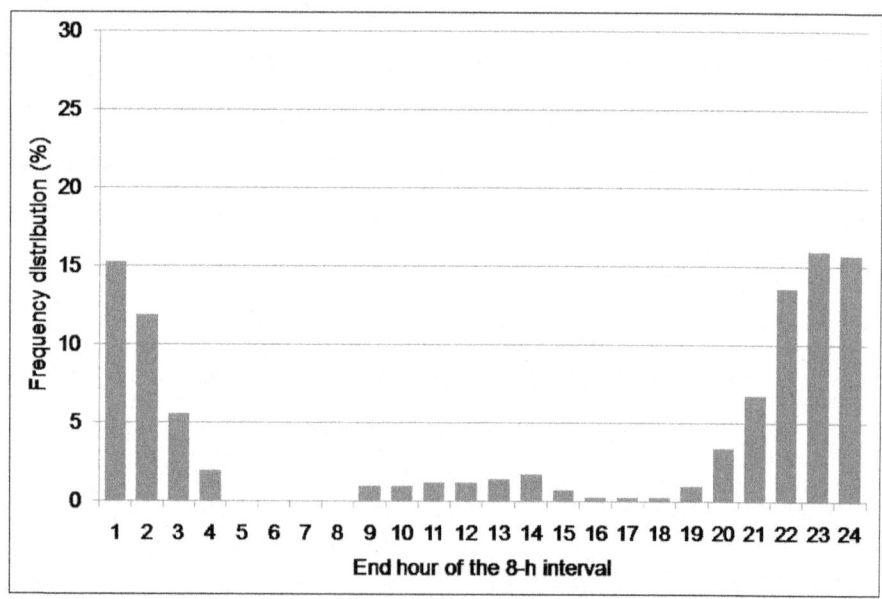

Figure 13: Daily distribution of estimated running 8-h average CO concentrations greater than 9ppm.

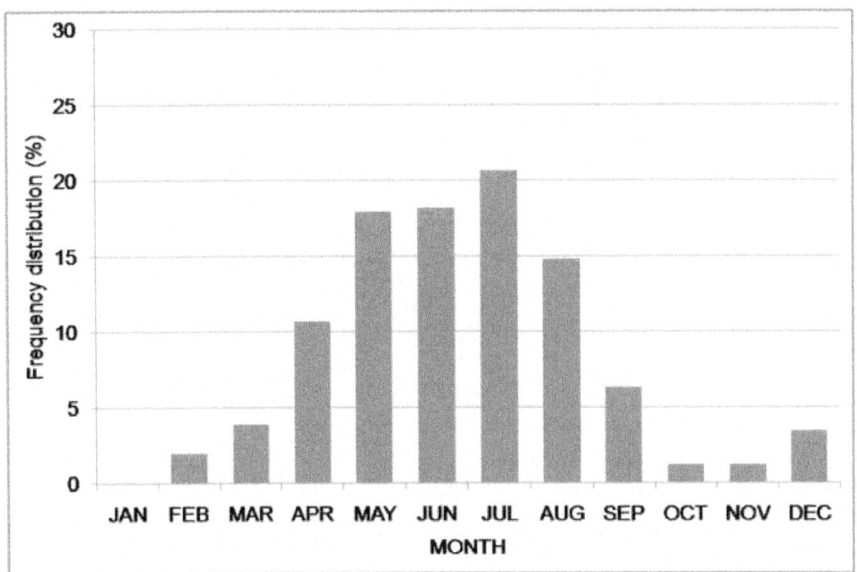

Figure 14: Monthly distribution of estimated running 8-h average CO concentrations greater than 9ppm.

Concentrations of No$_2$ in the MABA

Three years of hourly NO$_2$ ground level concentrations are estimated for the whole MABA. The higher NO$_2$ concentration values are obtained at downtown of the city of Buenos Aires and along the highways of the Metropolitan Area of Buenos Aires. The highest hourly NO$_2$ concentration estimated in the three years period is 184ppb, at downtown. Hourly concentrations are below the air quality standard (200ppb) for the city of Buenos Aires (Res. 198/06, city of Buenos Aires) and for the Greater Buenos Aires (Res.242/97, Province of Buenos Aires). The spatial distribution of the mean hourly NO$_2$ concentrations for the rush hour in the evening is shown in Fig. 15. These results are the hourly values obtained for 20:00 averaged over the three years.

The spatial distribution of annual mean NO$_2$ concentrations in the MABA is included in Fig. 16. The concentration distribution pattern shows a large spatial variability across the urban area. Different areas with high concentration values can be identified, as highways, areas with dense traffic and close to the airports. NO$_2$ annual concentration may reach 28ppb downtown. All values are below the air quality standard (53ppb) for the CBA (Res. 198/06, city of Buenos Aires) and for the GBA (Res. 242/97, Province of Buenos Aires).

Figure 15: Mean hourly ground level NO$_2$ concentrations for rush hour in the evening (20:00)

As mentioned above, model results indicate that NO$_2$ hourly background concentrations may exceed the air quality guideline proposed by the World Health Organisation (100ppb) (WHO, 2006b) at some places in the MABA. The mean annual number of hourly NO$_2$ concentrations that exceed 100ppb at each grid cell is shown in Fig. 17. Most exceedances occur in the city of Buenos Aires, where they may reach a maximum of 50 cases per year at one grid cell located downtown. In the Greater Buenos Aires, hourly NO$_2$ concentrations greater than 100ppb have been obtained in the Northern and Southern highways. These highways constitute two main entrances to the CBA. Also, exceedances are obtained close to the international airport located in the GBA, approximately 30km southwest the CBA.

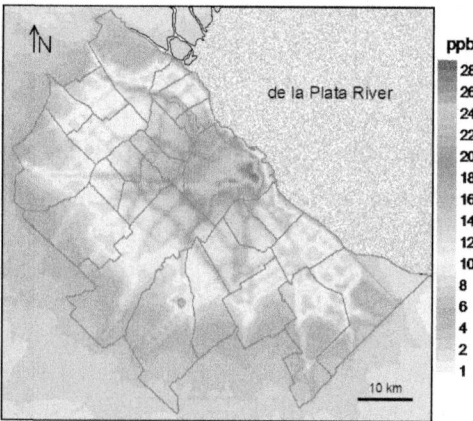

Figure 16: Annual mean ground level NO$_2$ concentrations.

Figure 17: Annual mean number of hourly NO_2 concentrations greater than 100ppb.

The analysis of the situations with hourly NO_2 concentrations greater than 100ppb obtained in the three years, reveals that the extension of the affected area is 1km² in 29% of the cases but it may reach a maximum of 43km² (1%) (Fig. 18). As shown in Fig. 19, high values of hourly NO_2 concentrations are obtained mainly during rush hours in the morning (07:00 to 09:00) and the evening (18:00 to 22:00). The frequency of values greater than 100ppb is higher in the evening/night than in the morning.

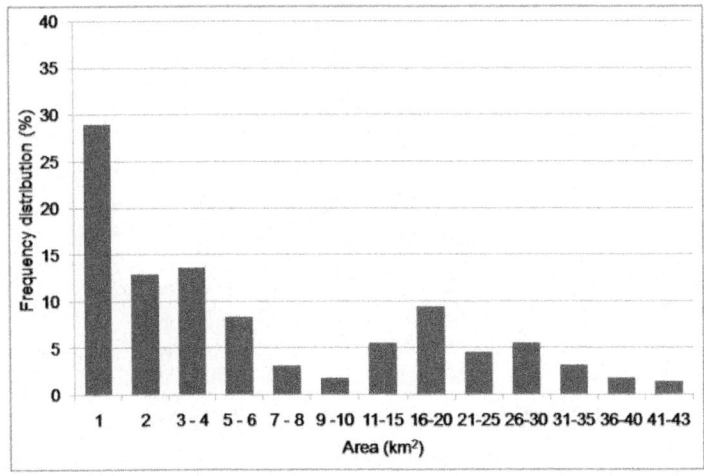

Figure 18: Frequency distribution of affected areas (km²) with hourly NO_2 concentrations greater than 100ppb.

High vehicle emissions and reduced atmospheric dispersion conditions are responsible for this result. Situations with hourly NO_2 concentrations greater than 100ppb in the MABA are more frequent from May to August (Fig. 20).

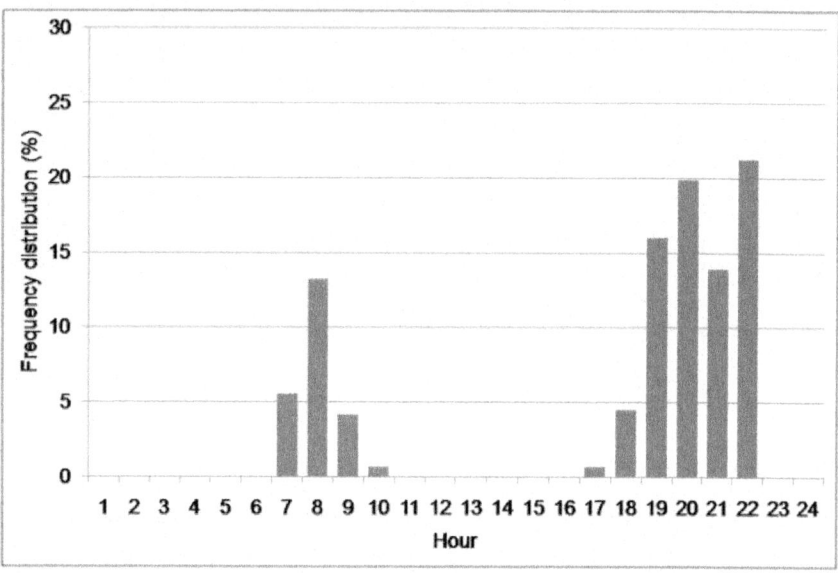

Figure 19: Daily distribution of hourly NO_2 concentrations greater than 100ppb.

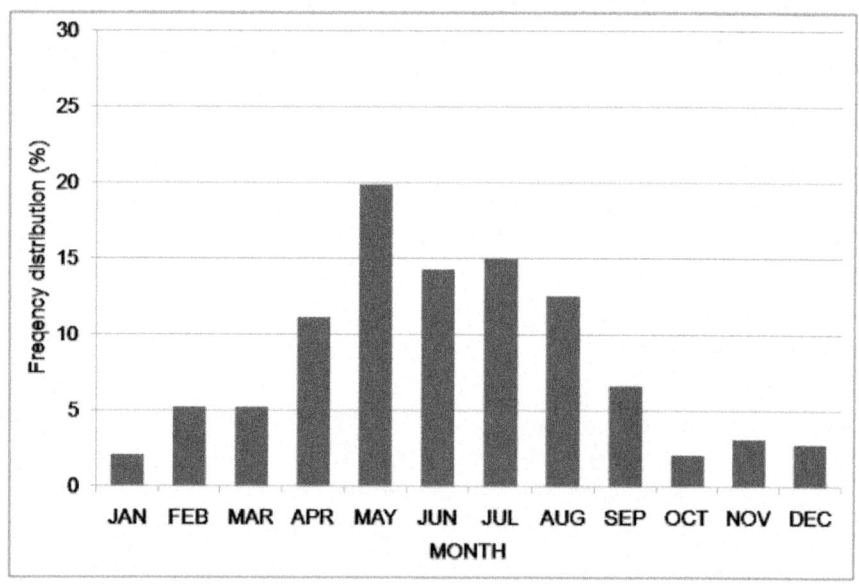

Figure 20: Monthly distribution of hourly NO_2 concentrations greater than 100ppb.

SUMMARY

This chapter presents the results of a high spatial and temporal resolution version of the area source emission inventory of carbon monoxide (CO) and nitrogen oxides (NO_x), and the evaluation of the air quality in the Metropolitan Area of Buenos Aires (MABA). The inventory includes mobile sources (passenger cars/taxis, buses and aircrafts) and fixed sources (emissions arising from residential, commercial and industrial buildings). Its main originality is that it deals as much as possible with distinctive on-road traffic features in order to describe more accurately the emission distribution at a scale comparable to that of the air quality models. The emission inventory for the Metropolitan Area of Buenos Aires can be characterised by the presence of an important contribution of CO and NO_x emitted from mobile sources. Mobile sources contribute with 99.4% of CO and 80.6% of NO_x annual emissions of area sources in the MABA. Annual area source emission rates estimated for the city of Buenos Aires are 324.7 Gg-CO yr^1 and 22.9 Gg-NO_x yr^1 and for the Greater Buenos Aires are 294.6 Gg-CO $yr^1$and 43.9 Gg-NO_x yr^1. Therefore, for the MABA annual area source emissions result 619.3 Gg-CO $yr^1$and 66.8 Gg-NO_x yr^1. Spatial distributions of carbon monoxide and nitrogen oxides emissions show an appreciable variation across the MABA. The urban atmospheric dispersion model DAUMOD is applied to evaluate the air quality in the MABA due to the contribution of area source emissions in the urban area. Estimations of horizontal distributions of CO and nitrogen dioxide (NO_2) background concentrations in the MABA are presented. Air quality regulations established for CO and NO_x in the MABA are accomplished. The analysis of running 8-h average CO concentrations greater than 9ppm reveals that 41% of these cases affect areas of 1km^2. Only in 10% of these cases the affected areas show extensions between 16-35 km^2. Model results indicate that NO_2 hourly background concentrations may exceed the air quality guideline proposed by the World Health Organisation at some places in the MABA. Most exceedances occur in the city of Buenos Aires, where they may reach a maximum of 50 cases per year at one grid cell located downtown. In the Greater Buenos Aires, hourly NO_2concentrations greater than the air quality guideline have been obtained in the Northern and Southern highways. The analysis of the situations with hourly NO_2 concentrations greater than the air quality guideline reveals that the extension of the affected area is 1km^2 in 29% of the cases but it may reach a maximum of 43km^2 (1%).

ACKNOWLEDGEMENTS

The authors kindly acknowledge the support given by ENARGAS, the National Secretary of Transportation, the Buenos Aires city Government and the Secretary of Transportation of the Province of Buenos Aires on providing valuable information on fuel and gas consumptions and traffic flow patterns. The support from the National Scientific and Technological Research Council of Argentina (CONICET) is also acknowledged

REFERENCES

1.   M. F. Andrade, R. M. Miranda, A. Fornaro, A. Kerr, B. Oyama, P. A. Andre, P. Saldiva, 2010 Vehicle emissions and PM2.5 mass concentrations in six Brazilian cities. Air Quality Atmosphere & Health. DOI 10.1007/s11869-010-0104-5.

2.   J. Ariztegui, J. Casanova, M. Valdes, 2004 A structured methodology to calculate traffic emissions inventories for city centres. The Science of the Total Environment, 334-335 , 101 109 .

3.   M. Arkouli, A. G. Ulke, W. Endlicher, G. Baumbach, E. Schultz, U. Vogt, M. Müller, L. Dawidowski, A. Faggi, U. Wolf-Benning, G. Scheffknecht, 2010 Distribution and temporal behaviour of particulate matter over the urban area of Buenos Aires, Atmospheric Pollution Research, 1 1 8 .

4.   S. P. Arya, 1999 Air Pollution Meteorology. Oxford University Press. New York.

5.   S. H. Azondèkon, J. M. Martel, 1999 "Value" of additional information in multi-criterion analysis under uncertainty. European Journal of Operational Research, 117 45 62 .

6.   D. J. Ball, S. W. Radcliffe, 1979 An inventory of sulfur dioxide emissions to London´s air. Research Report 23. Greater London Council, London.

7.   S. P. Beaton, G. A. Bishop, D. H. Stedman, 1992 Emission characteristics of Mexico City vehicles. Journal Air & Waste Management Association, 42 1424 1429 .

8.   R. Berkowicz, 2000 A simple model for urban background pollution. Environmental Monitoring and Assessment, 65 259 267 .

9.   B. Bocca, S. Caimi, P. Smichowski, D. Gómez, S. Cairoli, 2006 Monitoring Pt and Rh in urban aerosols from Buenos Aires, Argentina. The Science of the Total Environment, 358 255 264 .

10.   H. Bogo, R. M. Negri, Román. E. San, 1999 Continuous measurement of gaseous pollutants in Buenos Aires City. Atmospheric Environment, 33 2587 2598 .

11.   H. Bogo, D. R. Gómez, S. L. Reich, R. M. Negri, Román. E. San, 2001 Traffic pollution in downtown of Buenos Aires City. Atmospheric Environment, 35 1717 1727 .

12. H. Bogo, M. Otero, P. Castro, M. J. Ozafrán, A. Kreiner, E. J. Calvo, R. M. Negri, 2003 Study of atmospheric particulate matter in Buenos Aires city. Atmospheric Environment, 37 1135 1147 .

13. C. Borrego, O. Tchepel, A. M. Costa, J. H. Amorim, A. I. Miranda, 2003 Emission and dispersion modelling of Lisbon air quality at local scale. Atmospheric Environment, 37 5197 5205 .

14. T. M. Butler, M. G. Lawrence, B. R. Gurjar, J. van Aardenne, M. Schultz, J. Lelieveld, 2008 The representation of emissions from megacities in global emission inventories. Atmospheric Environment, 42 703 719 .

15. D. J. Carruthers, R. J. Holroyd, J. C. R. Hunt, W. S. Weng, A. G. Robins, D. D. Ashley, D. J. Thompson, F. B. Smith, 1994 UK-ADMS: a new approach to modelling dispersion in the earth's boundary layer. Journal of Wind Engineering, 52 139 153 .

16. CERC 2003 ADMS-Urban. An Urban Air Quality Management System. User Guide. Version 2.0. Cambridge Environmental Research Consultants Ltd., Cambridge.

17. A. D'Avignon, F. A. Carloni, E. L. L. Rovere, C. B. S. Dubeux, 2010 Emission inventory: An urban public policy instrument and benchmark. Energy Policy, 38 4838 4847 .

18. R. G. Derwent, D. R. Middleton, 1996 An empirical function for the ratio 2 NOx. Clean Air, 26 57 62 .

19. J. Dixon, D. R. Middleton, R. G. Derwent, 2001 Sensitivity of nitrogen dioxide concentrations to oxides of nitrogen controls in the United Kingdom. Atmospheric Environment, 35 3715 3728 .

20. D. Draper, 1995 Assessment and propagation of model uncertainty. Journal of Royal Statistical Society, 57 45 97 .

21. European Environment Agency. 2001 Joint EMEP/CORINAIR Atmospheric Emission Inventory Guidebook, Third Edition, Copenhagen.

22. L. A. Fagundez, V. L. Fernández, T. H. Marino, I. Martín, D. A. Persano, y. Rivarola, M. Benítez, I. V. Sadañiowski, Codnia, & Zalts A. 2001 Preliminary air pollution monitoring in San Miguel, Buenos Aires. Environmental Monitoring and Assessment, 71 61 70 .

23. H. C. Frey, J. Zheng, 2002 Quantification of variability and uncertainty in utility NOx emission inventories. Journal Air & Waste Management Association, 52 1083 1095 .

24. H. C. Frey, S. Li, 2003 Methods for quantifying variability and uncertainty in AP-42 emission factors: case studies for natural gas-fueled engines. Journal of Air & Waste Management Association, 53 1436 1447 .

25. GCBA 2006 Informes sobre Índice de Tránsito. (Período 2004-2006). Gobierno de la Ciudad de Buenos Aires. Buenos Aires. (in Spanish).

26. F. A. Gifford, 1970 Atmospheric Diffusion in an Urban Area, NOAA Research Lab. N° 33. Oak Ridge, N. C.

27. F. A. Gifford, S. R. Hanna, 1973 Modelling urban air pollution. Atmospheric Environment, 7 131 136 .

28. S. E. Gryning, A. A. M. Footslog, J. S. Irwin, B. Sivertsen, 1987 Applied dispersion modelling based on meteorological scaling parameters. Atmospheric Environment, 21 79 89 .

29. B. R. Gurjar, T. M. Butler, M. G. Lawrence, J. Lelieveld, 2008 Evaluation of emissions and air quality in megacities. Atmospheric Environment, 42 1593 1606 .

30. S. Hanna, R. Britter, P. Franzese, 2002 Simple screening models for urban dispersion. Proceedings of the 8th International Conference on Harmonisation within Atmospheric Dispersion Modelling for Regulatory Purposes, Sofia, Bulgaria, October 2002.

31. J. Harte, C. Holdren, R. Schneider, C. Shirley, 1991 Toxics A to Z. A guide to everyday pollution hazards. The Regents of the University of California. USA.

32. INDEC. 2010 Censo Nacional de Población, Hogares y Viviendas 2010: total del país, resultados provisionales. 1a edición. Buenos Aires. Instituto Nacional de Estadística y Censos. (in Spanish).

33. JICA-SAyDS, 2002 Estudio o línea de base de concentración de gases contaminantes en atmósfera en el área de Dock Sud en Argentina. Agencia de Cooperación Internacional del Japón en Argentina-Sec. de Desarrollo Sustentable y Política Ambiental. Informe Final. (in Spanish).

34. A. Karppinen, J. Kukkonen, T. Elolähde, M. Konttinen, T. Koskentalo, E. Rantakrans, 2000 A modelling system for predicting urban air pollution: model description and applications in the Helsinki metropolitan area. Atmospheric Environment, 34 3723 3733

35. Y. J. Kim, 1996 Preparation of Emissions Inventories and Establishment of the National Emission Inventory System of Air Pollutants in Korea. Proceedings of the Conference on the Emissions Inventory: Programs & Progress, Research Triangle Park, NC, Pittsburg, June 1996, 683 686 .

36. N. A. Mazzeo, L. E. Venegas, 1991 Air pollution model for an urban area. Atmospheric Research, 26 165 179 .

37. N. A. Mazzeo, L. E. Venegas, 2002 Estimation of cumulative frequency distribution for carbon monoxide concentration from wind-speed data in Buenos Aires (Argentina). Water, Air and Soil Pollution, Focus, 2 419

432 .

38. N. A. Mazzeo, L. E. Venegas, 2003 Carbon monoxide and nitrogen oxides emission inventory for Buenos Aires City (Argentina). Proceedings of the 4th International Conference on Urban Air Quality- Measurement, Modelling and Management, Prague, Czech Republic, March 2003. 159 162 .

39. N. A. Mazzeo, L. E. Venegas, 2004 Some aspects of air pollution in Buenos Aires City (Argentina). International Journal of Environment and Pollution, 22 365 378 .

40. N. A. Mazzeo, L. E. Venegas, 2008 Design of an air quality surveillance system for Buenos Aires city integrated by a NOx monitoring network and atmospheric dispersion models. Environmental Modelling & Assessment, 13 349 356 .

41. N. A. Mazzeo, L. E. Venegas, 2010 Chapter 2: Development and application of a methodology for designing a multi-objective and multi-pollutant air quality monitoring network for urban areas. In: Air Quality, A. Kumar (Ed.), 23 47 , Sciyo, Rijeka, Croatia. www.sciyo.com

42. N. A. Mazzeo, L. E. Venegas, H. Choren, 2005 Analysis of NO2 O3 and NOx concentrations measured at a green area of Buenos Aires City during wintertime, Atmospheric Environment, 39 3055 3068 .

43. N. A. Mazzeo, Rojas. A. L. Pineda, L. E. Venegas, 2010 Carbon monoxide emitted from the city of Buenos Aires and transported to neighbouring districts, International Journal of Latin American Applied Research, 40 267 273 .

44. C. A. Mc Hugh, D. J. Carruthers, H. A. Edmunds, 1997 ADMS-Urban: An air quality management system for traffic, domestic and industrial pollution. International Journal of Environment & Pollution, 8 437 440 .

45. D. R. Middleton, A. R. Jones, A. L. Redington, D. J. Thomson, R. S. Sokhi, L. Luhana, B. E. A. Fisher, 2008 Lagrangian modelling of plume chemistry for secondary pollutants in large industrial plumes. Atmospheric Environment, 42 415 427 .

46. C. A. Miller, G. Hidy, J. Hales, C. E. Kolb, A. S. Werner, B. Haneke, D. Parrish, H. C. Frey, L. Rojas-Bracho, M. Deslauriers, B. Pennell, J. D. Mobley, 2006 Air emission inventories in North America: a critical Assessment. Journal Air & Waste Management Association. 56 1115 1129 .

47. M. Mohan, L. Dagar, B. R. Gurjar, 2007 Preparation and Validation of Gridded Emission Inventory of Criteria Air Pollutants and Identification

of Emission Hotspots for Megacity Delhi. Environmental Monitoring & Assessment, 130 323 339 .

48. R. E. Morris, T. C. Myers, 1990 User's Guide to the Urban Airshed Model, Vol. I-V. U.S. Environmental Protection Agency, Research Triangle Park, NC.

49. Y. Nishikawa, A. Kannari, 2010 Atmospheric concentration of ammonia, nitrogen dioxide, nitric acid and sulphur dioxide by passive method within Osaka Prefecture and their emission inventory. Water, Air & Soil Pollution. 215 229 237 .

50. H. R. Olesen, 1995 Regulatory dispersion modelling in Denmark. International Journal of Environment and Pollution, 5 412 417 .

51. F. Pasquill, F. B. Smith, 1983 Atmospheric Diffusion, John Wiley & Sons, New York.

52. Rojas. A. L. Pineda, L. E. Venegas, N. A. Mazzeo, 2007 Emission inventory of carbon monoxide and nitrogen oxides for area sources at Buenos Aires Metropolitan Area (Argentina). Proceedings of the 6th International Conference on Urban Air Quality, Limassol, Cyprus, March 2007.

53. Rojas. A. L. Pineda, L. E. Venegas, 2008 Dry and wet deposition of nitrogen emitted in Buenos Aires city to waters of de la Plata river. PinedaRojas. A. L.VenegasL. E. (2008). Dry and wet deposition of nitrogen emitted in Buenos Aires city to waters of de la Plata river. Water, Air and Soil Pollution, Vol. 193, pp. 175-188. , Air and Soil Pollution, 193 175 188 .

54. Rojas. A. L. Pineda, L. E. Venegas, 2009 Atmospheric deposition of nitrogen emitted in the Metropolitan Area of Buenos Aires to coastal waters of de La Plata River, Atmospheric Environment, 43 1339 1348 .

55. Rojas. A. L. Pineda, L. E. Venegas, 2010 Interannual variability of estimated monthly nitrogen deposition to coastal waters due to variations of atmospheric variables model input, Atmospheric Research, 96 88 102 .

56. H. M. Regan, H. R. Akcakaya, S. Ferson, K. V. Root, S. Carroll, L. R. Ginzburg, 2003 Treatments of uncertainty and variability in ecological risk assessment of single-species populations. Human and Ecological Risk Assessment, 9 4 12 .

57. G. Rideout, D. Gourley, J. Walker, 2005 Measurement of in-service vehicle emissions in Sao Paulo, Santiago and Buenos Aires. ARPEL Environmental Report #25. Otawa. ESAA. Canada.

58. D. Romano, D. Gaudioso, R. De Lauretis, 1999 Aircraft emissions: a comparison of methodologies based on different data availability.

Environmental Monitoring & Assessment, 56 51 74 .

59. D. Romano, A. Bernetti, R. De Lauretis, 2004 Different methodologies to quantify uncertainties of air emissions. Environmental International, 30 1099 1107 .

60. S. Saija, D. Romano, 2002 A methodology for estimation of road transport air emissions in urban areas of Italy. Atmospheric Environment, 36 5377 5383 .

61. J. Sallés, J. Janischewski, A. Jaecker-Voirol, B. Martin, 1996 Mobile source emission inventory model. Application to Paris area. Atmospheric Environment, 30 1965 1975 .

62. M. Seika, N. Metz, R. M. Harrison, 1996 Characteristics of urban and state emissions inventories- a comparison of examples from Europe and the United States. The Science of the Total Environment, 189 190, 221 234 .

63. P. J. Sturm, Ch. Sudy, R. A. Almbauer, J. Meinhart, 1999 Updated urban emission inventory with a high resolution in time and space for the city of Graz. The Science of the Total Environment, 235 111 118 .

64. P. J. Sturm, 2003 Air Pollutants Emissions in Cities. In Moussiopoulos N. (Ed). Air Quality in Cities. Saturn. EUROTRAC-2. Subproject Final Report. Springer.

65. G. Tsilingiridis, T. Zachariadis, Z. Samaras, 2002 Spatial and temporal characteristics of air pollutant emissions in Thessaloniki, Greece: investigation of emission abatement measures. The Science of the Total Environment, 300 99 113 .

66. US.EPA. 1995 Compilation of Air Pollution Emission Factors, AP-42, 5th ed., United States Environmental Protection Agency, Office of Air Quality Planning and Standards, Research Triangle Park, NC.

67. L. E. Venegas, N. A. Mazzeo, 1999 Atmospheric stagnation, recirculation and ventilation potential of several sites in Argentine, Atmospheric Research, 52 43 57 .

68. L. E. Venegas, N. A. Mazzeo, 2000 Carbon monoxide concentrations in a street canyon at Buenos Aires City (Argentina). Environmental Monitoring & Assessment, 65 417 424 .

69. L. E. Venegas, L. E. Mazzeo, 2002 An Evaluation of DAUMOD Model in Estimating Urban Background Concentrations. Water, Air and Soil Pollution: Focus, 2 5-6, 433 443

70. L. E. Venegas, N. A. Mazzeo, 2003 Air quality in an area of Buenos Aires City (Argentina), Proceedings of the III Congresso Interamericano de Qualidade do Ar, Canoas, Brasil, July 2003. (in Spanish).

71. L. E. Venegas, N. A. Mazzeo, 2005 Application of atmospheric dispersion models to evaluate population exposure to 2 concentration in Buenos Aires. International Journal of Environment and Pollution, 25 224 238 .

72. L. E. Venegas, N. A. Mazzeo, 2006a Air Quality Monitoring Network Design to Control PM10 in Buenos Aires. International Journal of Latin American Applied Research, 36 241 247 .

73. L. E. Venegas, N. A. Mazzeo, 2006b Modelling of urban background pollution in Buenos Aires city (Argentina). Environmental Modelling & Software, 21 577 586 .

74. L. E. Venegas, N. A. Mazzeo, 2007 Influence of surrounding areas and wind on air quality of Buenos Aires City. Proceedings of the 11th International Conference on Harmonisation within Atmospheric Dispersion Modelling for Regulatory Purposes, 2 Cambridge, UK, July 2007, 327 331 .

75. L. E. Venegas, N. A. Mazzeo, 2010a An ambient air quality monitoring network for Buenos Aires city. International Journal of Environment and Pollution, 40 184 194 .

76. L. E. Venegas, N. A. Mazzeo, 2010b Air quality at different sites in the city of Buenos Aires. Proceedings of the A&WMA International Specialty Conference. Leapfrogging Opportunities for Air Quality Improvement, Xi'an, China, May 2010, 175 180 .

77. U. Vogt, W. Endlicher, G. Baumbach, E. Schultz, L. Dawidowski, M. Arkouli, M. Müller, U. Wolf-Benning, G. Ulke, 2007 Air quality and urban climate investigations in the megacity of Buenos Aires. Proceedings of the 6th International Conference on Urban Air Quality, Emissions Measurements and Modelling, Limassol, Cyprus, March 2007.

78. H. Wang, L. Fu, Y. Zhou, X. Du, W. Ge, 2010 Trends in vehicular emissions in China's mega cities from 1995 to 2005. Environmental Pollution, 158 394 400 .

79. WHO. 2006a Air quality guidelines. Global update 2005. World Health Organization.

80. WHO. 2006b WHO Air quality guidelines for particulate matter, ozone, nitrogen dioxide and sulfur dioxide. Global update 2005. World Health Organization. WHO/SDE/PHE/OEH/06.02. Geneve. 20pp

81. E. Zárate, L. C. Belalcázar, A. Clappier, V. Manzi, Bergh. H. Van den, 2007 Air quality modelling over Bogota, Colombia: Combined techniques to estimate and evaluate emission inventories. Atmospheric Environment, 41 6302 6318 .

Chapter 5

SPATIAL ANALYSIS ON THE CONCENTRATIONS OF AIR POLLUTANTS IN BASRA PROVINCE (SOUTHERN IRAQ)

Shukri I. Al-Hassen[1], Abdul Wahab A. Sultan[2], Adnan A. Ateek[2], Hamid T. Al-Saad[3], Salah Mahdi[3], Abdulzahra A. Alhello[3]

[1]Department of Geography, University of Basra, Basra, Iraq

[2]Technical College, Southern Technical University, Basra, Iraq

[3]Department of Environmental Chemistry, University of Basra, Basra, Iraq

ABSTRACT

This paper aims to analyze the geographic distribution of air pollutant concentrations in Basra Province, Southern Iraq, and to cartographically determine the spatial variation of air pollution levels as well as to recognize the hottest spots of air pollution within the study area, and conclude that the levels of air pollution in the study area are spatially varied, with an irregular spatial pattern and some hotspots.

INTRODUCTION

Air pollution may be defined as any atmospheric condition in which certain substances are present in such concentrations that they can produce harmful effects on man and his environment. An air pollutant, however, is any gas or substance (such as SO_x, NO_x, CO, and HCs) or particulate matter (such as smoke, dust, fumes, and aerosols) that leads to ambient air contamination. A pollutant may originate from natural or anthropogenic sources, or both. Pollutants occur throughout much of the troposphere; however, pollution close to the earth's surface within the boundary layer is of most concern because of the relatively high concentrations resulting from sources at the surface [1] [2] .

Air pollutant concentrations depend mainly on the total mass of pollution emitted into the atmosphere, together with the atmospheric conditions that affect its fate and transport. Obviously, air pollution has many and varied sources, including cars, smokestacks, and other industrial inputs into the atmosphere as well as wind erosion of soil. Large emissions from both anthropogenic and natural sources over long periods enhance concentrations, as do the chemical and physical properties of these pollutants. For example, when nitrogen oxides and hydrocarbons in car exhaust are emitted into warm, sunlit air, they readily form ozone molecules (O_3). Similarly, the solubility of a pollutant affects how efficiently it is removed by rainfall. In addition, atmospheric conditions have a major effect upon pollutants once these pollutants are emitted into (e.g., nitrogen oxides from car exhaust) or formed within (e.g., O_3) the atmosphere. Pollution dispersal is controlled by atmospheric motion, which is affected by wind, stability, and the vertical temperature variation within the boundary layer. Stability, in turn, influences both air turbulence and the depth at which mixing of polluted air takes place ([2] ; see also [3] -[7]).

In the study area (Basra province), located in southern Iraq, air pollution is of major public concern, which is currently the object of extensive scientific research. Studies of Al-Asadi [8] , Al-Mayahi [9] , Al-Imarah et al. [10] , Al-Saad et al. [11] , Garabedian [12] , Al-Hassen [13] , Qassim [14] , Douabul et al. [15] , Sultan et al. [16] , Karmalla et al. [17] , Abdullah and Hussien [18] , and Al-Hassen et al. [19] , are an example of some local-scale research in this respect. The study area is rich in petroleum and many existing industrial and human activities as the main sources of gaseous emissions around it (Figure 1). These are emission sources to contaminate the ambient air in Basra. According to above mentioned studies, Basra was recorded, in the last two decades, high quantities of gaseous emissions causing elevated levels of outdoor air pollution, and that concentration of some air pollutants was at risk to the public health. One of the most important driving forces is atmospheric conditions in this region.

(a)

(b)

(c)

(d)

Figure 1: Photography of some gaseous emission sources in the study area. (a) Petrochemical plant; (b) Najybia power station; (c) Emissions of petroleum exploitation near Burchesya; (d) Iranian Abadan refinery offshore seeba.

The previous studies, however, are focused on values of gaseous pollutant concentration and its implications on the human health in the study area; thus the objective of the present study is to spatially analyze the geographic distribution of air pollutant concentrations in Basra, and to cartographically determine the spatial variation of air pollution levels on the map of this region. The major aim is to recognize the hottest spot of air pollution within whole areas of Basra province. Basra has a hot desert climate (Köppen climate classification BWh), like the rest of the surrounding region, though it receives slightly more precipitation than inland locations due to its location near the coast. During the summer months, from June to August, Basra is consistently one of the hottest cities on the planet, with temperatures regularly exceeding 40°C (104°F) and approaching 45°C (113°F) in July. In winter Basra experiences mild weather with average high temperatures around 20°C (68°F). On some winter nights, minimum temperatures are below 0°C (32°F). High humidity—sometimes exceeding 90%—is common due to the proximity to the marshy Persian Gulf [20].

MATERIALS AND METHODS

As shown in Figure 2, seventeen sampling stations were chosen in the study area. The selected stations divided to cover the eastern and western region within the study area, and the geographic distribution of sampling stations was taken in consideration in the vicinity of human settlements and involved a variety of local environments (see Table 1).

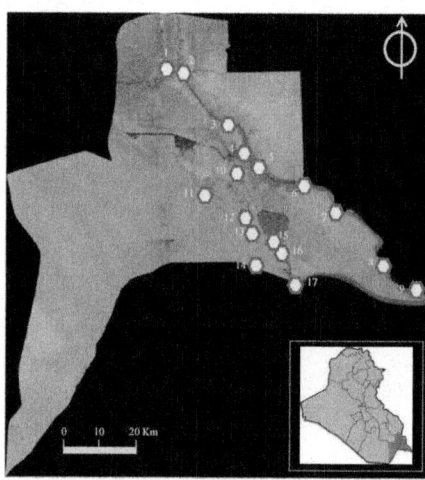

Figure 2: Map of the study area (Basra Province, Southern Iraq), showing the selected sampling stations. Note: Numbers of sampling stations correspondences their names listed in Table 1 and Table 2.

Table 1: Type of environment in the selected sampling stations, and the major sources of gaseous emissions in the study area

No.	Sampling station	Type of environment	Major source of air pollution
1	Midaina	Agricultural/industrial/urban	Petroleum exploitation and urban contaminates
2	Qurna	Agricultural/industrial/urban	Petroleum exploitation and urban contaminates
3	Dayer	Agricultural/industrial/urban	Petroleum exploitation and urban contaminates
4	Garmatt Ali	Agricultural/urban	Power station
5	Ashar	Urban	Urban contaminates
6	Abo Khaseeb	Agricultural/urban	Urban contaminates
7	Seeba	Agricultural/industrial	Petroleum industry and exploitation
8	Faw	Agricultural/urban	Urban contaminates
9	Ras Absha	Marine	Natural contaminates
10	Shatt Al-Basra	Riverine	Natural and urban contaminates
11	Burchesya	Industrial	Petroleum industry and exploitation
12	Petrochemical plant area	Industrial	Petroleum industry
13	South gas plant area	Industrial	Petroleum industry
14	Safwan	Urban	Urban contaminates
15	Khor Al-Zubayer	Industrial/urban	Industrial and urban contaminates
16	Gas terminal	Industrial/marine	Petroleum industry and shipping
17	Umm Qasr	Urban/marine	Urban contaminates

The selected stations were in order to monitor the concentrations of gaseous pollutants released into ambient air of the study area. A variety of gaseous pollutants such as carbon monoxide (CO), carbon dioxide (CO_2), sulfate oxides (SO_x), nitrogen oxides (NO_x), ozone (O_3), petroleum hydrocarbons (HCs), methane (CH_4), hydrogen sulfide (H_2S), and formaldehyde (HCHO), measured in this work. Concentrations of HCs, NO_x, SO_x, HCHO, O_3, and CO_2 measured using the portable detection instrument of Drager Chip-Measurement System, Germany, whereas the portable instrument of RK1 Gas Monitoring Eagle II, USA, detected the pollutants of CH_4, H_2S, and CO.

Fieldwork carried out during a one daytime of spring 2015, and all measurements were done at the same time in purpose of obtaining on readings approaching to reality. Therefore, the working team divided into two groups, one of them work in the eastern region (Midaina, Qurna, Dayer, Garmatt Ali, Ashar, Abo Khaseeb, Seeba, Faw, and Ras Absha), while the other group work in the western region (Shatt Al-Basra, Burchesya, Petrochemical plant area, South Gas Plant area, South Gas Plant area, Safwan, Khor Al-Zubayer, Gas terminal, Umm Qasr). The procedure of measuring was as described by Douabul et al. [15] .

RESULTS AND DISCUSSION

Table 2 lists the obtained results for concentrations of some air pollutants in the given study area and period. These concentrations recorded from the direct readings displayed on the screen of both employed detectors. The values were adjusted in a statistic form to make a more geographic mode. Thus, the values have been graphically and cartographically represented in Figure 3 and Figure 4, respectively. In this study, data analysis and explanation will conducted in the terms of each given element with the emphasis on spatial variation and geographic distribution of air pollution, as follows:

1)   Carbon monoxide (CO) is a colorless, odorless, and tasteless gas that is slightly less dense than air. It is toxic to humans when encountered in concentrations above about 35 ppm. In the atmosphere, it is spatially variable and short lived, having a role in the formation of ground-level ozone [21] .

Table 2: Concentrations of air pollutants measured in the study area based on the selected sampling stations, 2015

No.	Sampling station	CO ppm	CO_2 ppm	NO_x ppm	SO_x ppm	H_2S ppm	HCs ppm	CH_4 ppm	HCHO ppm	O_3 ppb	Index*
1	Midaina	4.16	215.12	0.72	0.54	1.53	1.62	8.52	0.42	0.02	25.85
2	Qurna	6.25	286.45	0.83	0.64	1.82	1.93	8.86	0.35	0.03	34.12
3	Dayer	8.92	260.21	0.86	0.63	1.92	3.25	10.21	0.46	0.02	31.83
4	Garmatt Ali	10.23	280.38	0.95	1.25	1.98	5.28	9.52	0.63	0.06	34.47
5	Ashar	12.32	225.32	0.65	0.92	1.21	12.21	13.28	0.92	0.12	29.66
6	Abo Khaseeb	20.63	250.12	0.83	1.28	1.26	24.28	14.25	1.23	0.14	34.89
7	Seeba	30.23	280.11	1.45	2.28	2.68	31.23	25.68	1.86	0.23	41.63
8	Faw	10.24	240.61	0.79	1.68	2.06	22.52	13.34	0.66	0.06	32.44
9	Ras Absha	2.52	180.32	0.42	0.43	1.25	1.74	5.54	0.24	0.01	21.38
10	Shatt Al-Basra	10.68	210.11	0.65	0.72	1.12	7.25	10.11	0.72	0.04	26.82
11	Burchesya	40.23	310.27	4.25	10.23	6.24	30.21	22.65	1.52	0.12	69.30
12	Petrochemical plant area	16.23	200.10	1.31	1.65	1.32	10.53	13.21	0.31	0.06	27.19
13	South gas plant area	20.53	220.31	1.45	3.21	3.1	18.23	10.25	0.62	0.16	30.87
14	Safwan	18.22	210.53	0.93	1.24	2.5	10.82	9.34	0.52	0.13	28.24
15	Khor Al-Zubayer	14.28	226.3	1.23	2.52	1.8	11.23	9.93	0.68	0.09	29.78
16	Gas terminal	18.34	228.23	1.86	4.38	2.1	22.31	16.83	0.84	0.11	32.77
17	Umm Qasr	16.12	180.98	0.98	3.49	3.5	24.63	12.46	0.75	0.07	26.99
	Mean	**15.30**	**235.61**	**1.15**	**2.18**	**2.19**	**14.01**	**12.58**	**0.74**	**0.08**	**31.49**

Data based on Fieldwork. *Index means a sum of values for the selected parameters divided by its number. It may be expressed, in the other meaning, the intensity of pollution.

Data based on Fieldwork. *Index means a sum of values for the selected parameters divided by its number. It may be expressed, in the other meaning, the intensity of pollution.

As listed in Table 2 and Figure 3 & Figure 4(a), the Burchesya sampling station records the highest value of CO is 40.23 ppm, while the lowest is 2.52 ppm in Ras Abasha station. The mean concentration of CO is 15.30 ppm within the all sampling stations.

2)   Carbon dioxide (CO_2) is a colorless, odorless gas vital to life on Earth. Carbon dioxide exists in the Earth's atmosphere as a trace gas at a concentration of about 0.04 percent (400 ppm) by volume. It is present in deposits of petroleum oil and natural gas [22] .

Table 2 and Figure 3 & Figure 4(b) report that the maximum concentration of CO_2 is 310.27 ppm recorded in the Burchesya sampling station, while the minimum concentration is 180.32 ppm in the Ras Abasha station. In general, the sampling stations in the eastern region of the study area registers values higher than those that in the western stations. The mean concentration of CO_2 is 235.61 ppm.

3)   Nitrogen oxide (NO_x) is a prominent air pollutant; it may refer to a binary compound of oxygen and nitrogen, or a mixture of such compounds. This reddish-brown toxic gas has a characteristic sharp, biting odor and is a prominent air pollutant [23] .

Table 2 and Figure 3 & Figure 4(c) indicate that the highest concentration of NO_x registered in the study area was 40.23 at the Burchesya sampling station, this may be a record value to compare with given in the previous studies yet. The lowest concentration was 0.42 in the Ras Abasha station. The mean concentration of NO_x is 1.15 ppm.

4)   Sulfur oxide (SO_x) refers to many types of sulfur and oxygen containing compounds such as SO, SO_2, SO_3, S_7O_2, S_6O_2, S_2O_2, etc. SO_x is a toxic gas with a pungent, irritating, and rotten smell [24] .

In this study, as shown in Table 2 and Figure 3 & Figure 4(d), the maximum concentration of SO_x was 10.23 ppm in the Burchesya sampling station, while the minimum concentration is 0.42 ppm in the Ras Abasha station. The mean concentration of SO_x is 2.18 ppm.

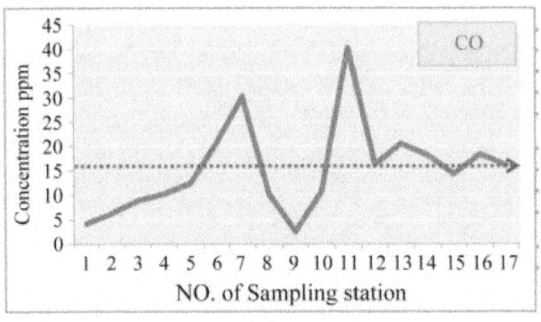

(a)

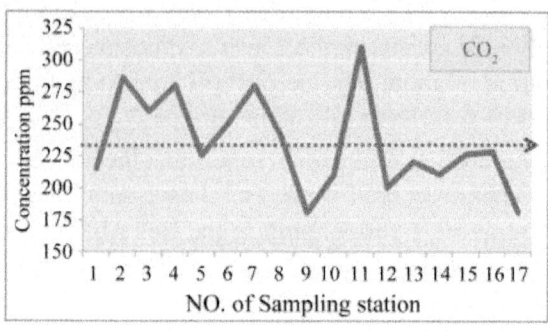

(b)

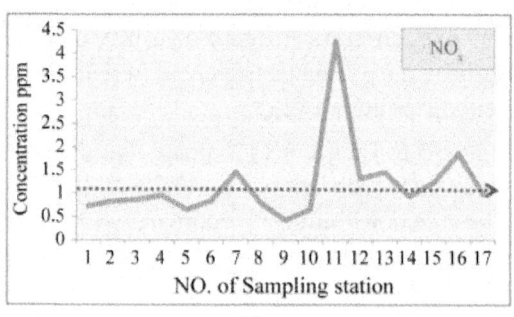

(c)

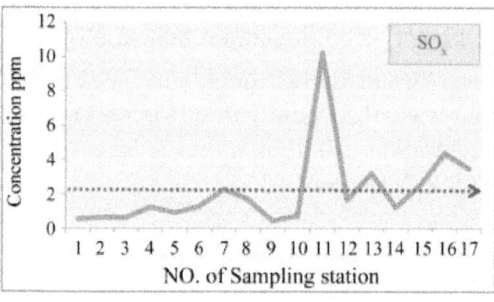

(d)

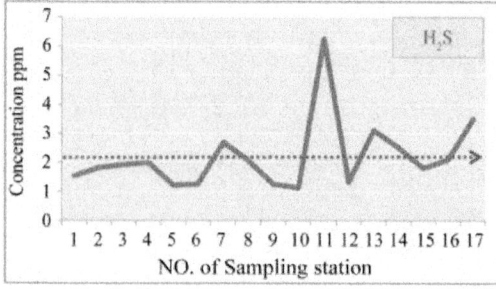

(e)

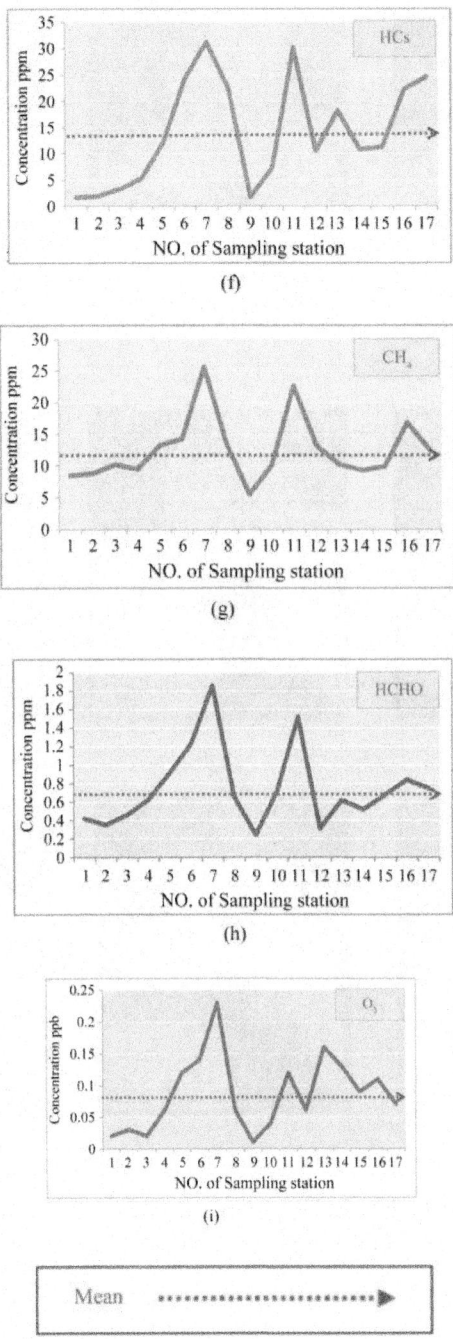

Figure 3: Graphic representation of the concentrations of air pollutants measured in the study area. Data based on Table 2.

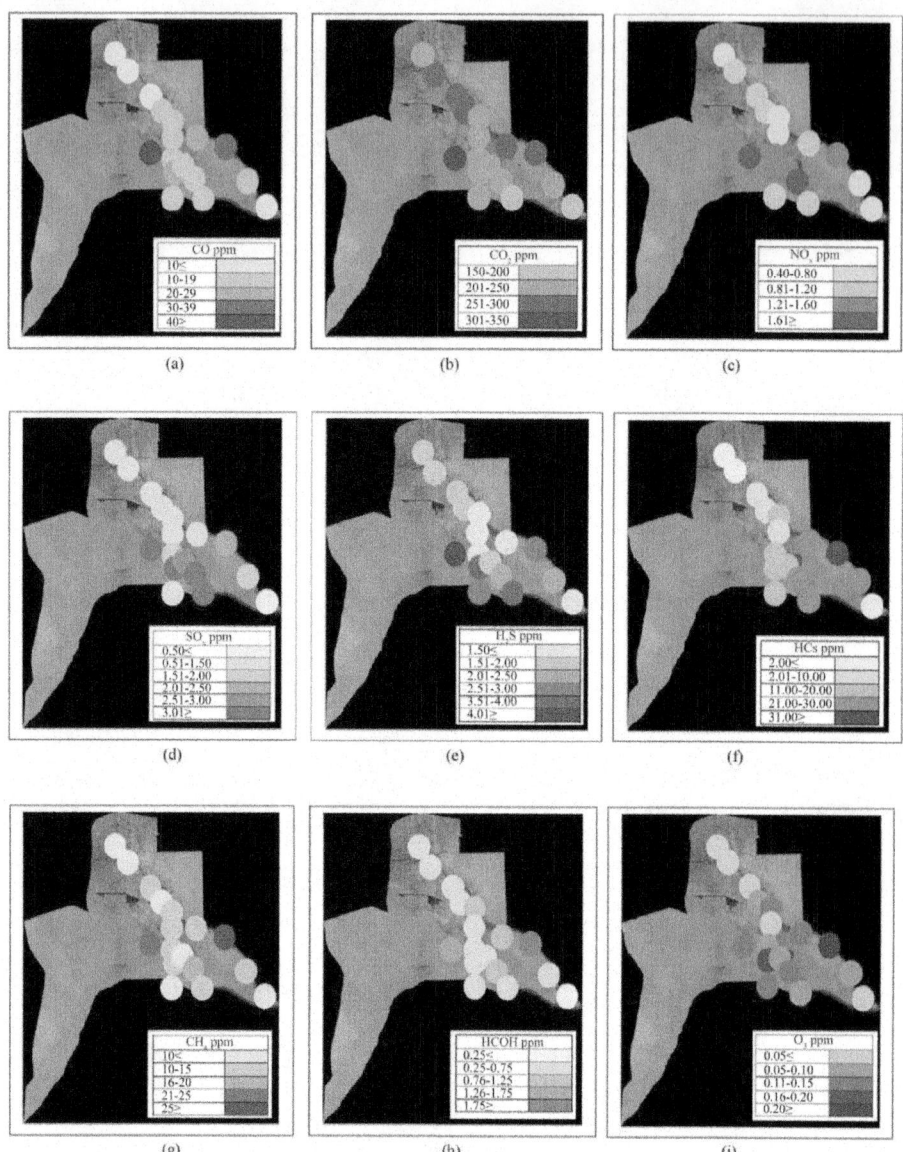

Figure 4: Cartographic representation of the concentrations of air pollutants measured in the study area. Data based on Table 1.

5)   Hydrogen sulfide is the chemical compound with the formula (H_2S). It is a colorless gas with the characteristic foul odor of rotten eggs; it is heavier than air, very poisonous, corrosive, flammable, and explosive [25].

Table 2 and Figure 3 & Figure 4(e) refer to the highest value of H_2S recorded in this study was 6.24 ppm in the Burchesya sampling station, while the lowest value was 1.21 ppm in the Ashar station. The mean concentration is 2.19 ppm.

6) Hydrocarbon (HCs) is an organic compound consisting entirely of hydrogen and carbon. Hydrocarbon poisoning such as that of benzene and petroleum usually occurs accidentally by inhalation or ingestion of these cytotoxic chemical compounds [22] .

   HCs gas concentrations in the study area were high, in general. The maximum concentration is 31.23 ppm registered at the Seeba sampling station, whereas the minimum concentration is 1.62 in the Midaina station. The mean concentration of HCs is 14.01 ppm, as shown in Table 2 andFigure 3 & Figure 4(f).

7) Methane is a chemical compound with the chemical formula (CH_4). It is the simplest alkane and the main component of natural gas. Methane is not toxic, yet it is extremely flammable and may form explosive mixtures with air [26] .

   The highest values of CH_4, as indicated in Table 2 and Figure 3 & Figure 4(g), were concentered in places with the petroleum industry and exploitation within the study area. The maximum concentration, however, is 25.68 ppm in the Seeba station, while the minimum concentration is 5.54 ppm in the Ras Abasha station. The mean concentration of CH_4 is 12.58 ppm.

8) Formaldehyde (HCHO) is a colorless, highly toxic, and flammable gas at room temperature that is slightly heavier than air. It has a pungent, highly irritating odor that is detectable at low concentrations, but may not provide adequate warning of hazardous concentrations for sensitized persons [27] .

   Table 2 and Figure 3 & Figure 4(h) show that the concentration of HCHO records the maximum value is 1.86 ppm in the Seeba station, while the minimum value is 0.24 ppm recorded in the Ras Abasha station. The mean concentration of HCHO is 0.74 ppm.

9) Ozone is an inorganic molecule with the chemical formula (O_3). It is a pale blue gas with a distinctively pungent smell. This same high oxidizing potential, however, causes ozone to damage mucous and respiratory tissues in animals, and also tissues in plants, above concentrations of about 100 ppb. This makes ozone a potent respiratory hazard and pollutant near ground level [28] .

In the study area, as shown in Table 2 and Figure 3 & Figure 4(i), the concentrations of ground level ozone (O_3) were largely varied within the selected sampling stations. The highest value is 0.23 ppb in the Seeba station, whereas the Ras Abasha station register the lowest value is 0.01 ppb. Thus, the mean concentration of O_3 is 0.08 ppb.

In general, the concentrations of air pollutants registered in the study area were spatially varied, it seems somehow a random pattern of distribution in the terms of each pollutant (see Figure 4). However, an overall spatial pattern may be drawn by using the index of pollution showing in Figure 5 based on Table 2. This index is a result of summing all values of each element dividing by its numbers, the resulting value is an approximate indicator of the spatial concentration of a pollutant. To simplify explaining the causes of spatial variation in pollutant concentrations, the mentionedTable 1 lists the major gaseous emission sources affecting air quality in the study area.

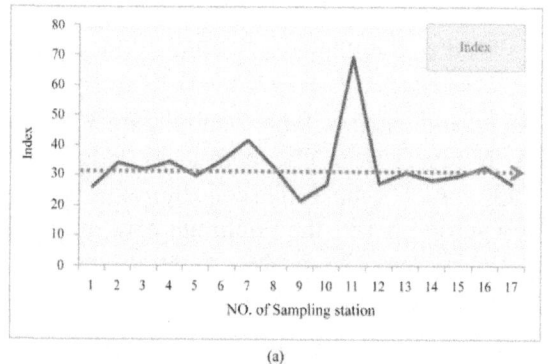
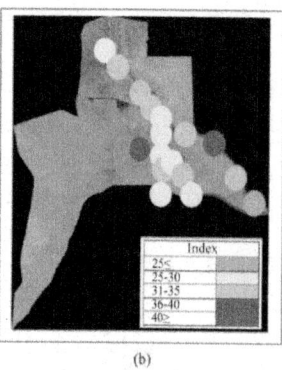

(a)   (b)

Figure 5: Geographical distribution of the levels of air pollution in the study area according to index of pollution. Data based on Table 2. Data based on Table 1.

As shown in Figure 5(a) & Figure 5(b), the elevated levels of air pollution focuses on the western region of the study area, this is due to existing industrial pollution sources. The sampling stations of Qurna, Dayer, and Garmatt Ali also report a significant index values with 34.12, 31.83 and 34.47 respectively, this refers to the petroleum exploitation existing in these places as well as the high values of CO_2, which, in turn, increase the index value in a comparison with the other sampling stations.

It is clear that, however, the highest values of the air pollution index with 69.30 and 41.63 have been recorded in the sampling stations of Burchesya and Seeba, respectively. For the former, this is due to the gaseous pollutants emitted from Abadan refinery offshore this site, in addition to there are a newer exploitation of petroleum within it; for the latter, this increase related to the gaseous emissions from dense works of petroleum exploitation on this site. The sampling station of Ras Abasha was the less polluted with an index value of 21.38, because of its marine nature and relatively distant from the influence of anthropocentric pollution sources.

CONCLUSIONS

In accordance to the spatial analysis of the geographical distribution of air pollutants, the present study concluded that the levels of air pollution in Basra province (the study area) were spatially varied, and that this spatial variation consequently correlated with the geographical distribution of gaseous emission sources existing in the study area. However, there is an irregular spatial pattern of air pollution concentration, although of some hotspots such as Burchesya and Seeba.

The authors recommend that mitigation to air pollution sources, particularly from stationary emission sources in the study area, is urgent action. Moreover, the installing of fixed stations designated to the air pollution monitoring is so necessary. Nevertheless, the accurate assessment of air pollution status in the study area needs more research and monitoring.

ACKNOWLEDGEMENTS

The authors acknowledge the following labs for the technical assistants: Environmental Analysis and Research Lab in Department of Geography, College of Arts, University of Basra; Gas Chromatography Lab in Department of Environmental Chemistry, Marine Science Center, University of Basra, and Air Pollution Lab in Department of Environmental and Pollution Engineering, Technical College, Southern Technical University.

REFERENCES

1. Admassu, M. and Wubeshet, M. (2006) Air Pollution: Lecture Notes for Environmental Health Science Students. University of Gondar Publications, Ethiopia, 5-6. http://www.cartercenter.org/resources/pdfs/health/ephti/library/lecture_notes/   nv_health_science_students/AirPollution.pdf

2. Matthias, A.D., Comrie, A.C. and Musil, S.A. (2006) Atmospheric Pollution. In: Peeper, I.L., Gerba, C.P. and Brusseau, M.L., Eds., Environmental and Pollution Science, 2nd Edition, Elsevier, San Diego, 377-394.

3.   Vallero, D.A. (2008) Fundamentals of Air Pollution. 4th Edition, Elsevier Inc., London, 3.

4.   Spellman, F.R. (1999) The Science of Environmental Pollution. Taylor & Francis Routledge, Pennsylvania, 245.

5.   Harrison, R.M. (2001) Air Pollution: Sources, Concentrations and Measurements. In: Harrison, R.M., Ed., Pollution: Causes, Effects and Control, 4th Edition, RSC, Cambridge, 169-192. http://dx.doi.org/10.1039/9781847551719-00169

6.   Gittins, M.J. (1999) Air Pollution. In: Bassett, W.H., Ed., Clay's Handbook of Environmental Health, 18th Edition, E & FN Spon, London, 729-776. http://dx.doi.org/10.4324/9780203016312.ch42

7.   Christoforou, C. (2004) Air Pollution. In: Stapleton, R.M., Ed., Pollution A to Z, Vol. 1, Macmillan Reference, New York, 30-38.

8.   Al-Asadi, K.A.W. (1998) The Influence of Climatic Factors on the Major Industries in Basra and Their Reflections on the Environmental Pollution. Ph.D. Thesis, College of Arts, University of Basra, 197. (In Arabic)

9.   Al-Mayahi, I.K. (2005) An Environmental Analysis on the Factors Affecting the Air Pollutants Quality at Basra Province. M.A. Thesis, College of Education, University of Basra, 240. (In Arabic)

10.   Al-Imarah, F.J.M., Al-Mohameed, R.S.J. and Ibraheem, S.I. (2007) Extent of Atmospheric Pollution by Some Industrial Emissions Released from Petrochemicals and Gas Liquefier Industries in Khor Al-Zubair. Journal of Kerbala, Special Issue on the Annual Environment of Meeting of Babylon University, 1-6.

11.   Al-Saad, H.T., Al-Imarah, F.J.M., Hassan, W.F., Jasim, A.H. and Hassan, I.F. (2010) Determination of Some Trace Elements in the Fallen Dust on Basra Governorate. Basrah Journal of Science, 28, 243-252. http://www.iasj.net/iasj?func=fulltext&aId=55085

12.   Garabedian, S.A.K. (2010) Study of the Main Pollutants of Air Caused by Transportation in Basra. Proceedings of the Scientific Conference of Marine Science Center and the National Specialized Workshop on Oceanography, 23-24 December 2008, 125-136. (In Arabic)

13.   Al-Hassen, Sh.I. (2011) Environmental Pollution in Basra City. PhD Thesis, College of Arts, University of Basra, Basra, 232. (In Arabic)

14.   Qassim, M.H. (2011) A Geographic Analysis for Air Pollution Problem in Al-Zubayr City and Its Healthy Effects. M.A. Thesis, College of Arts, University of Basra, Basra, 180. (In Arabic)

15. Douabul, A.A.Z., Al-Maarofi, S.S., Al-Saad, H.T. and Al-Hassen, Sh.I. (2013) Gaseous Pollutants in Basra City, Iraq. Air, Soil and Water Research, 6, 15-21. http://dx.doi.org/10.4137/ASWR.S10835

16. Sultan, A.W.A., Al-Hassen, Sh.I., Ateeq, A.A. and Al-Saad, H.T. (2013) Ambient Air Quality in the Industrial Area of Khor Al-Zubayr, Southern Iraq. Journal of Petroleum Research & Studies, 4, 1-11. http://www.iasj. net/iasj?func=fulltext&aId=87876

17. Karmalla, H.A., Al-Hassen, Sh.I., Adam, R.S. and Qassim, M.H. (2013) A Cartographic Representation to Levels and Impacts of Carbon Monoxide Pollutant in Basra City, Southern Iraq. Journal of Thi-Qar Science, 4, 105-115. (In Arabic) http://www.iasj.net/iasj?func=fulltext&aId=83346

18. Abdullah, A.S. and Hussien, H.H. (2014) Estimation of Gaseous Pollutants emitted from Al-Najybia Power Station in Basra. Journal of Thi-Qar Science, 4, 68-71. http://www.iasj.net/iasj?func=fulltext&aId=92673

19. Al-Hassen, Sh.I., Al-Qarroni, E.H., Qassim, M.H., Al-Saad, H.T. and Alhello, A.A. (2015) An Experimental Study on the Determination of Air Pollutant Concentrations released from Selected Outdoor Gaseous Emission Sources in Basra City (Southern Iraq). JIARM, 3, 88-98. http:// www.jiarm.com/FEB2015/paper19552.pdf

20. Wikipedia-Basra. https://en.wikipedia.org/wiki/Basra

21. ATSDR (Agency for Toxic Substances and Disease Registry), Toxic Substances Portal-Carbon Monoxide. http://www.atsdr.cdc.gov/toxfaqs/ TF.asp?id=1163&tid=253

22. WHO (World Health Organization) (2010) WHO Guidelines for Indoor Air Quality: Selected Pollutants. WHO Regional Office for Europe, Bonn, 454. http://www.euro.who.int/__data/assets/pdf_file/0009/128169/ e94535.pdf

23. ATSDR (Agency for Toxic Substances and Disease Registry), Toxic Substances Portal-Nitrogen Oxide. http://www.atsdr.cdc.gov/toxfaqs/ TF.asp?id=396&tid=69

24. ATSDR (Agency for Toxic Substances and Disease Registry), Toxic Substances Portal-Sulfate Oxide. http://www.atsdr.cdc.gov/toxfaqs/ TF.asp?id=252&tid=46

25. ATSDR (Agency for Toxic Substances and Disease Registry), Toxic Substances Portal-Hydrogen Sulfide. http://www.atsdr.cdc.gov/ substances/toxsubstance.asp?toxid=67

26. ATSDR (Agency for Toxic Substances and Disease Registry), Toxic Substances Portal-Methane. http://www.atsdr.cdc.gov/HAC/landfill/ html/ch4.html

27. ATSDR (Agency for Toxic Substances and Disease Registry), Toxic Substances Portal-Formaldehyde. http://www.atsdr.cdc.gov/mmg/mmg.asp?id=216&tid=39

28. WHO (World Health Organization) (2005) WHO Air Quality Guidelines for Particulate Matter, Ozone, Nitrogen Dioxide and Sulfur Dioxide: Global Update. Geneva, 20. http://whqlibdoc.who.int/hq/2006/who_sde_phe_oeh_06.02_eng.pdf

Chapter 6

ESTIMATION OF CITYWIDE AIR POLLUTION IN BEIJING

Jin-Feng Wang[1], Mao-Gui Hu[1], Cheng-Dong Xu[1], George Christakos[2], Yu Zhao[1,3]

[1] State Key Laboratory of Resources and Environmental Information System, Institute of Geographic Sciences and Natural Resources Research, Chinese Academy of Sciences, Beijing, China,

[2] Department of Geography, San Diego State University, San Diego, California, United States of America,

[3] School of Geosciences and Info-Physics, Central South University, Changsha, China

ABSTRACT

There has been a discrepancy between the daily air quality reports of the Beijing municipal government, observations recorded at the U.S. Embassy in Beijing, and Beijing residents' perceptions of air quality. This study estimates Beijing's daily area $PM_{2.5}$ mass concentration by means of a novel technique SPA (Single Point Areal Estimation) that uses data from the single $PM_{2.5}$ observation station of the U.S Embassy and the 18 PM_{10} observation stations of the Beijing Municipal Environmental Protection Bureau. The proposed technique accounts for empirical relationships between different types of observations, and generates best linear unbiased pollution estimates (in a statistical sense). The technique extends the daily $PM_{2.5}$ mass concentrations obtained at a single station (U.S. Embassy) to a citywide scale using physical relations between pollutant concentrations at the embassy $PM_{2.5}$ monitoring station and at the 18 official PM_{10} stations that are evenly distributed across the city. Insight about the technique's spatial estimation accuracy (uncertainty) is gained by means of theoretical considerations and numerical validations involving real data. The technique was used to study citywide $PM_{2.5}$ pollution during the 423-day period of interest (May 10, 2010 to December 6, 2011). Finally, a freely downloadable software library is provided that performs all relevant calculations of pollution estimation.

INTRODUCTION

Beijing, the capital city of China, is an international metropolis with a population of over 19 million. As in many big cities worldwide, air pollution is a major concern for city residents. Particulate matter (PM) is the air pollutant that most commonly affects people's health, where PM_{10} and $PM_{2.5}$ are the two main PM pollutants, i.e., PM consisting of particles with aerodynamic diameters ≤ 10 μm and ≤ 2.5 μm, respectively [1], [2]. The sources of PM10 consist of smoke, dirt and dust from factories, farming and roads, as well as mold, spores, and pollen. $PM_{2.5}$ is linked to toxic organic compounds, heavy metals (from smelting, processing, and others), burning of plant material, and forest fires.

$PM_{2.5}$ is a greater health threat than the PM_{10} particles. Laboratory studies have confirmed that the smaller the particle, the more likely it is to lodge in the lungs [3]. In situ studies have shown that these small particles can penetrate indoors, thus altering the home environment. The particles may cause an increase in cardiac and respiratory morbidity and mortality [4]. Indeed, significant increases in deaths from heart and lung disease occur during multi-day periods with high concentrations of fine particles [5]. More than 500,000 deaths per year have been reported worldwide due to $PM_{2.5}$ pollution [6].

In the case of Beijing, there is considerable discrepancy between air pollution levels in terms of PM_{10} records provided by the municipal government, $PM_{2.5}$ observations from individual unofficial stations, and perceptions among the local population. Rapid population growth, urbanization, and greater numbers of vehicles have inevitably caused a considerable increase in air pollution emissions throughout the city [7]–[12]. PM_{10} concentration is a mandatory air quality index that is routinely observed at several official PM_{10} monitoring stations and published daily by the Beijing Municipal Environmental Protection Bureau (BJ-EPB). The U.S. Embassy in Beijing has kept unofficial hourly $PM_{2.5}$ records since spring 2008, using a single monitoring station atop its building [13]. On the other hand, according to BJ-EPB the official stations monitoring Beijing's air quality are evenly distributed across the city in accordance with relevant scientific standards, whereas the U.S. Embassy data do not accurately represent the overall pollution level in the city [14]. As a result, in the last few years a serious disagreement has emerged between the daily air pollution assessments provided by the BJ-EPB [15], the U.S. Embassy, and those based on population's perceptions. For example, on October 23, 2011, a thick smog blanket over Beijing revealed a major discrepancy between the categorizations of "slightly polluted" air suggested by BJ-EPB data and "hazardous" air quality determined by U.S. Embassy monitoring [13], [16].

PM_{10} and $PM_{2.5}$ concentrations are related, since most of the PM_{10} is contributed by $PM_{2.5[17]}$–[19]. Therefore, evaluating the PM_{10}-$PM_{2.5}$ relationship can provides information on $PM_{2.5}$ concentrations in areas that are not monitored for it [20], [21]. In this study, we proposed a technique to estimate daily averages of $PM_{2.5}$ concentrations in Beijing, by integrating daily $PM_{2.5}$ observations at the single U.S. Embassy station and their physical correlations with PM_{10} data obtained at a spatially exhaustive monitoring network operated by BJ-EPB. The proposed technique, called SPA (Single Point Areal Estimation), takes advantage of the aforementioned physical link between $PM_{2.5}$ and PM_{10} concentrations to generate areal $PM_{2.5}$ pollution estimates over the entire city. In other words, the PM_{10} observations served as the key secondary information that can improve the estimation of Beijing's areal daily $PM_{2.5}$ concentration [22].

MATERIALS AND METHODS

MATERIALS

Daily PM_{10} concentration data were collected from May 10, 2010 to December 6, 2011 at the 18 authorized (BJ-EPB) observation stations, which are evenly distributed across the city. Daily $PM_{2.5}$ concentrations reported by the embassy monitoring station were also gathered for the same period. Days with long periods of missing $PM_{2.5}$ (hourly) data were discarded based on the following criterion: if during a day there were consecutive data gaps of more than 3 hours or the cumulative amount of missing data exceeded 12 hours, that day was not included in pollution estimation. The final result was a dataset covering a 423-day period. We also acquired information about the geographic locations of the U.S. Embassy and 18 BJ-EPB stations, as well as data on population density, main traffic routes, traffic flow volumes, daily mean wind direction and speed, and geomorphology. All data were stored in a Geographic Information System (GIS), and are represented in Figure 1.

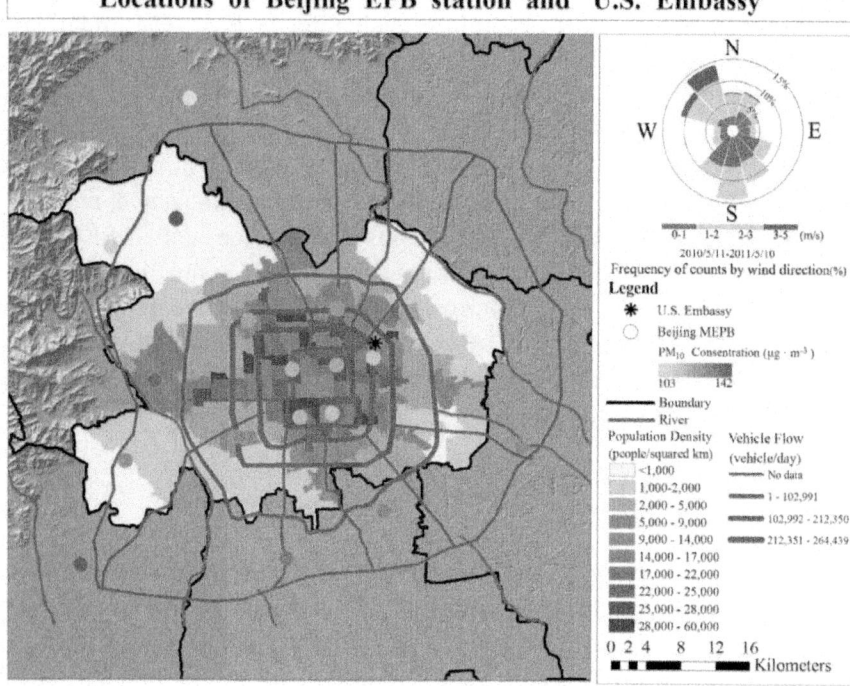

Figure 1: Location of BJ-EPB PM_{10} monitoring stations and U.S Embassy $PM_{2.5}$ station (Beijing, China).

The SPA Technique

We developed a technique, called Single Point Areal Estimation (SPA), which belongs to the category of biased areal estimation techniques [23]. SPA was used to extend the temporal $PM_{2.5}$ data recorded at a single (U.S. Embassy) monitoring station to areal-average $PM_{2.5}$ pollutant estimates, taking advantage of physical correlations between the $PM_{2.5}$ mass concentrations (U.S. Embassy station) and the PM_{10} data (18-station BJ-EPB network). This point-to-area transformation yields best linear unbiased estimates (BLUE) of $PM_{2.5}$ spatial averages over the entire city of Beijing. A formal derivation of the SPA technique is given in the following.

The objective of the SPA technique is to estimate citywide $PM_{2.5}$ pollution in the Beijing area. The estimate is based on $PM_{2.5}$ data from a single monitoring station at the U.S. Embassy in Beijing, and PM_{10} concentrations observations obtained at the official BJ-EPB monitoring network. Figure 2 outlines the SPA method.

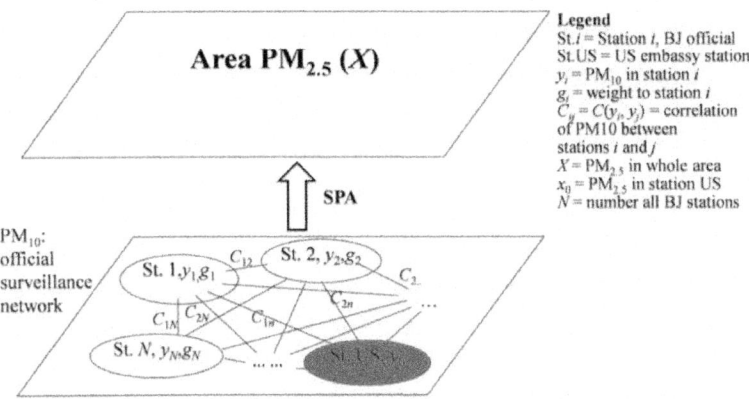

Figure 2: Relationship between stations and $PM_{2.5}$ areal concentration: y_i denotes PM_{10} concentration reported by station i, and X is areal $PM_{2.5}$ concentration for Beijing; St. US denotes the U.S. Embassy station at which daily $PM_{2.5}$ concentration x_0 is observed; X is estimated by x_0 using the SPA technique, based on observed $PM_{2.5}$ data at the embassy station, and their correlation with PM_{10} concentrations observed at the 18 (evenly distributed) stations operated by BJ-EPB.

The true average $PM_{2.5}$ concentration (X) over the entire area per time unit (e.g., daily) is calculated in theory by

$$X = \sum_{i=1}^{N} g_i x_i,$$

(1)

where x_i ($i=1, \ldots, 18$) denotes $PM_{2.5}$ concentration at the i-th station (which, in the present study, was not available from the official surveillance network); N denotes the total number of observation stations (18 in this case); g_i denotes the weight (contribution) of the i-th observation station to $PM_{2.5}$ estimation so that $\sum_{i=1}^{N} g_i = 1$ (unbiased estimation). There is only one $PM_{2.5}$ monitoring station (U.S. Embassy). Accordingly, the areal $PM_{2.5}$ concentration for Beijing is estimated by

$$\hat{X} = w_0 x_0,$$

(2)

where x0 denotes hourly PM2.5 concentration at the single monitoring station, as reported by the embassy and made available via the web site Twitter.com; w0 denotes the weight assigned to the embassy PM2.5 observation. This weight is estimated by minimizing

$$w_0 = \text{argmin}\left[v\hat{X} = E(w_0 x_0 - X)^2 \right],$$

(3)

where vX is the variance of the estimated area-averaged X (=PM2.5 concentration); and the $E(\cdot)$ denotes statistical mean.

At the same time, it is valid that

$$E\hat{X} = E(w_0 x_0),$$
(4)

i.e., the SPA technique generates an unbiased pollutant estimate that is also the best (in the minimum mean squared estimation error sense).

Derivation of the SPA Equations

The variance of $\hat{X}$ is derived as

$$vX = E(w_0 x_0 - X)^2 = E[(w_0 x_0 - X) - E(w_0 x_0 - X)]^2$$
$$= C(w_0 x_0, w_0 x_0) - 2C(w_0 x_0, X) + C(X, X),$$
(5)

where $C(\cdot)$ is the covariance between concentrations at any pair of points (the covariance provides a quantitative assessment of the spatial dependence between concentrations at these points).

The first term in Eq. (5) is

$$C(w_0 x_0, w_0 x_0) = w_0^2 C(x_0, x_0);$$
(6)

the second term is

$$2C(w_0 x_0, X) = 2w_0 C(x_0, \sum_{j=1}^{N} g_j x_j) = 2w_0 \sum_{j=1}^{N} g_j C(x_0, x_j),$$
(7)

and the third item is

$$C(X, X) = E(\sum_{j=1}^{N} g_j x_j - E\sum_{j=1}^{N} g_j x_j)^2$$
$$= E[\sum_{j=1}^{N} g_j(x_j - Ex_j)]^2$$
$$= \sum_{i=1}^{N} \sum_{j=1}^{N} g_i g_j C(x_j, x_j)$$
(8)

By substituting Eqs. (6)–(8) into Eq. (5), we obtain

$$v\hat{X} = w_0{}^2 C(x_0,x_0) - 2w_0 \sum_{j=1}^{N} g_j C(x_0,x_j)$$

$$+ \sum_{i=1}^{N} \sum_{j=1}^{N} g_i g_j C(x_j,x_j).$$

$$(9)$$

Taking into consideration the unbiased condition of Eq. (4), the Lagrange parameter μ is introduced into Eq. (9) in the following manner:

$$v\hat{X} = w_0^2 C(x_0,x_0) - 2w_0 \sum_{i=1}^{N} g_i C(x_0,x_i)$$

$$+ \sum_{i=1}^{N} \sum_{j=1}^{N} g_i g_j C(x_i,x_j) + 2\mu(\sum_{i=1}^{N} g_i - 1)$$

$$(10)$$

Minimization of Eq. (10) with respect to the g_i's, w_0 and μ is a standard optimization problem, leading to the system of equations (to be solved with respect to g_i, $i = 1,2,...,N$, w_0 and μ):

$$\begin{cases} \dfrac{\partial v\hat{X}}{\partial w_0} = w_0 C(x_0,x_0) - g_i \sum_{j=1}^{N} g_j C(x_0,x_j) = 0 \\[2mm] \dfrac{\partial v\hat{X}}{\partial g_i} = -w_0 C(x_0,x_i) - g_i C(x_i,x_j) + \sum_{j \neq i}^{N} g_j C(x_i,x_j) + \mu = 0 \\[2mm] \dfrac{\partial v\hat{X}}{\partial \mu} = \sum_{i=1}^{N} g_i - 1 = 0 \end{cases}$$

$$(11)$$

This system of equations can be written in matrix notation as

$$\begin{bmatrix} C(x_0,x_0) & C(x_0,x_1) & C(x_0,x_2) & \cdots & C(x_0,x_N) & 0 \\ C(x_1,x_0) & C(x_1,x_1) & C(x_1,x_2) & \cdots & C(x_1,x_N) & 1 \\ C(x_2,x_0) & C(x_2,x_1) & C(x_2,x_2) & \cdots & C(x_2,x_N) & 1 \\ \vdots & \vdots & \vdots & \vdots & \vdots & \\ C(x_N,x_0) & C(x_N,x_1) & C(x_N,x_2) & \cdots & C(x_N,x_N) & 1 \\ 0 & 1 & 1 & \cdots & 1 & 0 \end{bmatrix}$$

$$\begin{bmatrix} -w_0 \\ g_1 \\ g_2 \\ \vdots \\ g_N \\ \mu \end{bmatrix} = \begin{bmatrix} 0 \\ 0 \\ 0 \\ \vdots \\ 0 \\ 1 \end{bmatrix}$$

$$(12)$$

The solution of Eq. (12) yields w_0, g_i and μ, as appropriate.

Accuracy of the SPA Technique

A variety of studies have discussed the uncertainty sources affecting the accuracy of data-based air quality estimates [24], [25]. Generally, there is an inverse relationship between uncertainty and accuracy – the higher the data uncertainty, the lower the accuracy of a model or technique. Usually the accuracy of a technique is measured in terms of its estimation error. The theoretical background of the **SPA** technique considers both horizontal correlations between samples, and vertical correlations between samples and area populations. It subsequently produces pollutant estimates that satisfy two key criteria – unbiasedness and minimum estimation error. Accordingly, **SPA** is a network-based estimation technique that is resistant to shifts [26] such as dust storms, which are addressed by statistical autocorrelation parameters in the model.

In this study, the horizontal (spatial) correlation between $PM_{2.5}$ concentrations is approximated by that between spatial PM_{10} concentrations. The estimation error of this approximation is small due to various reasons:

The citywide $PM_{2.5}$ concentration estimated by SPA is defined as the weighted spatial $PM_{2.5}$ average from all 18 stations (for each station the weight was proportional to the associated Voronoi area). Note that spatial topology –which is a key determinant of horizontal (spatial) autocorrelation [27]– is identical for $PM_{2.5}$ and PM_{10} [28].

Both particulates vary in space and time, subject to the same weather conditions, providing a valuable determinant of horizontal correlation [29], [30]. Vertical correlations between $PM_{2.5}$ and PM_{10} concentrations were calibrated in terms of the observed data.

Empirical evidence has shown that $PM_{2.5}$ and PM_{10} concentrations are highly correlated, with values as high as 0.85 and 0.97, respectively [31], [32].

In the SPA technique, the correlation coefficients between $PM_{2.5}$ and PM_{10} are calibrated by the data so that they can correct for potential discrepancies (see section 2 in the *SI* text). Historical data have shown high correlations between the U.S. Embassy $PM_{2.5}$ concentrations and the 18 PM_{10} observation stations (Table 1). The maximum and minimum values of Pearson correlation efficient are 0.85 and 0.69, respectively.

Table 1: Pearson correlation coefficient between the U.S. Embassy PM2.5 concentration and 18 Beijing EPB PM_{10} concentrations

BJ-EPB Station	r	BJ-EPB Station	r
Aotizhongxin	0.81	Longquanzhen	0.82
Changpingzhen	0.72	Nongzhanguan	0.83
Dongsi	0.83	Tiantan	0.82
Fengtaihuanyuan	0.85	Tongzhouzhen	0.79
Gucheng	0.81	Wanliu	0.81
Guanyuan	0.83	Wanshouxigong	0.84
Haidingbeibuxinqu	0.69	Yizhuangkaifaqu	0.82
Huangcunzhen	0.80	Yungang	0.81
Liangxiang	0.82	Zhiwuyuan	0.77

doi:10.1371/journal.pone.0053400.t001

Estimation precision was further assessed by a validation study using an exhaustive PM_{10} dataset in the study area. In particular, daily areal PM_{10} concentrations were estimated by the SPA technique based on records at each of the 18 PM_{10} stations. The actual daily areal PM_{10} concentration is the weighted spatial PM_{10} average from all 18 stations (for each station, the weight was proportional to the associated Voronoi area; see *Supporting material*). Subsequently, the areal PM_{10} concentration estimated by each of the 18 PM_{10} monitoring stations and SPA was compared to the actual concentration value, resulting in good agreement (Table 2 and Figure S1 in SI text).

Table 2: Summary of R^2 values of the linear relationships between Beijing areal PM_{10} estimated on the basis of a single station using SPA and the true area

BJ-EPB Station	R^2	BJ-EPB Station	R^2
Aotizhongxin	0.961	Longquanzhen	0.921
Changpingzhen	0.862	Nongzhanguan	0.966
Dongsi	0.969	Tiantan	0.941
Fengtaihuanyuan	0.961	Tongzhouzhen	0.867
Gucheng	0.933	Wanliu	0.947
Guanyuan	0.964	Wanshouxigong	0.971
Haidingbeibuxinqu	0.764	Yizhuangkaifaqu	0.888
Huangcunzhen	0.896	Yungang	0.925
Liangxiang	0.849	Zhiwuyuan	0.901

doi:10.1371/journal.pone.0053400.t002

This result supports the reliability of the SPA technique when used to estimate areal pollution concentration based on a single monitoring station. An SPA software is provided that can be used to perform the data calculations of this study (www.sssampling.org/SPA). Readers can apply the SPA software to their own data.

RESULTS

Daily $PM_{2.5}$ mass concentrations observed at the embassy station ranged from 4 to 487 µg/m³ for the 423-day period. The annual average concentration (December 7, 2010–December 6, 2011) was 98.85 µg/m³, with high temporal variability. For the entire time series, the highest $PM_{2.5}$ concentrations (>300 µg/m³) occurred during 10 days: December 7 and November 18–19, 2010, February 21–24, October 23 and December 5, 2011; see Figure 3.

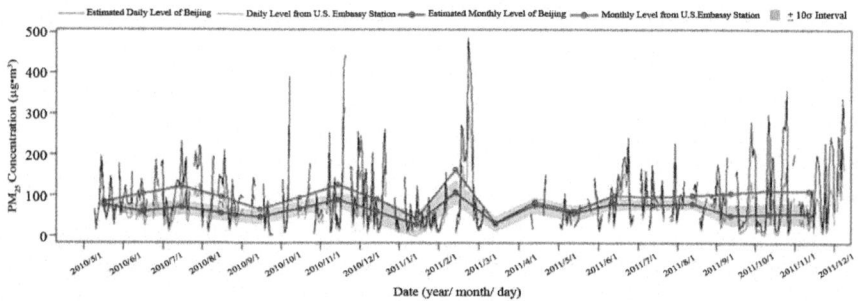

Figure 3: $PM_{2.5}$ concentration observed by a single station (U.S. Embassy), and estimated citywide $PM_{2.5}$ areal concentration (Beijing, China).

During the same period, estimated citywide $PM_{2.5}$ daily pollution in Beijing ranged from 2.86 to 318.29 µg/m³. The annual average pollution was 64.78 µg/m³. The highest concentrations (>300 µg/m³) occurred during two days, November 19, 2010 and February 21, 2011, as shown in Figure 3.

Discussion

It was found that the U.S. Embassy $PM_{2.5}$ observations exhibited approximately the same trend as citywide $PM_{2.5}$ areal concentrations estimated by the SPA technique, although the embassy's concentration values were clearly higher. The most important reason for this could be that the U.S. Embassy is at the city center, where population density and traffic volume are the highest in the city. The ratio between the embassy's $PM_{2.5}$ concentration and the estimated area-average concentration pollution varied with time. It is affected by the dynamic correlation between $PM_{2.5}$ and PM_{10}, caused by variation in local emission and atmospheric conditions between the embassy and the entire city.

Estimated area-average $PM_{2.5}$ concentrations varied on a daily and monthly basis. The lowest concentrations occurred during January and March 2011, owing to the large number of windy days (refer to Figure S2 in *SI* text for monthly wind speeds). Estimation uncertainty is high for March 2011, because of serious data gaps. The highest concentrations occurred during July and November 2010, and during February and July–September 2011. During November, formation of a temperature inversion layer was observed over Beijing, which is a meteorological condition that plays an important role in the accumulation of $PM_{2.5}$. The $PM_{2.5}$ mass concentration peak during February was most likely due to emissions from coal consumption for heating purposes [33], [34]; this was the month with the lowest temperatures and slowest winds during 2011. July–September was the hottest period during a year. Long and intense solar irradiation during summer favors photochemical formation of aerosol particles [35], [36], which benefits the synthesis of $PM_{2.5}$. This caused the high $PM_{2.5}$ levels observed during that season. As regards seasonal variation, winter and summer had higher $PM_{2.5}$ levels, with concentrations 68.74 $\mu g/m^3$ and 70.42 $\mu g/m^3$, respectively. Spring and fall concentrations were 63.59 $\mu g/m^3$ and 61.54 $\mu g/m^3$, respectively.

In sum, $PM_{2.5}$ pollution in Beijing remained relatively high during the study period (Figure 3). Daily and annual interim target-1 standards recommended by the World Health Organization (WHO) are 75 $\mu g/m^3$ and 35 $\mu g/m^3$, respectively [37]. As mentioned earlier, the annual (December 7, 2010– December 6, 2011) average concentration in Beijing was 64.78 $\mu g/m^3$. During that period, daily concentrations during 93 out of 259 days exceeded the WHO standard. Compared to the Beijing $PM_{2.5}$ levels of five years ago reported in previous studies[33], [34], this level has dropped significantly. The situation may be attributed to a policy of prioritizing development of public transport, displacement of heavy industrial factories away from the city, and other efforts associated with the 2008 Beijing Olympics. Yet, the number of cars in the city has grown, from 2.6 million in 2005 to 5 million in 2010. Furthermore, air quality remains dependent on weather conditions, which means that considerable willingness and effort are needed to eliminate $PM_{2.5}$ sources, thereby clearing the sky over the city.

CONCLUSION

PM air pollution is a severe problem for Beijing city, as is demonstrated by both the official PM_{10} and the estimated $PM_{2.5}$ concentrations. The areal $PM_{2.5}$ concentration estimated by the proposed SPA technique was found to be a little lower than that observed at the U.S. Embassy monitoring station that is located at the city center and near a traffic junction. Validation results showed that the SPA technique is a useful tool in the estimation of areal PM2.5 concentration, even when only one PM2.5 observation station is available. Concerning the in situ implementation of SPA, (i) the key input to the technique is the correlation (covariance) between the $PM_{2.5}$ and PM_{10} stations calculated from historical data, (ii) the estimation weight of the PM2.5 station was obtained by solving a linear equation (equation (12)) and, subsequently, (iii) the areal PM2.5 concentration was calculated from equation (2). Concluding, given the prohibitive costs of measurement campaigns and monitoring networks, the proposed SPA technique can be an effective and accurate pollution estimation tool, especially in cases in which, due to limited monitoring stations or in remote areas or in the past, other sources of information need to be used.

AUTHOR CONTRIBUTIONS

Conceived and designed the experiments: JFW GC. Performed the experiments: MGH CDX YZ. Analyzed the data: JFW MGH CDX YZ. Contributed reagents/materials/analysis tools: MGH YZ GC. Wrote the paper: CDX JFW

REFERENCES

1.   Bayraktar H, Turalioglu FS, Tuncel G (2010) Average mass concentrations of TSP, PM10 and PM2.5 in Erzurum urban atmosphere, Turkey. Stochastic Environmental Research and Risk Assessment 24: 57–65. doi: 10.1007/s00477-008-0299-2

2.   Wiwanitkit V (2008) PM10 in the atmosphere and incidence of respiratory illness in Chiangmai during the smoggy pollution. Stochastic Environmental Research and Risk Assessment 22: 437–440. doi: 10.1007/s00477-007-0149-7

3.   Kaiser J (2000) Air pollution - Evidence mounts that tiny particles can kill. Science 289: 22–23. doi: 10.1126/science.289.5476.22

4.   Ostro B (2004) Outdoor Air Pollution: Assessing the Environmental Burden of Disease of Outdoor Air Pollution at National and Local Levels. Geneva: World Health Organization, (WHO Environmental Burden of Disease Series, No. 5).

5.   Pyne S (2002) Air pollution - Small particles add up to big disease risk. Science 295: 1994–1994. doi: 10.1126/science.295.5562.1994a

6.   Nel A (2005) Air pollution-related illness: Effects of particles. Science 308: 804–806. doi: 10.1126/science.1108752

7.   Akimoto H (2003) Global air quality and pollution. Science 302: 1716–1719. doi: 10.1126/science.1092666

8. Lelieveld J, Berresheim H, Borrmann S, Crutzen PJ, Dentener FJ, et al. (2002) Global air pollution crossroads over the Mediterranean. Science 298: 794–799. doi: 10.1126/science.1075457

9. Kaiser J (1997) Particulate matter - Getting a handle on air pollution's tiny killers. Science 276: 33–33. doi: 10.1126/science.276.5309.33

10. Guzman F, Ruiz ME, Vega E (1996) Air quality in Mexico City. Science 271: 1040–1041. doi: 10.1126/science.271.5252.1040

11. Fu BJ (2008) Blue skies for China. Science 321: 611–611. doi: 10.1126/science.1162213

12. Marshall JD, Nethery E, Brauer M (2008) Within-urban variability in ambient air pollution: Comparison of estimation methods. Atmospheric Environment 42: 1359–1369. doi: 10.1016/j.atmosenv.2007.08.012

13. US-Embassy-in-Beijing (2011) U.S Embassy Beijing Air Quality Monitor, Available at:http://eng.embassyusa.cn/070109air.html.

14. XinHuaNet (2011) Chinese and U.S. Experts Explain Beijing Air Quality Monitoring Data, Available at: http://news.xinhuanet.com/society/2011-11/08/c_111153803.htm.

15. BJ-EPB (2011) Beijing Municipal Environmental Protection Bureau Daily Air Quality, Available at: http://www.bjepb.gov.cn/air2008/Air.aspx.

16. MEP-PRC (1996) People's Republic of China National Standard. Ambient air quality standard GB3095–1996: 2.

17. Brook JR, Dann TF, Burnett RT (1997) The relationship among TSP, PM(10), PM(2.5), and inorganic constituents of atmospheric particulate matter at multiple Canadian locations. Journal of the Air & Waste Management Association 47: 2–19.

18. Gomiscek B, Hauck H, Stopper S, Preining O (2004) Spatial and temporal variations Of PM1, PM2.5, PM10 and particle number concentration during the AUPHEP-project. Atmospheric Environment 38: 3917–3934. doi: 10.1016/j.atmosenv.2004.03.056

19. Lundgren DA, Hlaing DN, Rich TA, Marple VA (1996) PM(10)/PM(2.5)/PM(1) data from a trichotomous sampler. Aerosol Science and Technology 25: 353–357. doi: 10.1080/02786829608965401

20. Li CS, Lin CH (2002) PM1/PM2.5/PM10 characteristics in the urban atmosphere of Taipei. Aerosol Science and Technology 36: 469–473. doi: 10.1080/027868202753571287

21. Maraziotis E, Sarotis L, Marazioti C, Marazioti P (2008) Statistical analysis of inhalable (PM10) and fine particles (PM2.5) concentrations

in urban region of Patras, Greece. Global Nest Journal 10: 123–131.

22. van de Kassteele J, Koelemeijer RBA, Dekkers ALM, Schaap M, Homan CD, et al. (2006) Statistical mapping of PM10 concentrations over Western Europe using secondary information from dispersion modeling and MODIS satellite observations. Stochastic Environmental Research and Risk Assessment 21: 183–194. doi: 10.1007/s00477-006-0055-4

23. Wang JF, Reis BY, Hu MG, Christakos G, Yang WZ, et al. (2011) Area Disease Estimation Based on Sentinel Hospital Records. PLoS ONE 6: 1–8. doi: 10.1371/journal.pone.0023428

24. Christakos G, Vyas VM (1998) A composite space/time approach to studying ozone distribution over Eastern United States. Atmospheric Environment 32: 2845–2857. doi: 10.1016/s1352-2310(98)00407-5

25. Jerrett M, Newbold KB, Burnett RT, Thurston G, Lall R, et al. (2007) Geographies of uncertainty in the health benefits of air quality improvements. Stochastic Environmental Research and Risk Assessment 21: 511–522. doi: 10.1007/s00477-007-0133-2

26. Reis BY, Kohane IS, Mandl KD (2007) An epidemiological network model for disease outbreak detection. Plos Medicine 4: 1019–1031. doi: 10.1371/journal.pmed.0040210

27. Isaaks EH, Srivastava RM (1989). Applied Geostatistics. New York: Oxford University Press.

28. Kumar N (2009) An optimal spatial sampling design for intra-urban population exposure assessment. Atmospheric Environment 43: 1153–1155. doi: 10.1016/j.atmosenv.2008.10.055

29. Tiwari S, Chate DM, Pragya P, Ali K, Bisht DS (2012) Variations in Mass of the PM10, PM2.5 and PM1 during the Monsoon and the Winter at New Delhi. Aerosol and Air Quality Research 12: 20–29. doi: 10.4209/aaqr.2011.06.0075

30. Pandey P, Khan AH, Verma AK, Singh KA, Mathur N, et al. (2012) Seasonal Trends of PM2.5 and PM10 in Ambient Air and Their Correlation in Ambient Air of Lucknow City, India. Bulletin of Environmental Contamination and Toxicology 88: 265–270. doi: 10.1007/s00128-011-0466-x

31. Kumar R, Joseph AE (2006) Air pollution concentrations of PM2.5, PM10 and NO2 at ambient and kerbsite and their correlation in Metro City - Mumbai. Environmental Monitoring and Assessment 119: 191–199. doi: 10.1007/s10661-005-9022-7

32. Marcazzan GM, Vaccaro S, Valli G, Vecchi R (2001) Characterisation of PM10 and PM2.5 particulate matter in the ambient air of Milan

(Italy). Atmospheric Environment 35: 4639–4650. doi: 10.1016/s1352-2310(01)00124-8

33. Wang HL, Zhuang YH, Wang Y, Sun Y, Yuan H, et al. (2008) Long-term monitoring and source apportionment of PM(2.5)/PM(10) in Beijing, China. Journal of Environmental Sciences-China 20: 1323–1327. doi: 10.1016/s1001-0742(08)62228-7

34. Zheng M, Salmon LG, Schauer JJ, Zeng L, Kiang CS, et al. (2005) Seasonal trends in PM2.5 source contributions in Beijing, China. Atmospheric Environment 39: 3967–3976. doi: 10.1016/j.atmosenv.2005.03.036

35. Song Y, Tang XY, Xie SD, Zhang YH, Wei YJ, et al. (2007) Source apportionment of PM2.5 in Beijing in 2004. Journal of Hazardous Materials 146: 124–130. doi: 10.1016/j.jhazmat.2006.11.058

36. Wang Y, Zhuang GS, Tang AH, Yuan H, Sun YL, et al. (2005) The ion chemistry and the source of PM2.5 aerosol in Beijing. Atmospheric Environment 39: 3771–3784. doi: 10.1016/j.atmosenv.2005.03.013

Chapter 7

MEANDERING DISPERSION MODEL APPLIED TO AIR POLLUTION

Gervásio A. Degrazia, Andréa U. Timm, Virnei S. Moreira, Débora R. Roberti
Universidade Federal de Santa Maria/UFSM Brazil

INTRODUCTION

Generally in stable conditions, during situations of low wind speed $(\bar{u} \leq 1-2\text{ms}^{-1})$, low-frequency horizontal wind oscillations (meandering) are observed in a nocturnal Planetary Boundary Layer (PBL). The study of low wind speed conditions is of interest, partly because the simulation of airborne pollutant dispersion in these conditions is rather difficult. In fact, most of the existing regulatory dispersion models become unreliable as $\bar{u}$ approaches zero, so that their application is generally limited to $(\bar{u} > 2\text{ms}^{-1})$. The meandering movements are clearly distinct from those associated to a full developed turbulence, which are responsible for the pollutants dispersion in a PBL. Even when the stability reduces the vertical dispersion and the instantaneous plume may be thin, meandering disperses the plume over a rather wide angular sector. As a consequence, any air pollution operational dispersion model to be reliable must take into account the transport effect provocated by the meandering. Transport phenomenon in turbulence, including the diffusion of passive scalars and the dispersion of pollutants in the PBL, are controlled by the advection processes associated with the action of stochastic velocity fluctuations in time and space. As a consequence, a Lagrangian description following the movement of infinitesimal fluid particles, as they are carried by the velocity turbulent fluctuations, is conceptually correct and from practical point of view useful for describing turbulent transport (Yeung, 2002). Lagrangian stochastic particle models are powerful computational tools for the investigation of the atmospheric dispersion process (Rodean, 1996). In these models, the fluid particle displacements are produced by stochastic velocities and the movement evolution of a particle can be considerate a Markov process (Wang, 1945), in which past and future are statistically independent when the present is known.

This method is based on Langevin equation, which is derived from the hypothesis that the velocity is given by the combination between a deterministic and a stochastic term (Chandrasekhar, 1943). Each fluid particle moves taking into account the transport due to the mean wind velocity and the turbulent fluctuations of the wind velocity components. From the spatial distribution of the particles it is possible to determine the pollutant concentrations. The implementation of the Lagrangian stochastic dispersion model in air pollution problems permits to take into account complex situations such as turbulent flows generated above inhomogeneous topo-graphy (different terrains) (Carvalho et al., 2002), in non-stationary situations associated with the evolutionary transition periods of the PBL (sunset transition period) and low wind speed conditions which for many places in the world occur for a substantial percentage of time (Oettl et al., 2001). Concerning this last complexity, it is important to note that in low wind velocity situations, particularly during stable conditions, the turbulent dispersion in the PBL is poorly described. As a consequence, the occurrence of low wind speed is generally considered the most critical situation associated with the air pollution dispersion problem. The aim of this chapter is to report a turbulence parameterization that can be employed in Lagrangian stochastic dispersion models to describe the air pollution dispersion in the situation of low wind velocity stable conditions. This specific parameterization employs an observational value for the meandering period and turbulent velocity variances and decorrelation time scales varying with height for a stable boundary layer. Therefore, are used two classical approaches to obtain the turbulent velocity variances and the decorrelation time scales. The first turbulence approach used in this study was derived by Degrazia et al. (2000) utilizing Taylor statistical diffusion theory while the second one was developed by Hanna (1982). Degrazia et al. (2000) turbulence approach is based on the observed turbulent velocity spectra while Hanna (1982) approach is based on analyses of field experiments, theoretical considerations and second-order closure model. An additional aim is to incorporate this new parameterization for the meandering dispersion phenomenon in a numerical Lagrangian stochastic particle model to simulate the dispersion of air pollution in a low wind velocity stable boundary layer. The Lagrangian particle model employed in the present investigation is constituted by a system of two coupled Langevin equations describing the meandering dispersion associated to the lateral and longitudinal components of the wind velocity fluctuations. This dispersion model is based on the so-called Thomson simplest solution and can be applied to more general case of inhomogeneous turbulence. The horizontal coupling, occurring between the lateral and longitudinal velocity components, reproduces the meandering enhanced air pollution transport. Finally, the observed concentration data employed in the comparison with the coupled Lange-vin equations model, incorporating Degrazia et al. (2000) and Hanna (1982) approaches, were obtained from the low wind speed experiment performed in a stable boundary layer from the series of field observations conducted at the Idaho National Engineering Laboratory - INEL.

THE SCLE DISPERSION MODEL

Simulating the dispersion of contaminants in low wind speed stable conditions is a difficult physical task. In such situations, the airborne contaminants are dispersed over rather wide angular sectors and therefore it is no longer possible to establish a definite mean wind direction since low-frequency horizontal wind oscillations start to dominate and diffusion of contaminants in the PBL becomes controlled by these degrees of freedom, characterized by low frequencies (large characteristic time associated with the meandering period). The horizontal meandering of a flow occurs when the wind speed presents a threshold low value and the low-frequency horizontal wind oscillations generate autocorrelation functions of the horizontal wind components showing a looping behavior evidenced by the presence of accentuated negative lobes (Anfossi et al., 2005; Oettl et al., 2005). This oscillatory character associated with the meandering phenomenon and consequently the presence of large negative lobes in the observed autocorrelation functions were recently explained as an intrinsic property of atmospheric flows occurring in weak horizontal turbulent momentum flux conditions (Goulart et al., 2007). The low wind speed autocorrelation functions of the horizontal wind components displaying an oscillation behavior and the presence of large negative lobes can be very well fitted by the following relationship

$$R(\tau) = e^{-p\tau}cos(q\tau)$$

(1)

where

$$p = \frac{1}{(m^2 + 1)T_{Lu,v}}$$

(2)

and

$$q = \frac{m}{(m^2 + 1)T_{Lu,v}}$$

(3)

The functional form as given by the Eqs. (1), (2) and (3) is composed of the product of the classical exponential function (representing the autocorrelation function for a fully developed turbulence) by the cosine function (describes the meandering phenomenon associated with the observed low frequency horizontal wind oscillations) (Frenkiel, 1953). The Frenkiel function (Eq. (1)) is a hybrid formula described in terms of TLu,v, the local horizontal Lagrangian time scale for a fully developed turbulence, and m, the loop parameter which controls the meandering oscillation frequency associated with the horizontal wind. It is worth noting that large m values characterize the dominant presence of the meandering phenomenon in comparison with the fully developed turbulence. Anfossi et al. (2006) proposed a system of two coupled Langevin equations (SCLE) to describe the contaminants dispersion in meandering conditions. Therefore, the following system of equations describes the dispersion in low wind speed conditions

$$du = \{-p(u - \bar{u}) - q(v - \bar{v})\}\, dt + \sqrt{2pdt}\sigma_u \xi u \tag{4}$$

and

$$dv = \{q(u - \bar{u}) - p(v - \bar{v})\}\, dt + \sqrt{2pdt}\sigma_v \xi v \tag{5}$$

where u and v are the horizontal components of the wind velocity fluctuations, ξu and ξv are random Gaussian variables having zero mean and unit variance, σ_u and σ_v are standard deviations of the horizontal wind components. It is important to note that Eqs. (4) and (5) are valid assuming horizontal homogeneous conditions (Anfossi et al., 2006). For the vertical component of the velocity fluctuation w we solve the Langevin equation accor-ding to the usual LAMBDA model (Anfossi et al., 2006; Thomson, 1987)

$$dw = a_i(z, w) + b_0(z)dW_j \tag{6}$$

where dW_j is the incremental Gaussian Wiener process (with zero mean and variance dt), $b_0(z) = \sqrt{2\sigma_w^2/T_{Lw}}$ (σ_w^2 is the vertical turbulent velocity variance and T_{Lw} is the vertical local Lagrangian time scale) and ai(z, w) is computed by solving the Fokker-Planck equation associated with Eqs. (4) and (5) using a PDF of Gram-Chalier type truncated to the third-order (Ferrero & Anfossi, 1998).

The position of each particle, at each time step, is obtained by the numerical integration of Eqs. (4) and (5) and (6) and the following equation:

$$dx_i = u_i dt, \tag{7}$$

where i = 1, 2, 3, x_i is the position vector of each particle, u_i is its corresponding Lagrangian velocity vector. Therefore, to describe the diffusion of passive scalars in the PBL the Langevin equation is integrated according to the rules of the Ito calculus (Gardiner, 1997), which was developed to get solutions of the stochastic differential equations.

TURBULENCE PARAMETERIZATION FOR MODELING MEANDERING EFFECTS IN THE STABLE PBL

The fundamental parameters to reproduce meandering transport effects, employing the Eqs. (4) and (5), are the quantities m, σ_u,v and $T_{Lu^{,}v}$ which define the Eqs. (1), (2), (3), (4) and (5). In this section the values of these physical parameters will be computed. In PBL turbulent dispersion models the selection of an adequate parameterization plays a fundamental role to evaluate the contaminants concentration in the atmosphere. Thusly, the efficiency of each approach to reproduce correctly the contaminants concentration field depends on the manner turbulent parameters are related to physical properties of the PBL. In the specific case of parameterization of the enhanced horizontal diffusion of passive scalars, controlled by the meandering phenomenon, the variables p and q need to be introduced into Eqs. (4) and (5). Analyzing the Eqs. (1), (2) and 3) we can define the following meandering period (Anfossi et al., 2005).

$$T_* = \frac{2\pi(m^2 + 1)T_{Lu,v}}{m},$$

(8)

from which can be obtained a relation for m given by (Carvalho et al., 2006)

$$m = \frac{T_* + \sqrt{T_*^2 - 16\pi^2 T_{Lu,v}^2}}{4\pi T_{Lu,v}}$$

(9)

As seen in Eq. (9), this formula for the loop parameter, defines m as a quantity that can be estimated from the meandering period T^* and of the local horizontal time scale associated with a fully developed turbulence $T_{Lu,v}$. On the other hand, analyzing Eq. (9) it is possible to notice that the presence of large values for the local turbulent time scales impede the increasing of m and consequently the meandering reinforced transport tends to vanish in the PBL. Time series of sonic anemometer wind speeds were analyzed by Anfossi et al. (2005). These observational data suggest that the mean magnitude of the meandering period is of the order of $T_* \cong 2000s.$

Turbulence Parameterization Derived by Degrazia et al. (2000) and Hanna (1982)

It is possible to relate turbulent parameters (wind velocity standard deviations $\sigma_{u,v,w}$ and Lagrangian decorrelation time scales $T_{Lu,v,w}$) to spectral distribution of turbulent kinetic energy (TKE). Following this approach, Degrazia et al. (2000) developed expressions for the wind velocity variances and Lagrangian decorrelation time scales. The velocity variances were obtained directly from the integration of the turbulence velocity spectra (Caughey & Palmer, 1979). On the other hand, the Lagrangian decorrelation time scales were derived from the peak wavelength of the turbulent velocity spectra (Caughey, 1982). Therefore, the turbulent velocity variances ($\sigma_{u,v,w}$) and the local turbulent time scales ($T_{Lu,v,w}$) for a stable boundary layer are given respectively by the following expression (Degrazia et al., 2000)

$$\sigma^2_{u,v,w} = \frac{2.32 c_{u,v,w} \phi_\epsilon^{2/3} u_*^2}{[(f_m^*)^s_{u,v,w}]^{2/3}}$$

(10)

and

$$T_{Lu,v,w} = \frac{z}{\sqrt{c_{u,v,w}}} \left\{ \frac{0.059}{[(f_m^*)^s_{u,v,w}]^{2/3} (\phi_\epsilon^s)^{1/3} u_*} \right\}$$

(11)

where z is the height above the surface, $c_{u,v,w} = \alpha_{u,v,w}\alpha_u(2\pi k)^{-2/3}$ with $k = 0.4$ (von karman constant), $\alpha_u = 0.5 \pm 0.05$ and $\alpha_{u,v,w} = 1, 4/3, 4/3$, respectively (Champagne et al., 1977; Sorbjan, 1989). For a shear-dominated stable PBL the ad-imensional dissipation rate $\phi_\epsilon^s = \epsilon_s k z / u_*^3$ can be written as $\phi_\epsilon^s = 1.25(1 + 3.7z/\Lambda)$ where Λ is the local Obukhov length given by $\Lambda = L(1 - z/h)^{(1.5\alpha_1 - \alpha_2)}$, with h being the height of the turbulent stable PBL and L is the surface Obukhov length. Furthermore, for a shear-dominated stable PBL, the local friction velocity is defined by $u_* = (u_*)_0(1 - z/h)^{\alpha_1/2}$, where $(u_*)_0$ is the surface friction velocity and $\alpha_1 = 1.5$ and $\alpha_2 = 1.0$ (Nieuwstadt, 1984). Finally, the reduced frequency of the stable horizontal spectral peaks is provided by the following relation $(f_m^*)^s_{u,v,w} = (f_m)^n_{(u,v,w)s}(1 + 3.7z/\Lambda)$ where $(f_m)^n_{us} = 0.045$, $(f_m)^n_{vs} = 0.16$ and $(f_m)^n_{ws} = 0.33$ are the frequencies of the spectral peaks in the surface for neutral conditions (Olesen et al., 1984; Sorbjan, 1989). Based on analyses of field experiments (Hanna, 1981; 1968; Kaimal, 1976), theoretical consi-derations (Irwin, 1979; Panofsky et al., 1977) and second-order closure models (Wyngaard et al., 1974), Hanna (1982) proposed the following turbulence parameterization for the turbulent velocity variances $(\sigma_{u,v,w})$ and the local turbulent time scales $(T_{Lu,v,w})$

$$\frac{\sigma_u}{(u_*)_0} = 2\left(1 - \frac{z}{h}\right)$$

(12)

And

$$\frac{\sigma_w}{(u_*)_0} = \frac{\sigma_v}{(u_*)_0} = 1.3\left(1 - \frac{z}{h}\right),$$

(13)

$$T_{Lu} = 0.15 \frac{h}{\sigma_u} \left(\frac{z}{h}\right)^{0.5},$$

(14)

$$T_{Lv} = 0.07 \frac{h}{\sigma_v} \left(\frac{z}{h}\right)^{0.5}$$

(15)

And

$$T_{Lw} = 0.10 \frac{h}{\sigma_w} \left(\frac{z}{h}\right)^{0.8}$$

(16)

From Eqs. (10) and (11), it can be seen that the turbulent velocity variances and the local Lagrangian time scales associated with the fully developed turbulence in the shear-dominated stable turbulent flow in PBL are formulated in terms of a similarity theory and expressed by three fundamental parameters associated with the turbulence in a PBL; the friction velocity, the adimensional dissipation rate and the reduced frequency of the horizontal spectral peaks. On the other hand, Eqs. (12), (13), (14), (15) and (16) are expressed in terms of a similarity theory and described by two fundamental scales, the friction velocity (u_*) and the height of the turbulent stable PBL. Therefore, using Eq. (9), with $T_* \cong 2000s$ and $T_{Lu,v}$ given by the Eqs. (11), (14) and (15), the low-frequency horizontal wind oscillation effects can be parameterized and introduced into Eqs. (4) and (5) to simulate the observed dispersion of passive scalars, caused by the meandering transport in the stable PBL.

MEANDERING DISPERSION SIMULATION

The results of the proposed model are compared with the concentration data collected under stable conditions in low wind speeds over flat terrain at the Idaho Engineering Laboratory (INEL). These observed results have been published in a U.S. National Oceanic and Atmos-pheric Administration (NOAA) report (Sagendorf & Dickson, 1974). Because of wind direction large variability associated with the meandering phenomenon, a full 360o sampling grid was implemented. Arcs were laid out at radii of 100, 200 and 400 m from the emission point source. Samplers were placed at intervals of 6o on each arc for a total of 180 sampling locations. The receptor height was 0.76 m. The tracer SF6 was released at a height of 1.5 m. The 1 h average concentrations were determined by means of an electron capture gas chromato-graphy. Wind speeds measured at levels 2, 4, 8, 16, 32 and 61 m were used to calculate the coefficient for the exponential wind vertical profile. According to Brusasca et al. (1992) and Sharan & Yadav (1998) the roughness length used was $z_0 =$ 0.005m. The Monin-Obukhov length L and the friction velocity u_* were not available for the INEL experiment but can be roughly estimated by different formulations. Then, L may be calculated from an empirical formulation suggested by Zannetti (1990) and the stable turbulent PBL height h was determined according to expression derived by Zilitinchevick (1972). INEL observed concentrations $\chi_m\,(m^{-2})$ were normalized according to the following relation (Sagendorf & Dickson, 1974)

$$\chi_m = C_m \frac{U_4}{Q},$$

(17)

where C_m is the dimensional concentration expressed in gm^{-3}, U_4 is the mean wind speed at 4 m and Q is tracer emission rate (gs^{-1}). Consequently, predicted concentrations are for the INEL experiments expressed in (m^{-2}). For the simulations, the turbulent flow field is considered as inhomogeneous in the vertical direction and the transport is performed by the longitudinal component of the mean wind velocity. In the simulations the horizontal domain was determined according to sampler distances and the vertical domain was set equal to the observed PBL height h. The emission point source was localized at the domain centre. The time step was maintained constant and equal to $\Delta t = 0.5s$ during the simulations. The magnitude of this time step is of the order of the time scales of Kolmogorov's turbulent energy spectrum inertial subrange. Furthermore, this value of Δt performs the following inequality $\Delta t << T_L$. This condition ensures that the turbulent velocities can be considered a Markov process (Rodean, 1996). For each simulation, the number of particles released was 10^6. In the case of the INEL experiments, the cells of concentration at ground-level have a vertical dimension of $\Delta z = 3m$ (Anfossi et al., 2006). On the other hand, the horizontal dimensions were computed from the following relation

$$\Delta x = \Delta y = \frac{2\pi r}{N_{ang}},$$

(18)

where r is the arc radius and $N_{ang} = 60$ is the number of samplers per arc. This way of computing the cell size covers all the compass at the three radii without significant overlapping. From these criteria results dx = 10.47m, dy = 20.93m and dz = 41.87m for the three arcs res-pectively. The simulated concentrations were obtained by counting the number of particles in volumes generated from the vertical and horizontal dimensions above presented.

RESULTS AND DISCUSSION

The performance of the SCLE model, employing the meandering parameterization for u and v components as given by the Eqs (9), (10), (11), (12), (13), (14) and (15) is shown in Figu-res 1 to 6 and in Table 1. Figs. 1 and 2 show some typical simulation results obtained with the SCLE model employing Degrazia et al. (2000) turbulence parameterization. On the other hand, Figs. 3 and 4 show these same results reproduced with the SCLE model utilizing Hanna (1982) turbulence approach. Figs. 1 and 3 refer to experiment 8, which is characterized by a plume that spreads horizontally over a wide angular sector (one having the widest horizontal plume spread), meaning that the tracer is collected at all angles. The results of the simulations show that the SCLE model containing (Degrazia et al., 2000) and Hanna (1982) turbulence parameterization is able to reproduce the dispersion of the contaminants plume over all the 360o . Concerning to the environmental effects of air contaminants released in the PBL, the estimation of the location in which the maximum concentration occurs is a fundamental data. Regarding this information, the maximum concentration for experiment 8 at 400 m is approxi-mately well reproduced by the SCLE model using Degrazia et al. (2000) and Hanna (1982) parameterization. Observing Figs. 2 and 4, we can see that the SCLE model with the Degrazia et al. (2000) and Hanna (1982) turbulence approach simulates fairly well the observed maximum concentrations for experiment 6 at 100, 200 and 400 m. In addition, to obtain a global evaluation about of the quality of the simulations that were made using the meandering parameterization, the following statistical indices have been computed at each arc, for the ten INEL experiments: concmax, top5 and Sy. Concmax (m^{-2}) is the maximum ground-level concentration, top5 (m^{-2}) refers to the mean of the 5 highest measured and computed ground-level concentration and $S_y = \sqrt{\sum_{i=1}^{N}(\theta_i - \overline{\theta})^2 / \sum_{i=1}^{N} \chi_i}$, where θ_i are the sampler angles and $\overline{\theta}$ their average value (weighted with the concentrations). Considering the INEL experiments, the Figs. 5 and 6 show the results of these statistical indices (concmax, top5 and S_y) obtained from the Eqs. (4) and (5) employing the meandering parameterization and respectively Degrazia et al. (2000) and Hanna (1982) turbulence approach. Figs. 5 (a,b and c) (Degrazia et al. (2000) turbulence approach) and Figs. 6 (a,b and c) (Hanna (1982) turbulence approach) show respectively the scatter diagram between observed and predicted concmax, top5 and S_y values. Observing these figures it is possible to notice that there is a certain data spread, however this scattering is not sufficient to avoid a reasonable alignment in relation to the straight of perfect agreement.

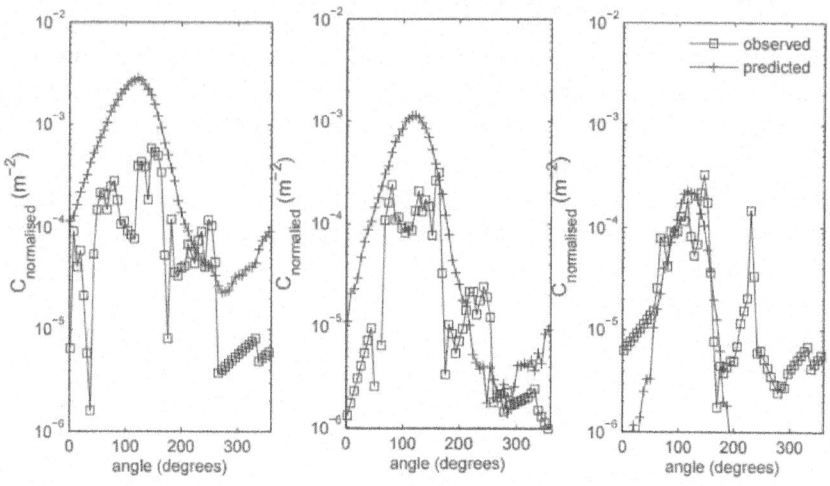

Figure. 1: Normalized ground level concentration (m^{-2}) for experiment 8 at 100, 200 and 400 m as a function of the sampler angles for (Degrazia et al., 2000) turbulence parameterization. Open squares indicate observed concentrations and crosses indicate simulated concentrations.

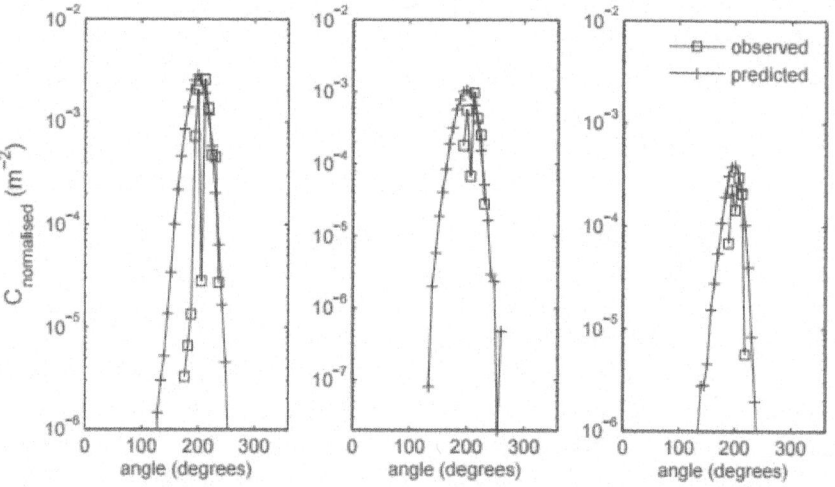

Figure. 2: As Fig. 1 but for experiment 6.

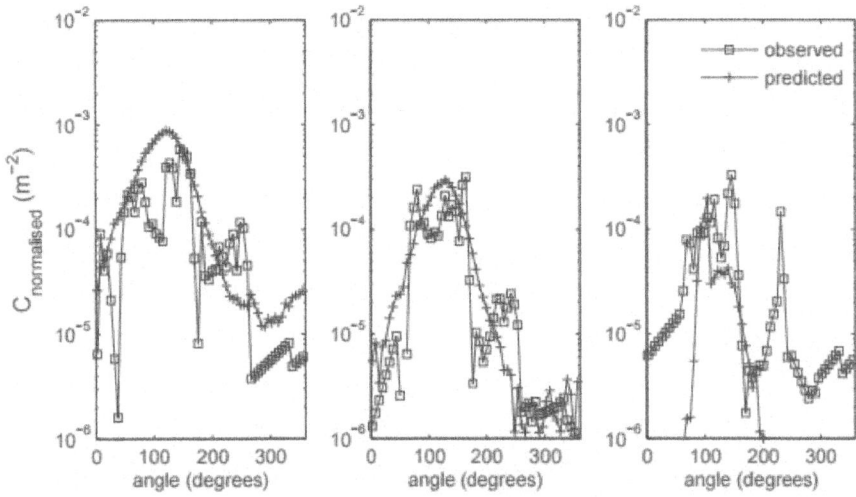

Figure. 3: Normalized ground level concentration (m^{-2}) for experiment 8 at 100, 200 and 400 m as a function of the sampler angles for Hanna (1982) turbulence parameterization. Open squares indicate observed concentrations and crosses indicate simulated concentrations.

Table 1 shows the results of the statistical analysis made with observed and predicted values of peak concentration (n = 30). Furthermore, this Table presents a comparison between SCLE

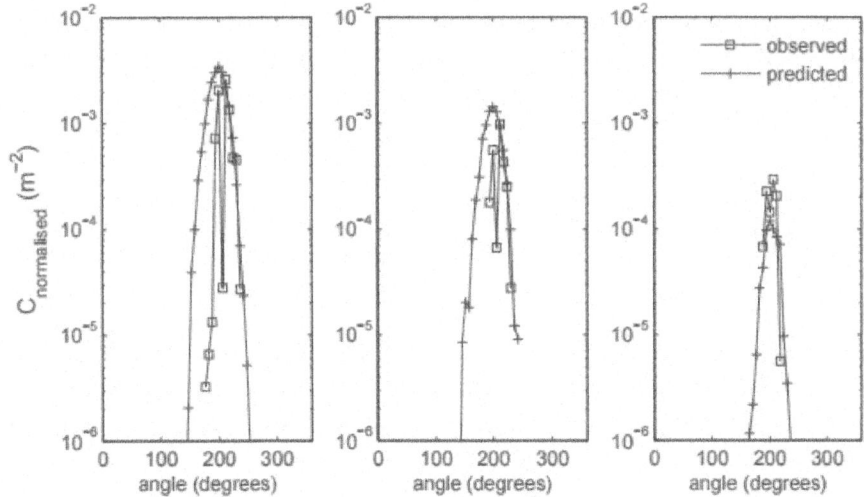

Figure. 4: As Fig. 3 but for experiment 6.

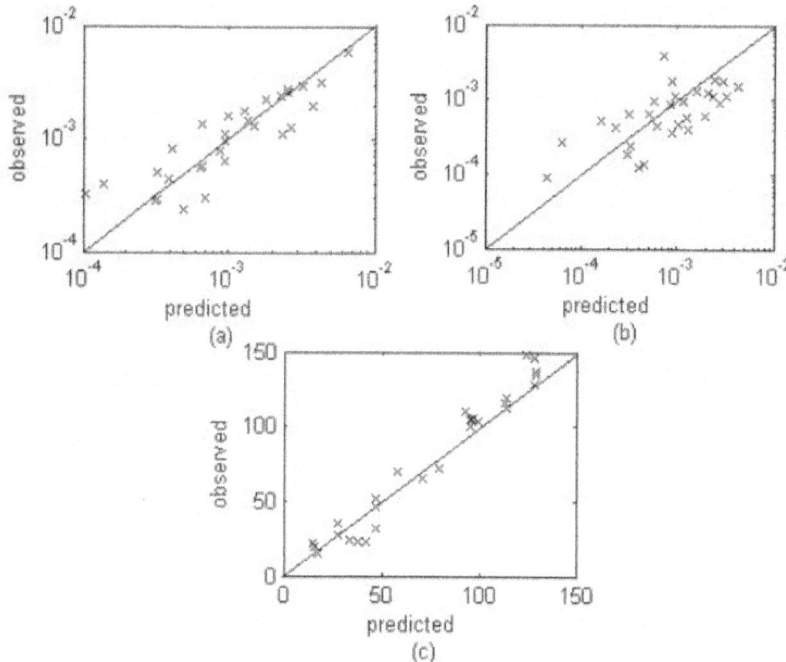

Figure. 5: Plot of (a) concmax, (b) top5 and (c) S_y calculated with SCLE model utilizing the Degrazia et al. (2000) turbulence parameterization. X-axis shows predicted values whereas observed values are on the Y-axis.

model employing the parameterization (Eqs. (9), (11), (14) and (15)) and other three models (Oettl et al., 2001; Sagendorf & Dickson, 1974; Sharan & Yadav, 1998). The statistical indices are suggested by Hanna (1989):

$NMSE = \overline{(C_o - C_p)^2}/\overline{C_o}\overline{C_p}$ (Normalized Mean Square Error)
$FB = (\overline{C_o} - \overline{C_p})/(0.5(\overline{C_o} + \overline{C_p}))$ (Fractional Bias)
$FS = 2(\sigma_o - \sigma_p)/(\sigma_o + \sigma_p)$ (Fractional Standard Deviation)
$R = \overline{(C_o - \overline{C_o})(C_p - \overline{C_p})}/\sigma_o\sigma_p$ (Correlation Coeficient)
$FA2 = 0.5 \leq C_o/C_p \leq 2$ (Factor 2)

where C is the analyzed quantity (concentration) and the subscripts "o" and "p" represent the observed and the predicted values, respectively. The overbars in statistical indices indicate averages. The statistical index FB indicates if the predicted quantity underestimates or overestimates the observed one. The statistical index NMSE represents the quadratic error of the predicted quantity in relation to the observed one. The statistical index FS indicates the measure of the comparison between predicted and observed plume spreading. The statistical

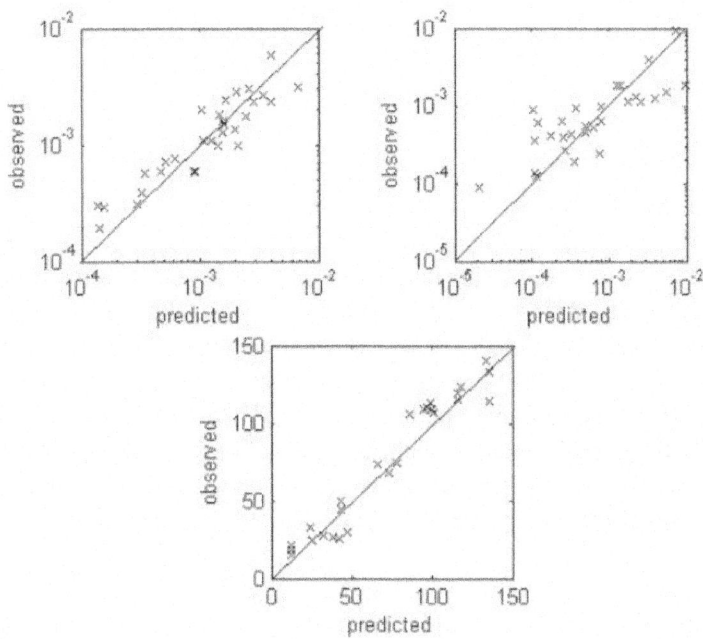

Figure. 6: Plot of (a) concmax, (b) top5 and (c) S_y calculated with SCLE model utilizing the Hanna (1982) turbulence parameterization. X-axis shows predicted values whereas observed values are on the Y-axis.

index FA2 provides the fraction of data for which $0.5 \leq C_o/C_p \leq 2$. As nearest zero are the NMSE, FB and FS and as nearest one are the R and FA2, better are the results. Analyzing the results shown in Table 1, it is possible to infer that the SCLE model employing the meandering parameterization developed in this study reproduces well the experimental data in stable meandering conditions. Furthermore, the SCLE model with these meandering enhanced dispersion parameterizations presents results comparable or even better than ones obtained by other models. Therefore, the proposed parameterizations for the meandering phenomenon reported in the present analysis generate magnitudes of the statistical indices that are within acceptable ranges, with NMSE, FB and FS values relatively near to zero and R and FA2 relatively near to 1.

Table 1: Statistical evaluation considering other for the INEL experiment

	NMSE	R	FA2	FB	FS
SCLE model Degrazia et al. (2000)	0.19	0.88	0.89	-0.054	-0.21
SCLE model Hanna (1982)	0.35	0.78	0.90	-0.038	-0.22
Sagendorf and Dickson (1974)	0.60	0.42	0.80	0.06	-
Sharan and Yadav (1998)	0.53	0.55	0.60	-0.02	-
Oettl et al. (2001)	0.21	0.86	0.87	-0.13	-

CONCLUSIONS

The investigation deals with contaminants dispersion associated with the low wind speed cases. In low wind velocity conditions in the stable PBL, the meandering horizontal of the wind is a physical mechanism that dominates the horizontal spread of contaminants. This means that as the average wind speed decreases, large horizontal low frequency oscillations start to control the flow field, surpassing both the transport and the small-scale diffusion generated by the fully developed turbulence. Therefore, based on meandering phenomenon observational evidences and Degrazia et al. (2000) and Hanna (1982) turbulence approach, which allows to calculate local (at distinct-heights) horizontal Lagrangian time scales, para-meterizations for the reinforced meandering diffusion were derived and presented. The parameters p and q (Eqs. 2 and 3) are important quantities describing the meandering phenomenon. They are defined in terms of m, which controls the magnitude of negative lobes in the meandering observed autocorrelation functions, and of the local horizontal time scales $T_{Lu,v}$ for a fully developed turbulence. Therefore, Eq. (9) is a formulation for the meandering parameterization that is described in terms of the meandering period T_* and of the time scales $T_{Lu,v}$. Concerning to the T_*, an observed representative mean value of the meandering period of the order of $T_* = 2000s$, obtained from a large number of experimental data, was employed in this study. This phenomenological choice for the value of T_*, allows that the present approach can be applied to meandering distinct cases. On the other hand, the parameterization of $T_{Lu,v}$ was obtained from Degrazia et al. (2000) and Hanna (1982) turbulence descriptions. Thusly, the equations that provide $T_{Lu,v}$ (Eqs. 11, 14 and 15) are expressed in terms of a similarity theory describing the shear dominated stable turbulence. The meandering parameterizations above discussed were evaluated and tested through the comparison with observational data and other different meandering dispersion models. The results generated by the simulations using the two coupled Langevin equations (SCLE), employing the new parameterizations, agree well with the experimental data, pointing that the present approaches reproduce the contaminants meandering spread process adequately in low wind speed stable conditions. Finally, considering the good agreement between the results of the proposed model with the experimental ones, the new parameterizations for the meandering phenomenon employed in the SCLE are found to be suitable to simulate meandering enhanced dispersion of contaminants in a low wind speed stable PBL.

ACKNOWLEDGEMENT

The authors acknowledgement the financial support provided by CAPES (Coordenação de Aperfeiçoamento de Pessoal de Nível Superior) and CNPq (Conselho Nacional de Desenvolvimento Científico e Tecnológico).

REFERENCES

1. Anfossi, D.; Alessandrini, S.; Trini Castelli, S.; Ferrero, E.; Oettl, D. & Degrazia, G. (2006). Tracer dispersion simulation in low wind speed conditions with a new 2D Langevin equation system, Atmospheric Environment, Vol. 40, pp. 7234-7245.

2. Anfossi, D.; Oettl, D.; Degrazia, G. & Goulart, A. (2005). An analysis of sonic anemometer observations in low wind speed conditions, Boundary-Layer Meteorology, Vol. 114, pp. 179-203.

3. Brusasca, G.; Tinarelli, G. & Anfossi, D. (1992). Particle model simulation of diffusion in low wind speed stable conditions, Atmospheric Environment, Vol. 26A, pp. 707-723.

4. Carvalho, J.C.; Degrazia, G.A.; Vilhena, M.T.; Magalhães, S.G.; Goulart, A.; Anfossi, D.; Acevedo, O.C. & Moraes, O.L.L. (2006). Parameterization of meandering phenomenon in a stable atmospheric boundary layer, Physica A, Vol. 368, pp. 247-256.

5. Caughey, S.J. (1982). Observed characteristics of the atmospheric boundary layer. In: Nieuwstadt, F.T.M., van Dop, H. (Eds.), Atmospheric Turbulence and Air Pollution Modelling, Reidel, pp. 107-158, Dordrecht.

6. Caughey, S.J. & Palmer, S.G. (1979). Some aspects of turbulence structure thought the depth of the convective boundary layer, Quarterly Journal of the Royal Meteorological Society, Vol. 105, pp. 811-827.

7. Chandrasekhar, S. (1943). Stochastic Problems in Physics and Astronomy, Review of Modern Physics, Vol. 15, pp.1-89.

8. Champagne, F.H.; Friehe, J.C.; La Rue, J.C. & Wyngaard, J.C. (1977). Flux measurements, flux estimation techniques, and fine-scale turbulence

measurements in the unstable surface layer over land, Journal of Atmospheric Science, Vol. 34, pp. 515-530.

9.   Carvalho, J.C.; Anfossi, D.; Castelli, S.T. & Degrazia, G.A. (2002). Application of a model system for the study of transport and diffusion in complex terrain to the TRACT experiment, Atmospheric Environment, Vol. 36, pp. 1147-1161.

10. Cirillo, M.C. & Poli, A.A. (1992). An intercomparison of semiempirical diffusion models under low wind speed, stable conditions, Atmospheric Environment, Vol. 26A, pp. 765-774.

11. Csanady, G.T. (1992). Turbulent diffusion in the environment, Geophysics and Astrophysics Monographs, Reidel, 248pp, Boston.

12. Degrazia, G.A.; Anfossi, D.; Carvalho, J.C.; Mangia, C.; Tirabassi, T. & Campos Velho, H.F. (2000). Turbulence parameterization for PBL dispersion models in all stability conditions, Atmospheric Environment, Vol. 34, pp. 3575-3583.

13. Dosio, A.; Guerau de Arellano, J.V. & Holtslag, A.A.M. (2005). Relating Eulerian e Lagrangian statistics for the turbulent dispersion in the Atmospheric Convective Boundary Layer, Journal of the Atmospheric Sciences, Vol. 62, No. 4, pp. 1175-1191.

14. Ferrero, E. & Anfossi, D. (1998). Comparison of FDPs, closure schemes and turbulence parameterizations in Lagrangian stochastic models, International Journal of Environment and Pollution, Vol. 9, pp. 384-410.

15. Frenkiel, F.N. (1953). Turbulent diffusion: mean concentration distribution in a flow field of homogeneous turbulence, Advances in Applied Mechanics, Vol. 3, pp. 61-107.

16. Gardiner, C.W. (1997). Handbook of stochastic methods, Springer, 442pp, Berlin. Goulart, A.; Degrazia, G.A.; Acevedo, O.C. & Anfossi, D. (2007). Theoretical considerations of meandering winds in simplified conditions, Boundary-Layer Meteorology, Vol. 125, pp. 279-287.

17. Hanna, S.R. (1989). Confidence limits for air quality model evaluations, as estimed by Bootstrap and Jacknife Resampling Methods, Atmospheric Environment, Vol. 23, pp. 1385-1398.

18. Hanna, S.R. (1982). Applications in air pollution modelling, In: Atmospheric Turbulence and Air Pollution Modelling, Nieuwstadt, F.T.M., van Dop, H., Reidel, page numbers (275-310), Dordrecht.

19. Hanna, S.R. (1981). Lagrangian and Eulerian time-scale relations in the daytime boundary layer, Journal of Applied Meteorology, Vol. 20, pp. 242-249.

20. Hanna, S.R. (1968). A method of estimating vertical eddy transport in the planetary boundary layer using characteristics of the vertical velocity spectrum, Journal of the Atmospheric Sciences, Vol. 25, pp. 1026-1033.

21. Irwin, J.S. (1979). Estimating plume dispersion-arecommended generalized scheme. Preprints fourth symposium on turbulence, diffusion and air pollution, American Meteorological Society, 45 Beacon Street, Boston, Mass. 02108, pp. 62-69.

22. Kaimal, J.C. (1976). Turbulence structure in the convective boundary layer, Journal of Atmospheric Science, Vol. 33, pp. 2152-2169.

23. Nieuwstadt, F.T.M. (1984). The turbulent structure of the stable, Nocturnal Boundary Layer, Journal of the Atmospheric Sciences, Vol. 41, No. 14, pp. 2202-2216.

24. Oettl, D.; Almbauer, R.A. & Sturm, P.J. (2001). A new method to estimate diffusion in stable, low-wind conditions, Journal of Applied Meteorology, Vol. 40, pp. 259-268.

25. Oettl, D.; Goulart, A.; Degrazia, G.A. & Anfossi, D. (2005). A new hypothesis on meandering atmospheric flows in low wind speed conditions, Atmospheric Environment, Vol. 39, pp. 1739-1748.

26. Olesen, H.R.; Larsen, S.E. & Hojstrup, D. (1984). A new hypothesis on meandering atmospheric flows in low wind speed conditions, Atmospheric Environment, Vol. 39, pp. 1739-1748.

27. Panofsky, H.A.; Tennekes, H.; Lenschow, D.H. & Wyngaard, J.C. (1977). The characteristics of turbulent velocity components in the surface layer under convective conditions, Boundary-Layer Meteorology, Vol. 11, pp. 355-361.

28. Rodean, H.C. (1996). Stochastic Lagrangian Models of Turbulence Diffusion, American Meteorological Society, 84pp, Boston.

29. Sagendorf, J.F. & Dickson, C.R. (1974). Diffusion under low wind speed, inversion conditions, NOAA Technical Memorandum ERL ARL-52, 89pp.

30. Sharan, M. & Yadav, K. (1998). Simulation of diffusion experiments under light wind, stable conditions by a variable k-theory model, Atmospheric Environment, Vol. 32, pp.3481-3492.

31. Sharan, M.; Yadav, K. & Sing, M.P. (1995). Comparison of sigma schemes for estimation of air pollutant dispersion in low winds, Atmospheric Environment, Vol. 29, pp. 2051-2059.

32. Sorbjan, Z. (1989). Structure of the Atmospheric Boundary Layer, Prentice-Hall, Englewood Cliffs, NJ, 317pp.

33. Thomson, D.J. (1987). Criteria for the selection of stochastic models of particle trajectories in turbulent flows, Journal of Fluids Mechanics, Vol. 180, pp. 529-556.

34. Wang, M.C. (1945). Influence of magnetic field on the Brownian motion of charged particle, Journal of the Physical Society of Japan, Vol. 28, pp. 559-564.

35. Wilson, J.D. & Sanford, B.L. (1996). Review of Lagrangian stochastic models for trajectories in turbulent atmosphere, Boundary-Layer Meteorology, Vol. 78, pp. 191-210.

36. Wilson, R.B.; Start, G.E.; Dickson, C.R. & Ricks, N.R. (1976). Diffusion under low wind speed conditions near Oak Ridge, Tennessee, NOAA Technical Memorandum ERL ARL-61,83pp.

37. Wyngaard, J.C.; Coté, O.R. & Rao, K.S. (1974). Modelling of the atmospheric boundary layer Advances in Geophysics, Academic Press, Vol. 18A, pp. 193-212, New York.

38. Yeung, P.K. (2002). Lagrangian investigations of turbulence, Annual Review of Fluid Mechanics, Vol. 34, pp. 115-142.

39. Zannetti, P. (1990). Air Pollution Modeling. Teories, Computational Methods and Available Software, Kluwer Academic Publisher, 444pp, New York.

40. Zilitinchevick, S.S. (1972). On the determination of the height of the Ekman boundary layer, Boundary-Layer Meteorology, Vol. 3, pp. 141-145.

Chapter 8

AIR POLLUTION AND CULTURAL HERITAGE: SEARCHING FOR "THE RELATION BETWEEN CAUSE AND EFFECT"

Eleni Metaxa

School of Chemical Engineering, National Technical University of Athens, Greece

INTRODUCTION

Pollution of the natural environment is largely unintended and unwanted consequences of human activities in manufacturing, transportation, agriculture and waste disposal. High levels of pollution are largely a consequence of industrialization, urbanization and the rapid increase of human population in modern times. Pollutants are commonly classified according to the part of the environment primarily effected by them, either by air, water or land. Sub-grouping depends on characteristics of the pollutants themselves: chemical, physical, thermal and others. Many pollutants affect more than one resource. The substances that pollute the atmosphere are either gases, finely divided soils, or finely dispersed liquids aerosols. Five major classes of pollutants are discharged into the air: carbon monoxide, sulphur oxides, hydrocarbons, nitrogen oxides and particulates (dust, ash). The principle source of air pollution is the burning of fossil fuels, e.g., coal, oil and derivatives of the latter, such as gasoline, in internal combustion engines or for heating or industrial purposes. The term heritage was used for first time from experts in the early seventies, to declare all the human creation with artistic features, which have been delivered to us as hereditary asset, namely as heritage. At the end of the same decade, the term heritage acquired collective sense and it was used to talk about European Heritage or later about Universal Heritage; in any case to indicate monuments, objects and places. If in a sense culture is the evolution of human life in space and time, the "monumentsremnants" of the human creation of all the times form the prints, the signs, the evidences, the strides of the human-beings progress within the time: "past narrates its history...". Thus, monuments form an undivided entirety with time and place, with man, his surroundings and his history. These unique and unprecedented fingerprints of human civilization form the natural and cultural heritage of a place, of a country, of a people, the peculiar features of a nation which characterize its identity. Cultural heritage is continuously undergoing numerical strains: anthropogenic and natural ones, from which the former can be anticipated or/and prevented, whereas the latter not. The result of these strains is the deterioration of all the materials. In fact, there is no material which is not to be downgraded. The Second Law of Thermodynamics inevitably intervenes and finally results in the deterioration of all the materials. For this reason, materials' deterioration is independent, in practice, on their surroundings and it is taking place in any environment, even without the direct contact of the materials with the constituents of a corrosive environment. Of course, the environment impacts quantitatively the deterioration or corrosion phenomenon taking place, by means of the impact on the rate of the deterioration process(es) and the kind of the produced substances. Air pollution as an anthropogenic reason for materials' deterioration forms a problem of a great importance, because it has catastrophic consequences, universally, in health, in the environment and in the cultural heritage monuments and artifacts. The most famous kind of atmospheric pollution is the photochemical cloud, whose components are complicated chemical reactions in atmosphere, which have as principal reactants the hydrocarbons, nitrogen oxides, sulphur oxides, ozone and ultraviolet radiation. The conservation of works of art and antiquities is intended to: (a) the preservation of cultural heritage, (b) the deceleration of their deterioration processes and (c) the restoration, in some cases, of their form in order to be comprehensible from the public. All of these purposes can be achieved with: (i) control of the environment, (ii) saving static interferences on the monument (i.e., structural conservation), which restore the static sufficiency of the monument, so it does not collapse; (iii) saving interferences on the surface of the monument (i.e., surface conservation), since all decay actions start from the surface of the monument. As the Nobel prize-winners Wolfgang Pauli and Enrico Fermi have felicitously worded: "if God made solids, surfaces were work of Devil"! Indeed, solid surfaces are not uniform, namely homogeneous, but they present heterogeneity, which in general arises from the existence of "imperfections" of various origins. These imperfections are distributed randomly on the surface of the solid material influencing its potential. An absolutely serious scientific approach of the problem of confrontation of historic buildings and monuments decay because of air pollution presupposes the finding of the relation between "cause and effect", namely of "how and why air-pollutants interact with each other and with the solid surfaces". Then someone could interfere and inhibit a corruptive action on them, by restricting even minimizing the conditions are being responsible for. A scientific answer in the previous question presupposes knowledge of the mechanism of materials surfaces deterioration due to polluted surrounding atmosphere.

A SCIENTIFIC APPROACH TO THE PROBLEM OF CULTURAL HERITAGE DETERIORATION DUE TO AIR POLLUTION

In order to study the action of air pollutants on cultural heritage monuments is important not only to obtain results by pure chemical analysis of monuments but also to clarify the mechanism of this action. This mechanism may consist of various steps in series, which are usually rate processes, with the deposition as the first step, or sometimes equilibrium states, such as the distribution of air pollutant(s) between the solid surface and the nearby atmospheric environment through adsorption-desorption phenomena. Thus, a simulation of various physicochemical actions of air pollutant(s) on the solid surface must be done followed by the experimental determination of various physicochemical parameters pertaining to the adsorption-desorption phenomena and possible surface heterogeneous reactions constants as well. A schematic representation of the possible physicochemical actions taking place between air pollutant(s) and monuments surface could be the following one:

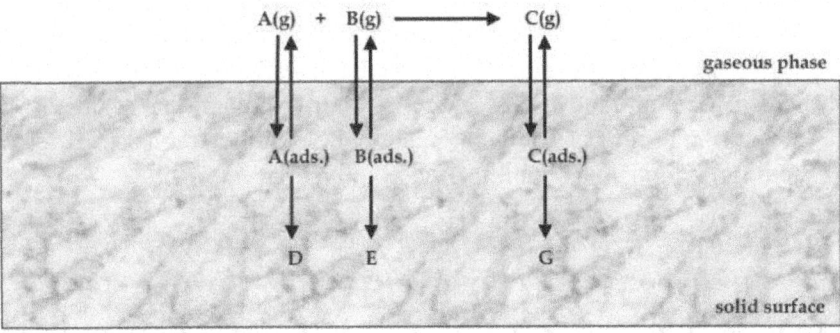

Figure. 1: A model describing the action of air pollutants with the surface of the monument

The model in Fig. 1 is based on the general concept of an open system, consisting of the exposed solid surface, above the which convection currents and diffusion currents as well are causing the transport of the gaseous pollutants

A(g) and B(g) parallel and perpendicular to the solid surface, while a possible simultaneous interaction between them may produce another gaseous pollutant C(g), which may be also adsorbed onto the solid surface or/and desorbed back to the gaseous phase, or to be undergone a surface heterogeneous reaction, e.g. dissociation or isomerization. As soon as gaseous pollutants A(g) and B(g) are nearing the solid surface, adsorption phenomena are taking place, followed either by a surface chemical reaction between the adsorbed species producing D and E, or a desorption of them back to the gaseous phase. Therefore, the rate processes describing the above phenomena are the following ones: (i) diffusion of the pollutants from the gaseous to the solid surface, (ii) adsorption of them onto the solid surface, (iii) a possible surface heterogeneous reaction and (iv) desorption of the pollutants back to the gaseous phase. Therefore an estimation of the crucial relations between environmental factors and materials' deterioration cannot only based on simple measurements of various physicochemical quantities which are validating the materials' decay, but also "timeresolved measurements" are necessary to be done, since only the latter can give information about the actual mechanism of materials' decay. The latter has, in fact, a "local" character, in the sense that it depends on the active sites of the solid surface which are available for adsorption at any particular time t. The achievement of this purpose could be done by using a dynamic experimental methodology, which could supply us with "real-time" measurements concerning the whole physicochemical phenomena taking place. To this direction, the novel method of the Reversed-Flow Inverse Gas Chromatography (RF-IGC) has already been successfully applied for various interacting systems gas−solid material or/and gas1/gas2−solid (e.g., gas=HCs, NO_x, SO_x, O_3, etc. and solid=a marble sample, a ceramic, a pigment, etc.). The results of these applications of RF-IGC in the investigation of the deterioration mechanism of cultural heritage caused by air-pollutants have already been published in high impact factor International Scientific Journals and reported in Scientific Symposiums both in Greece and abroad as well.

A BRIEF OVERVIEW OF VARIOUS METHODS AND TECHNIQUES USED FOR STUDYING ENVIRONMENTAL IMPACTS ON CULTURAL HERITAGE

Cultural heritage is comprised of a great variety of materials including buildings, monuments, pigments and art objects. Thus analytical data are essential for determining the state of conservation of the object, as well as the causes and mechanisms of its deterioration. The analytical methods used in this field of research are identical with those used at the cutting edge of modern science. Techniques developed for advanced physics and chemistry can apply to both of ancient and modern materials, since problems encountered in both the advanced technology and cultural heritage areas are similar. However, there is one essential difference between the analysis of ancient and modern materials, since an art or ancient object cannot be replaced and the consumption or damaging of even a small part of it for analytical purposes must be undertaken only where vital data cannot otherwise be obtained. Thus, a significant number of different modern instrumental methods for cultural heritage characterization are available and they have already been used for the investigation of the weathering effects of air pollution on them, supplying us with information on morphology, chemical composition and structure of the materials present in the monument, archaeological artifact, or art object. Depending on the information required and the procedure involved, the analysis can be considered destructive or nondestructive and it can be carried out on the bulk or the object surface. In addition, the obtained data can be panoramic or sequential and the measurements can be directly performed on the work itself or on a sample, depending on the instrumental technique used. In any case, however, one should aim at the maximization of information and the minimization of the consumed volume of the cultural object.

Materials characterization generally includes determination of chemical composition, of crystalline and molecular structure and of morphology of the object under investigation (A. Doménech-Carbó et al., 2009). The major instrumental methods used for characterizing the chemical composition of the object either in layers or/and in its bulk include: (i) spectroscopic (e.g., XRF, AAS, ICP-AES, Mössbauer spectroscopy) or/and spectrometric techniques (e.g., ICP-MS, LA-ICP-MS), which have been widely used in the identification and determination of major, minor and trace-elements composing either inorganic or organic type cultural objects. The provided information and the application of each specific technique depends on the range of electromagnetic radiation and the phenomenon involved in its interaction with the materials present in the analyzed object (A. Doménech-Carbó et al., 2009; Jenkins, 2000; Putzig et al., 1994); (ii) activation methods (e.g., NAA, PAA), which are based on the interaction of the object material with (fast) neutrons or protons and provide information about the major, minor and trace element composition of the art and archaeological object, which, in turn, can be used to establish their provenance and temporal origin (A. Doménech-Carbó et al., 2009). Concerning the characterization of the crystalline and molecular structure of cultural goods, the analytical techniques most frequently used are grouped into diffraction methods (XRD), spectroscopic (e.g., UV-VIS, FTIR, DRIFT, ATR, FTIR-PAS, Raman, NMR, EPR) and spectrometric methods (e.g., MS, DTMS, DPMS, MALDI), chromatographic methods and thermoanalytical methods (e.g., TG, DTA, DSC) (A. Doménech-Carbó et al., 2009; Jenkins, 2000; Putzig et al., 1994). The majority of instrumental methods which yield morphological, topological and textural information of objects are mostly microscopy techniques (e.g., light microscopy (LM), electron microscopy (SEM, ESEM, TEM) and atomic force microscopy (AFM)) (A. Doménech-Carbó et al., 2009). By using light microscopy (either the low-magnification or the high-magnification technique), characteristics of materials such as the percentage of aggregates, pores, temper or specific minerals, pore or grain size and grain shape as well can be determined, allowing for a better analysis and interpretation of composition, technology, provenance, deterioration and conservation. In addition, the use of electrons instead of light in these instruments permits the characterization of the finest topography of the object surface and additional analytical information can be obtained. The AFM maps the topography of a substrate by monitoring the interaction force between the sample and a sharp tip attached to the end of a cantilever, so that the morphology of the surface of the studied solid sample can be reproduced at nanometer resolution (A. Doménech-Carbó et al., 2009). In addition, whenever a more elaborate surface analysis is pursued, methods based on the interaction of the incident energy provided by a microbeam of photons, electrons, or particles with the atoms or molecules located in the surface of the object sample are used. In such studies, the concept of "surface" should not considered in a strict sense, since the investigation concerns a depth in the range of a few μm on the solid surface. Such surface analysis techniques most frequently used in the characterization of cultural objects include high-resolution spatially resolved microspectroscopes, such as micro-FTIR (μFTIR), microRaman (μRaman), laser-induced breakdown spectroscopy (LIBS), micro-XRF (μXRF), XPS, PIXE, etc (A. Doménech-Carbó et al., 2009; Giakoumaki et al., 2007; Jenkins, 2000; Putzig et al., 1994). It is worthy of noting that the time-resolved versions of the previous spectroscopic methods (TRS), although it is not new, has opened up a wide range of nascent application areas, including test and measurement in materials characterization. Though the basic technique differs little from the traditional spectroscopic methods, it allows us to measure the temporal dynamics and the kinetics of photophysical processes. The advantage of TRS over traditional spectroscopy is that it enables scientists to make more exact measurements of a sample's properties (Bhargava and Levin, 2003; Isnard, 2006; Miliani et al., 2010; Osticioli et al., 2009; Putzig et al., 1994; Quellette, 2004). In what follows some representative examples of various analytical techniques commonly used in this field are reported for a better understanding of the particular contribution of each method used.

FTIR-studies in materials decay: Infrared radiation is usually defined as that electromagnetic radiation whose frequency is between ~ 14300 and 20 cm^{-1}

(namely, ~ 0.7 and 500 μm). Within this region of the electromagnetic spectrum, chemical compounds absorb IR-radiation providing there is a dipole moment change during a normal molecular vibration, molecular rotation, molecular rotation/vibration, or a lattice mode or from combination, difference and overtones of the normal molecular vibrations. The frequencies and intensities of the IR-bands exhibited by a chemical compound uniquely characterize the material and its IR-spectrum can be used to identify and quantify the particular substance in an unknown sample. Thus, FTIR and μFTIR-spectroscopy is useful for the study of degradation forms of cultural heritage, as it permits to identify the degradation phases and to establish the structural relationship between them and the substrate. A representative example of application of this method concerns the results obtained on marble from a Roman sarcophagus, located in the medieval cloister of St. Cosimato Convent in Rome (Italy) and on oolitic limestone from the façade of St. Giuseppe Church in Syracuse (Sicily). The IR-spectra of these samples showed the presence of degradation products composed of calcium sulphate hydrate, commonly called gypsum ($CaSO_4 \cdot 2H_2O$) and calcium oxalate, as well as the presence of organic matter probably due to conservation materials. The qualitative distribution maps of degradation products, obtained by means of micro-FTIR (μFTIR) operating in ATR-mode, revealed that the degradation process is present deep inside the stones also if it is not visible macroscopically (La Russa et al., 2009).

- SEM-studies in deterioration of glass: Deterioration of glass includes both chemical and structural changes. The initial stage of attack is a process that involves ion-exchange between alkali ions, which are present in the silicate structure of the glass, such as Na, K, and hydrogen from the environment. This leads to the formation of a leached or socalled "gel layer" in which alkaline elements are depleted. In case of atmospheric attack, the leached ions will interact with components from the ambient air such as carbon dioxide and sulphur dioxide which will lead to a crust formation including products such as a calcite ($CaCO_3$) and gypsum ($CaSO_4 \cdot 2H_2O$) (Adriaens, 2005).

- A combination of stereo-microscope, XRD and ICP-OES techniques was used (Elgohary, 2008) for the investigation of stone degradation due to air-pollution in Amman citadel of Liwān. The whole investigation and specific measurements showed that the damage produced on the surfaces of various calcareous stone samples of this region, either being physical or chemical, such as crustation, crystallization, dirties accumulations and other deteriorating forms, was essentially the result of the synergistic action of rain water and the various gaseous pollutants at prevailed in the region under study.

GAS CHROMATOGRAPHIC INSTRUMENTATION FOR STUDYING THE IMPACTS OF AIR POLLUTION ON CULTURAL HERITAGE

A Brief Overview of Gas Chromatographic Techniques

Chromatography is a separation method that combines separation and analysis. It is wellknown that chromatographic separations are based on physicochemical processes such as diffusion, adsorption and chemical equilibrium of the studied solutes distributed among the mobile and the stationary phase. Gas chromatography (GC) is a technique that is used not only to separate substances from each-other, but also to study physicochemical properties. Some of these properties measured are concerned with the moving gaseous phase, giving emphasis on the determination of the properties of the solutes; for instance, diffusion coefficients of solutes into the carrier gas. Gas chromatographic analysis suffers from the socalled broadening factors, the majority of which is related to non-fulfilment of the assumptions under which the central chromatographic equation embraced by Van Deemter is derived; namely, the non-negligible axial diffusion of the solute gas in the chromatographic column, the non-linearity of the distribution (e.g. adsorption) isotherm and the non-instantaneous equilibration of the solute distribution among the mobile and the stationary phase. However, through these broadening factors gas chromatography is capable of making physicochemical measurements, which lead to very precise and accurate results, by using relatively cheap instrumentation and very simple experimental arrangements. Among the most widely used gas chromatographic methods for physicochemical measurements are the traditional techniques of elution development, frontal analysis and displacement development under constant gas flow-rate (Cazes, 2009).

The majority of gas chromatographic physicochemical measurements has been done by the inverse gas chromatography (IGC) technique, which uses the same experimental procedures employed in direct gas chromatography, but it focuses its interest on the stationary phase and its behavior towards known probe solutes; for instance, the catalytic properties of the solid stationary phase for reactions between gases. As in direct GC, the results used in IGC to derive information about the physicochemical properties of the stationary phase are based on net retention volumes, broadening of elution peaks and further on the analysis of the statistical moments of the peaks. The usual inverse gas chromatography (IGC), having the stationary phase of the system as the main object of investigation, is an integration method and not a time-resolved chromatography, since it totally ignores the heterogeneity of the adsorbing solid surface, it does not take into account the non-linearity of isotherms, the non-negligible axial diffusion in the chromatographic column and the kinetics of mass transfer across the gas/solid boundary (Cazes, 2009; Katsanos & Karaiskakis, 2004; Thielmann, 2004). All the afore-mentioned chromatographic systems are not usually in true equilibrium during the retention period, so that extrapolation to infinite dilution and zero carrier-gas flow-rate is required to approximate true equilibrium parameters. Moreover, they have not a time-resolved character of the experimental procedure, since they provide measurements for physicochemical properties statistically weighed over time and enclosed by the chromatographic elution peaks; some of these properties are indeed independent of time, but there are other properties strongly dependent on the time variable. A new version of IGC is a flow perturbation method, the so-called Reversed-Flow Inverse Gas Chromatography (RF-IGC), which has been introduced in 1980 by N. A. Katsanos et al., and since then it is extensively used as a tool to study various physicochemical processes (Katsanos, 1988; Katsanos & Karaiskakis, 2004). It is a differential method depending neither on retention times and net retention volumes, nor on broadening factors and statistical moments of the elution bands. In addition, the results of RF-IGC do not need extrapolation to infinite dilution and zero carrier gas flow rate to approximate true physicochemical parameters. All the determinations achieved by RF-IGC are based on rate measurements over an extended period of time, thus constituting a time-resolved chromatography (Katsanos & Karaiskakis, 2004).

The Novel Method of RF-IGC: Physical Description and Experimental Setup

The Reversed-Flow Inverse Gas Chromatography (RF-IGC) method: (i) abandons the main role of carrier-gas in classical gas chromatography and substitutes it with gaseous diffusion currents inside a new diffusion column perpendicular to the conventional chromatographic current (sampling column), the latter being a little far from the solid bed in which all the desired physicochemical phenomena take place in the absence of gas running; (ii) by means of a four or six port valve the direction of carrier-gas flow is reversed from time to time for short time intervals, thus creating extra narrow chromatographic peaks which are deposited onto the conventional chromatographic signal. All the above described are schematically presented in Figs. 2 and 3. By introducing these modifications, the carrier gas flow does not intervene with the measurement of the desired physicochemical quantities, which describe step by step the entire physicochemical phenomena taking place inside the diffusion column where no carrier gas flows but only a static pressure of it exists.

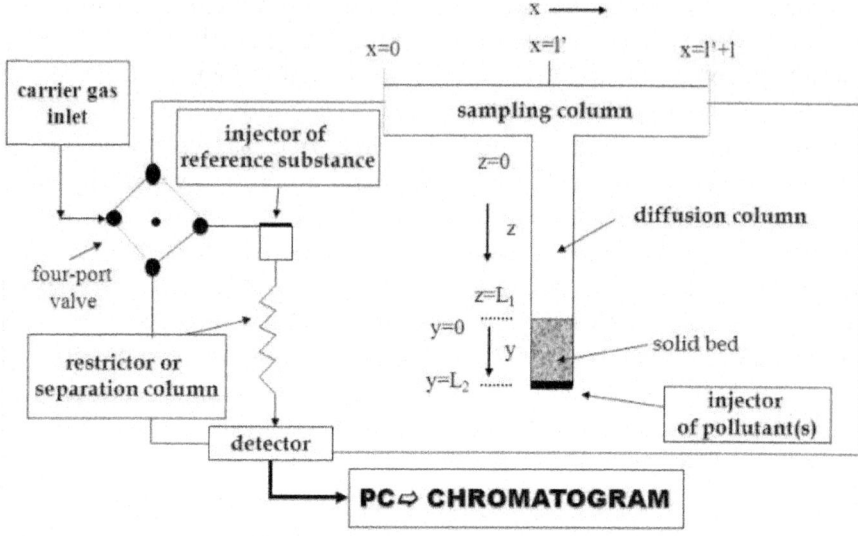

Figure. 2: Experimental setup of RF-IGC

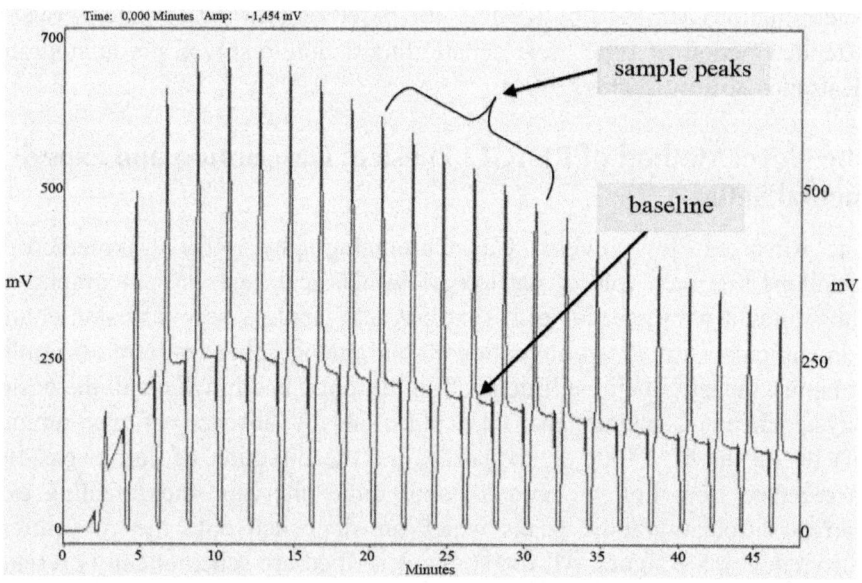

Figure. 3: A typical chromatogram obtained by RF-IGC

The extra chromatographic peaks (Fig. 3) obtained by repeatedly reversing the carrier gas direction for short time intervals are termed sample peaks, because they constitute samples of the phenomena taken from the region of their occurrence at various times, like small samples taken from a reaction occurring in a usual chemical flask containing the reactants. They have different heights depending on the time at which each flow reversal was made. Since this happens at various chosen times, it constitutes a time-resolved experiment like those in chemical kinetics. The experimental details by means of which the reversals are effected are shown in Fig. 2. From the series of the sample peaks obtained under various conditions, several physicochemical quantities have been determined and published (Agelakopoulou et al., 2009; Arvanitopoulou et al., 1994; Bakaoukas et al., 2005; Floropoulou et al., 2009; Katsanos et al., 1998,2003,2004; Metaxa et al., 2009a,2009b,2009c; RoubaniKalantzopoulou, 2004,2009; Roubani-Kalantzopoulou et al., 1996; Sotiropoulou et al., 1995). The sample peaks are predicted theoretically by the so-called chromatographic sampling equation (1), which describes the concentration-time curve of the sample peaks created by the flow reversals and has been derived using mass balances, rates of change, etc., and integrating the resulting partial differential equations under given initial and boundary conditions. It gives the concentration of the solute at the junction of the sampling and the diffusion column x=l' or z=0 of Fig. 2, for different values of the time variable. The sampling equation predicts the sample peaks theoretically and its predictions coincide with the experimental sample peaks shown in Fig. 3, the only difference being that the peaks predicted are square, whereas those actually found are not square owing obviously to nonideality. In fact, the experimental peaks can be made as narrow as we want, since the width at their half-height is equal to the duration of the carrier-gas flow reversal. The equation describing the height, H, of the sample peaks as a function of time, t, when each flowreversal was made has the form:

$$H^{1/M} = g \cdot c(l',t) = \sum_i A_i \exp(B_i,t)$$

(1)

where i runs from 1−4, M is the response factor of the detector used (M=1 for a flame ionization detector), g is the calibration factor of the detector (in cm^4 mol), c(l',t) is the measured sampling concentration of the gaseous analyte (in $mol \cdot cm^{-3}$) at x=l' or z=0 of Fig. 2, and A_i, Bi are functions of the physicochemical quantities pertaining to the various phenomena occurring in the solid bed region. The detailed content of Ai and Bi, as found from a non-linear least-square analysis of the plot of $H^{1/M}$ versus time t, leads to the clear determination of the physicochemical quantities of the mathematical model used; for instance, catalytic reaction constants, adsorption-desorption rate constants, gas and surface diffusion coefficients, local adsorption isotherms, local adsorption energies, local adsorption energy probability density functions, local lateral molecular interactions and adsorption rates, as will be explained for the action of various gaseous pollutants on calcareous stones, marbles and statues in the next sections. In addition, a brief account of the general principles which construct the mathematical model of the method of RF-IGC will be given in the following section.

Mathematical Model

The theoretical analysis for the measurement of the time resolved physicochemical parameters by RF-IGC is based on the following equations:

i.   two mass balances for the gaseous concentration of the analyte in the regions y and z, c_y (mol·cm^{-3}) and c_z (mol·cm-3), respectively;

$$\frac{\partial c_y}{\partial t} = D_2 \frac{\partial^2 c_y}{\partial y^2} - k_{-1} \frac{a_S}{a_y}(c_S^* - c_S) - k_{app}c_y$$

(2)

$$\frac{\partial c_z}{\partial t} = D_1 \frac{\partial^2 c_z}{\partial z^2} - k_{app}c_z$$

(3)

ii.   one rate of change of the adsorbed concentration c_S (mol· g^{-1}) of the analyte in the region y and

$$\frac{\partial c_S}{\partial t} = k_{-1}\left(c_S^* - c_S\right) - k_2 c_S$$

(4)

iii. one local adsorption isotherm, which correlates the adsorbed equilibrium concentration c_S^* (mol· g-1) on the solid surface with the non-adsorbed concentration c_y:

$$c_S^* = \frac{m_S}{a_S}\delta(y - L_2) + \frac{a_y}{a_S}k_1\int_0^t c_y(\tau)d\tau$$

(5)

k_{app}: is the apparent rate constant of a first- or pseudofirst-order reaction of the gaseous adsorbate in the gas phase (in s^{-1}).

D_2 : is the diffusion coefficient of this gaseous adsorbate into the gas phase in section y (in cm$_2$ s $^{-1}$).

k^{-1} : is the rate constant for desorption of the solute from the solid bulk (in s^{-1}).

c_S : is the adsorbed concentration of the gaseous adsorbate adsorbed on the solid at time t (in mol g^{-1}).

k_2 : is the rate constant of a possible first-order or pseudofirst-order surface reaction of the adsorbed solute (in s^{-1}).

With the initial conditions $c_y(0,y) = \frac{m}{a_y}\delta(y-L_2)$, and $c_S(0,y) = 0$, m being the amount (mol) of the gaseous adsorbate introduced as a pulse at $y=L_2$, all the required adsorption parameters are calculated from the experimental data – pairs (H, t) and various geometrical characteristics of the diffusion column and the solid bed – on the basis of the following equations, by means of a suitable PC-program based on non-linear least-squares regression analysis (Agelakopoulou et al., 2009; Arvanitopoulou et al., 1994; Bakaoukas et al., 2005; Floropoulou et al., 2009; Katsanos et al., 1998,2003,2004; Metaxa et al., 2009a,2009b,2009c; Roubani-Kalantzopoulou, 2004,2009; Roubani-Kalantzopoulou et al., 1996; Sotiropoulou et al., 1995):

a.   local adsorption energies, ε

$$\varepsilon = RT\left[\ln(KRT) - \ln(RT) - \ln K^0\right]$$
(6)

b.   local adsorption equilibrium concentrations, c_s * and c_y

$$c_s^* = \frac{a_y}{\alpha_s}k_1\frac{vL_1}{gD_z}\sum_{i=1}^{3}\frac{A_i}{B_i}\left[\exp(B_it)-1\right]$$
(7)

$$c_y = \frac{vL_1}{gD_1}\sum_{i=1}^{3}A_i\exp(B_it)$$
(8)

c.   local adsorption isotherm, θ_t

$$\theta_t = \frac{c_s^*}{c_{s\,max}^*}$$
(9)

$$\theta = 1 - \frac{1}{c_{s\,max}^*}\cdot\frac{1}{KRT}\cdot\frac{\partial c_s^*}{\partial c_y}$$
(10)

d.   local monolayer capacity, $c_{s\,max}^*$:

$$c_{s\,max}^* = c_s^* + \frac{\partial c_s^*/\partial c_y}{KRT}$$
(11)

e.   probability distribution function for adsorption energy, $\varphi(\varepsilon; t)$:

$$\varphi(\varepsilon;t) = \frac{\theta}{c_{s\,max}^* RT}\left[\frac{KRT\left(\partial c_s^*/\partial t\right)+\partial^2 c_s^*/\partial c_y\partial t}{\partial(KRT)/\partial t} - \frac{\partial c_s^*/\partial c_y}{KRT}\right]$$
(12)

f.   dimensionless parameter, β, for lateral interactions:

$$z\omega = \beta RT$$
(13)

where: ω is the lateral interaction energy, z the number of neighbors for each adsorption site and $^{\perp}$ is a dimensionless parameter. Thus, the "$\theta z\omega$" is the added to ε "differential energy of adsorption due to lateral interactions", namely:

$$\theta z\omega = \beta\theta RT \qquad (14)$$

All these relations are based on the Jovanovic local isotherm Eq. (15):

$$\theta(p,T,\varepsilon) = 1 - \exp(-Kp) \qquad (15)$$

where

$$K = K^0(T)\exp(\varepsilon/RT) \qquad (16)$$

R being the gas constant, and

$$K^0 = \frac{h^3}{(2\pi m)^{3/2}(kT)^{5/2}} \cdot \frac{v_S(T)}{b_g(T)} \qquad (17)$$

where: m is the molecular mass of the adsorbate; k is the Boltzmann's constant; h the Planck's constant; and the ratio $u_s(T)/b_g(T)$ of two partition functions, namely that of the adsorbed molecule, $u_s(T)$, and that for rotations-vibrations in the gas phase $b_g(T)$. This ratio is taken as a unity, approximately, as was done before (Agelakopoulou et al., 2009; Arvanitopoulou et al., 1994; Bakaoukas et al., 2005; Floropoulou et al., 2009; Katsanos et al., 1998,2003,2004; Metaxa et al., 2009a,2009b,2009c; Roubani-Kalantzopoulou, 2004,2009; Roubani-Kalantzopoulou et al., 1996; Sotiropoulou et al., 1995). The contribution of lateral molecular interactions in the overall phenomenon of adsorption and desorption is taken into account, by correcting Eq. (16) to include this type of energy:

$$K' = K^0 \exp\left(\frac{\varepsilon}{RT} + \beta\theta_i\right) = K\exp(\beta\theta_i) \qquad (18)$$

Thus, Jovanovic isotherm described in Eq. (15), is modified accordingly, as well as any other equation based on it.

Applications of RF-IGC in Studying Physicochemical Phenomena Being Responsible for Materials' Decay

The differential time-resolved RF-IGC method provides a new pathway for solids characterization, by supplying us with experimental local values of important physicochemical quantities, such as adsorption energy, adsorption isotherm, monolayer capacity, non-adsorbed concentration of gaseous analyte in equilibrium with the solid surface, probability density function and energy owing to lateral molecular interactions, which pertain to the particular surfaces. It is reminded that the term "local" means "with respect to time t", namely it regards only adsorption sites active at time t. In the following sections, some representative results are described, which have already published in high impact International Scientific Journals and announced in Scientific Symposiums, concerning cultural heritage monuments in Greece (Agelakopoulou et al., 2009; Arvanitopoulou et al., 1994; Bakaoukas et al., 2005; Floropoulou et al., 2009; Katsanos et al., 1998,2003,2004; Metaxa et al., 2009a,2009b,2009c; Roubani-Kalantzopoulou, 2004,2009; Roubani-Kalantzopoulou et al., 1996; Sotiropoulou et al., 1995).

The Local Character of Adsorption in Materials Decay: Chemisorption or Physisorption?

Due to their different crystallographic properties, natural stones exhibit different types of structural surfaces—each one with its own adsorption energy distribution function—which determines the variations in the observed weathering, deterioration patterns and processes of these materials. The adsorption of atoms or molecules onto these heterogeneous surfaces may occur either by the formation of strong chemical bonds (chemisorption) or via weaker physical attachment (physisorption) of the adsorbed molecules. It should be noted that although the majority of the reported results regards experiments done at relatively low temperatures, where physisorption is more expected, the structural materials of samples used (e.g., statues from Museums) are energetically upgraded, as a consequence of the whole process they have been undergone from the moment of their natural formation until the moment of their use as materials for creating statues. Thus, in the presence of an aggressive environment a chemisorption process is favorable; by means of RF-IGC experiments, chemisorption is observed taking place in the beginning of the experiments (Agelakopoulou et al., 2009; Floropoulou et al., 2009; Katsanos et al., 1998,2003,2004; Metaxa et al., 2009a,2009b,2009c; Roubani-Kalantzopoulou, 2004,2009). Furthermore, any kind of adsorption process (chemisorption or/and physisorption) takes place in different extent, because of the development of molecular lateral interactions (attractive or/ and repulsive) between the adsorbed species, in addition to the adsorbent-adsorbate interactions, the latter being the predominant type of forces characterizing chemisorption. In case of chemisorption, the lateral interactions between adsorbates are mostly repulsive, resulting in random topography. The adsorbent-adsorbate interactions leading to chemisorptions are chemical bond forces. The active sites, where chemisorption takes place, are included in the region A and correspond on the minima of potential energy of the surface, namely the maxima of adsorption energy. The distribution of the adsorbed molecules on these active sites is random. These sites correspond to high values of coverage θ and adsorption energy ε. The regions B and C include active centers that correspond to adsorption at sites of lower energy, ε and surface coverage, θ. They are characterized by a patchwise or/and an island topography of the adsorbed molecules and created by weak forces of Van der Waals type between the admolecules. Adsorption sites in B and C are not on a free surface, but they are created through the lateral interactions with molecules have been already adsorbed. When chemisorption is taking place, high adsorption energy values are observed, since they correspond to the minima of the potential energy of the surface. Contrary to this, physisorption entails low adsorption energy values, since it usually corresponds to the maxima of the surface potential (saddle points) [Rudzinski & Everett, 1992]. The decrease of adsorption entropy values verifies that adsorption occurred. All the above are depicted in Figs. (4a)-(4d). The "topography" of the solid surface is described experimentally through the time-resolved analysis of the energy distribution function $\varphi(\varepsilon;t)$ versus t, where distinguished regions corresponding to different kinds of active sites produced at different times are observed, as the three types of active sites recorded in Figs. (4c) and (5c). The number of distinct peaks recorded corresponds to different kinds of active sites appear and the area under these peaks corresponds to the collection of adsorbed species onto active sites with a definite mean energy value; in fact, this reflects or/and explains the "local" character of adsorption. It is worthy of noting that these results coincide with those extracted with simulation methods for the submonolayer adsorption of argon on the surface of crystalline rutile (Bakaev& Steele, 1992). In the absence of SO2, an adsorption-induced surface reconstruction creates new adsorption sites, as it is indicated in Figs. 4 (a, b, d) and Figs. 5 (a, b, d). The above observed surface reconstruction [Agelakopoulou et al., 2009; Floropoulou, 2009; Metaxa et al., 2009a, 2009b] may be attributed to one or more of the following reasons:

- An entropy increase owing to the close approach of adsorbates in various configurations which increase the surface entropy and unveil new active sites for adsorption on the solid surface [Jansen, 2008].
- The creation of sub-surface states which does not affect the surface free-energy [Christmann, 1995].
- A co-variation in chemisorption and physisorption so that an equilibrium has been established between adsorption and desorption.

The most possible reason in this case seems to be the entropic one, because of the momentary increment of the adsorption entropy observed in Fig. (4d) as a local maximum, which is followed by a new slight decrease in the adsorption entropy thus confirming the readsorption induced. Obviously, the third kind of active sites observed for the adsorption of C_2H_2 on the surface of the statues

(L291 of Kavala and L1991 of Philippi), in the absence of SO_2, resulted from this surface-reconstruction.; in the presence of SO_2, only two different types of active sites are observed, as Fig. (4c) indicates. Although the above-mentioned explanation for this surface reconstruction and readsorption induced concerns, in fact, the formation of islands without attractive interactions [Jansen, 2008], another surface reconstruction process which favours the formation of islands due to attractive lateral interactions have been proposed by other researchers (Velasco & Rezzano, 1999). The latter explanation could be also accepted in our case, in the sense that this island-formation occurs in higher time-values and after θ=1, at saddle-points positions (Roubani-Kalantzopoulou, 2004, 2009; Agelakopoulou et al., 2009; Metaxa et al., 2009a).

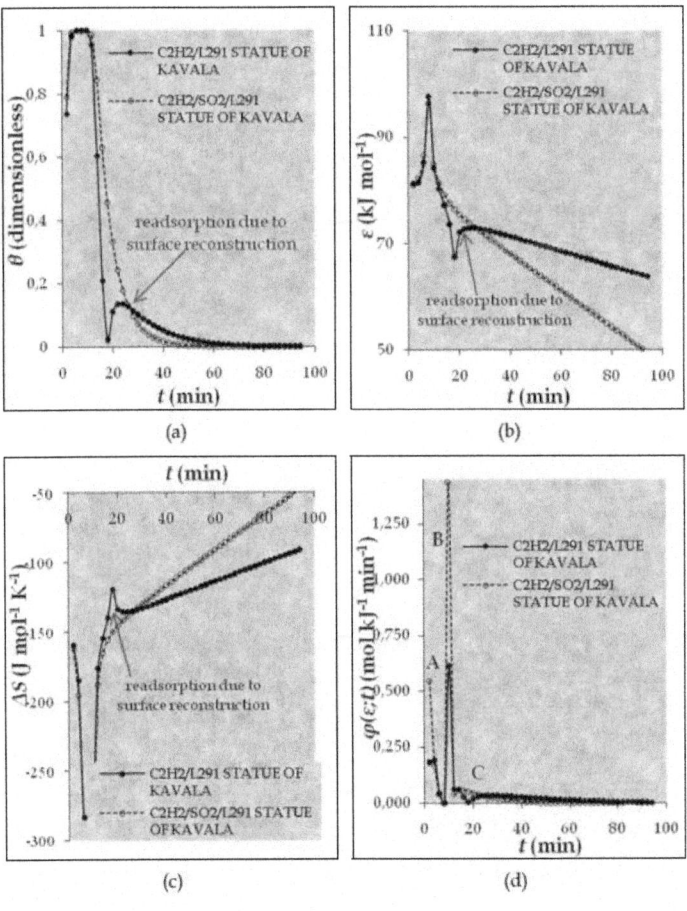

Figure. 4: Time-resolved analysis for the adsorption isotherms (4a), adsorption energies (4b), distribution energy functions (semi-logarithmic plot) (4c), and adsorption entropies (4d), concerning the systems $C_2H_{2(g)}/(SO_{2(g)})/L291$ Statue of Kavala, Greece.

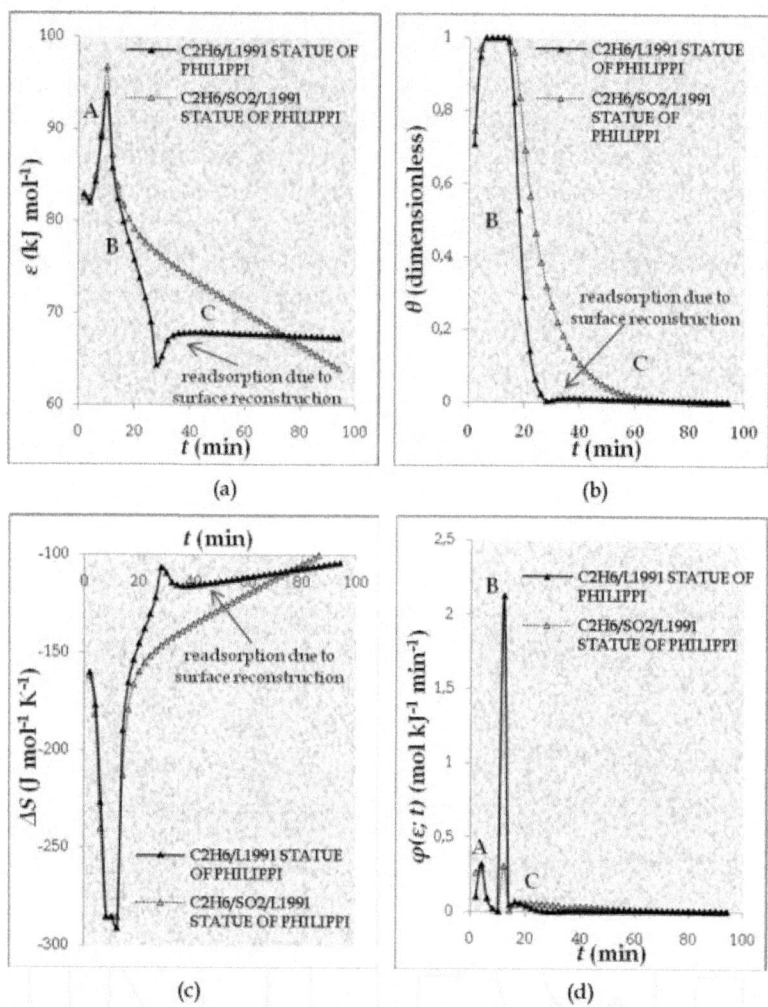

Figure. 5: Time-resolved analysis for the adsorption isotherms (4a), adsorption energies (4b), distribution energy functions (semi-logarithmic plot) (4c), and adsorption entropies (4d), concerning the systems $C_2H_{2(g)}/(SO_{2(g)})/L1991$ Statue of Philippi, Greece.

The role of Synergy in the Adsorption Phenomena

The synergistic effect of a second pollutant has also been examined and is very obvious how it operates in each case. For example, in the presence of SO_2 lower values for c^*_{ssmax} are determined for the adsorption of ethane on the surface of the ancient statue L1991 taken from the interior of the Museum of Philippi, near Salonica, in Greece. This fact could be

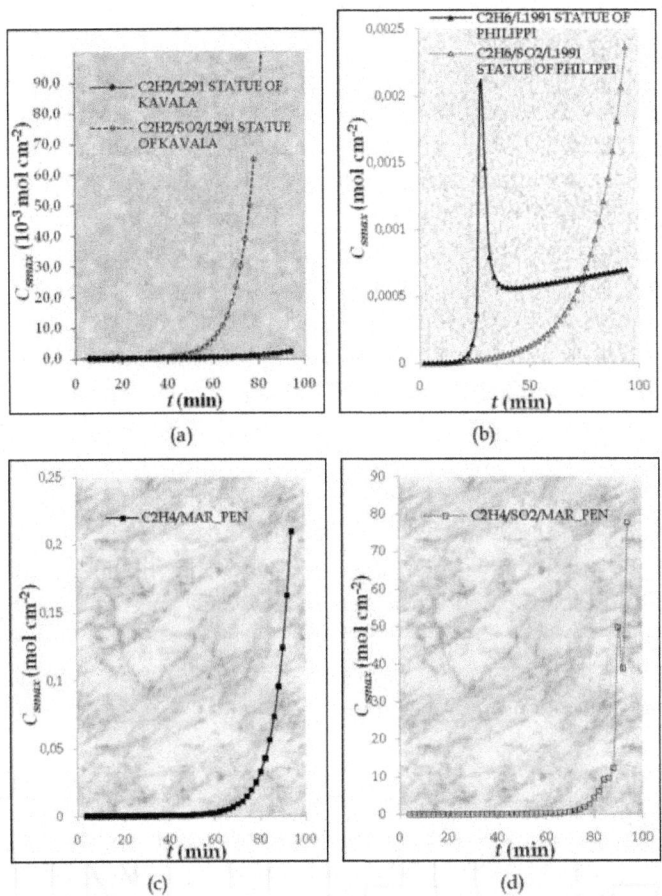

Figure. 6: Time-resolved analysis of local molecular capacity, c^*_{ssmax}, for the systems: (a) $C_2H_2/(SO_2)/L291$ statue of Kavala museum, (b) $C_2H_6/(SO_2)/L1991$ statue of Philippi museum and (c,d) $C_2H_4/(SO_2)/$Pentelic marble.

attributed either to an oxidation of the hydrocarbon molecules from SO_2 in the gaseous phase before adsorption takes place or to a competitive adsorption of SO_2-molecules towards ethane molecules on the active sites of the statue surface. This fact is confirmed by Fig. 5d, where the number of active sites on the statue surface dramatically decreases in the presence of sulphur dioxide. On the other hand, an opposite behavior is observed concerning the synergistic action of sulphur dioxide on the adsorption of acetylene on the surface of the ancient statue L291 - a pure calcite - from the exterior of Kavala Museum and on the adsorption of ethylene on the surface of a recently cut sample from Penteli mountain ore in Dionysos, Greece. The latter is also confirmed by Fig. 4d, where the number of active sites available for the adsorption of acetylene on the surface of L291 statue increases significantly in the presence of sulphur dioxide. Analogous observations have been drawn for the adsorption of ethane, ethene and ethyne (acetylene) on the surface of another ancient statue sample (L351), which was taken from the interior of the Kavala Museum and it was a pure dolomite, as an X-Ray diffraction analysis of this sample showed. Finally, the amounts of the hydrocarbons which totally adsorbed (c^*_s) on both of the statues from Museum of Kavala was calculated, with or without the presence of sulphur dioxide (Agelakopoulou et al, 2009). The results show that with the presence of SO_2 an increment of the estimated total adsorbed amount of acetylene is noted for the calcite's statue (L291) of Kavala, contrary to the other hydrocarbons where the presence of sulphur dioxide causes only a negligible or no effect. As regards the dolomite's statue (L351) of Kavala, the synergistic effect of SO_2 is more profound. First of all, the totally adsorbed amount of each hydrocarbon on L351 statue, either with or without the presence of sulphur dioxide, is higher than in case of L291 statue, a fact that is ascribable to the higher porosity of the former, which is dolomite, whereas the latter is calcite. Secondly, the higher adsorbed amount found for ethane in the absence of SO_2, something is reversed in the presence of SO_2 and concerns acetylene. In addition, the synergistic effect of SO_2 decreases from acetylene to ethene and ethane; obviously, the order of the bond of hydrocarbon has a significant role in the adsorption phenomenon.

The Influence of the Hydrocarbon's Bond on the Adsorption Phenomena

For the same solid adsorbent (statue, pure oxide, etc.), the influence of the type of the bond in the molecule of the hydrocarbon is related with its molecular weight which reflects on the diffusion coefficient of the molecule.

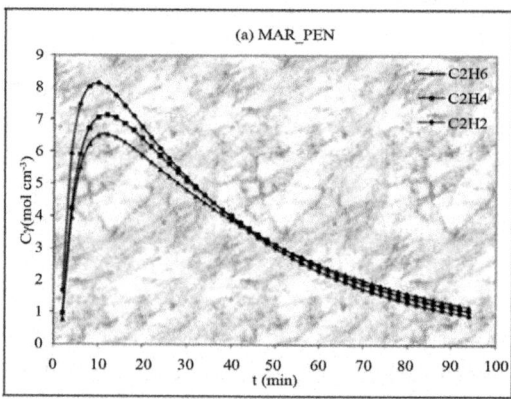

Figure. 7(a): Time-resolved analysis of the local non-adsorbed equilibrium concentration, c_y, of each hydrocarbon adsorbed on the various solid substrates: (a) marble of

Penteli, (b) L1991 statue of Philippi, (c) L291 statue of Kavala and (d) L351 statue of Kavala.

As the molecular weight of the hydrocarbon increases, the corresponding diffusion coefficient decreases. This becomes obvious in the following diagram (Fig. 7) which depicts the non-adsorbed concentration, c_y, of the hydrocarbon as a function of time t:

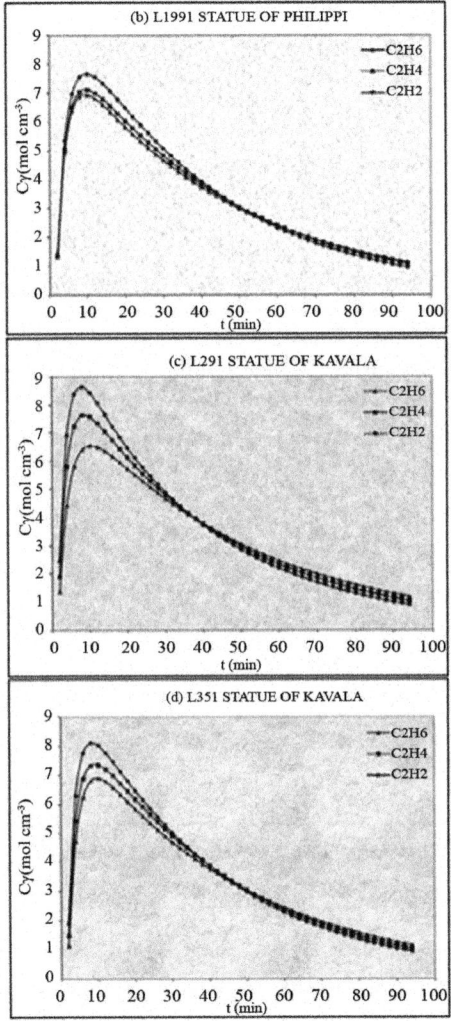

Figure. 7(b, c, d): Time-resolved analysis of the local non-adsorbed equilibrium concentration, c_y, of each hydrocarbon adsorbed on the various solid substrates: (a) marble of Penteli, (b) L1991 statue of Philippi, (c) L291 statue of Kavala and (d) L351 statue of Kavala.

Characterization of Cultural Heritage Deterioration by Means of XRD, SEM and RAMAN analysis

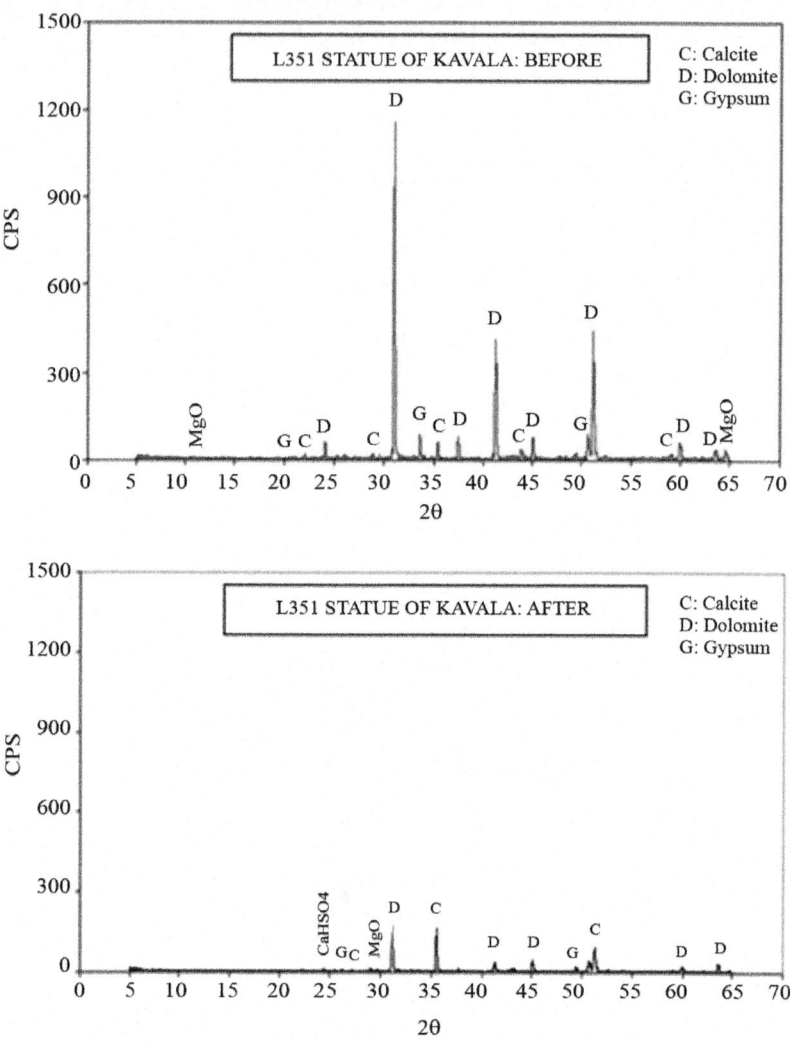

Figure. 8: XRD-diagram for the L351 statue from the interior of the Museum of Kavala: (on the top) before the injection of gaseous pollutants and (at the bottom) after the injection of gaseous pollutants.

The characterization of the above-mentioned cultural heritage materials through diffraction (XRD), spectroscopic (Raman) and microscopy techniques (SEM) was achieved following to the RF-IGC experiments. For example, the XRD-analysis for the L351 statue from the interior of the Museum of Kavala showed that it is a marble mostly composed of dolomite and quite less of calcite; the opposite stands for the L291 statue from the exterior of the same Museum. As it can also be revealed from the XRD-analysis (c.f. Fig. 8), some gypsum was detected before the exposure of the statue sample to the gaseous pollutants, so either the gypsum was a minor component of the marble or it was formed during a previous exposure of the statue to an atmosphere that permitted this formation. After the exposure to the pollutants, no gypsum was detected by any method used. This is rather expected because there was no humidity during the contact of the injected sulphur dioxide onto the marble. More, an improvement in the organization of the crystallites of the sample is observed after the injection of the gaseous pollutants. In the Raman spectra (c.f. Fig. 9) for the same statue after the exposure an accumulation of weak peaks at 1970-2200 cm^{-1} is distinguished owing to acetylene (C_2H_2). More, the main peak existed in 1095 cm^{-1}, before the exposure to the gaseous pollutants, which concerns the carbonate ion, is absent in the spectrum after the injection of the pollutants. The SEM-images for both statues (L351: dolomite and L1991: calcite) are shown in Fig. 10, where is evident the coarse and porous structure of dolomite. Calcite seems more tight and compact. The combination of Raman and SEM-EDAX analysis showed that SO_2 was adsorbed (Metaxa et al., 2009b).

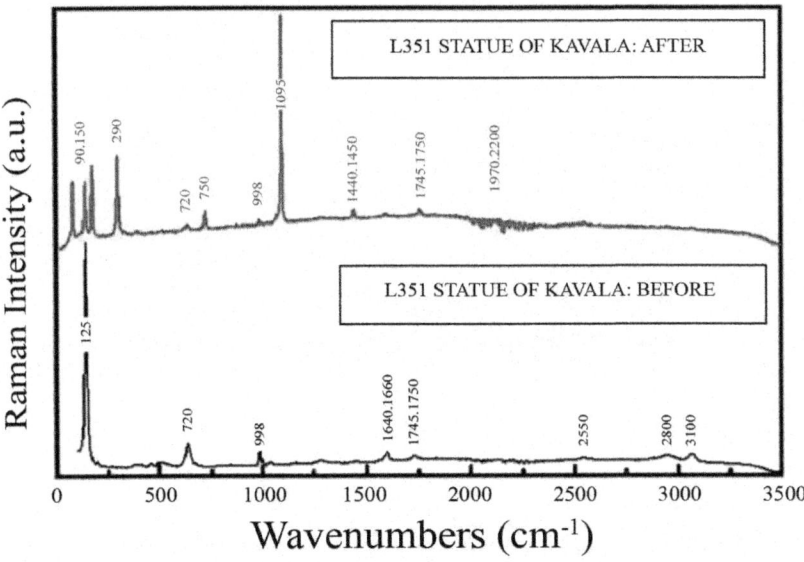

Figure. 9: Raman-spectra for the L351 statue from the interior of the Museum of Kavala: (on the top) before the injection of gaseous pollutants and (at the bottom) after the injection of gaseous pollutants.

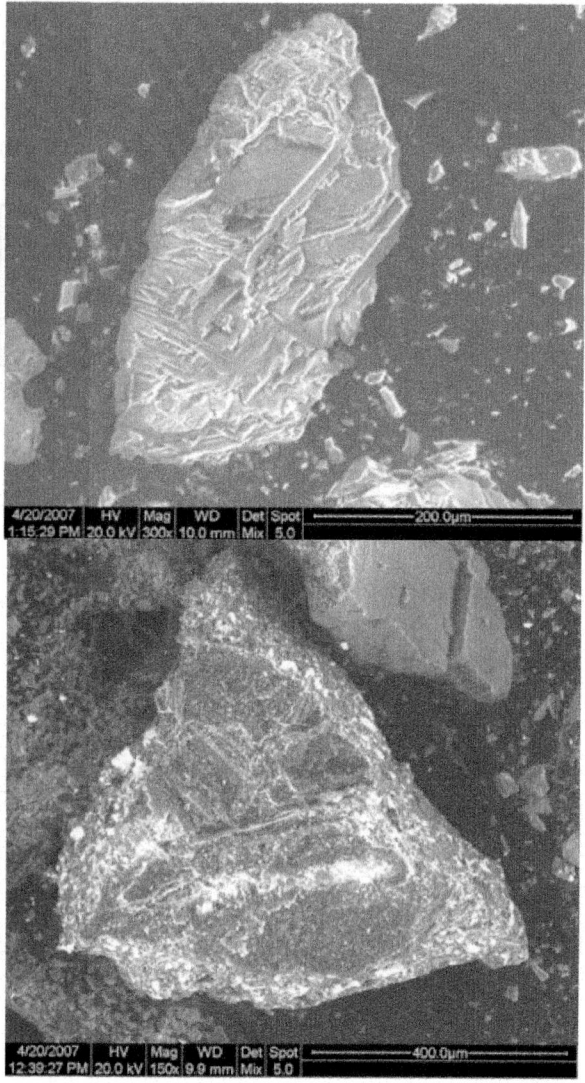

Figure. 10: SEM-images showing the shapes of calcite (on the top) and dolomite (at the bottom).

Extracting Information about Surface Heterogeneity

The "non-ideal" behavior of a solid surface in the direction of the adsorption of a gaseous substance on it consists of an inherent energetic heterogeneity component and an adsorption-induced heterogeneity component originated from the lateral interactions between the adspecies. Surface heterogeneity is responsible for the time-variation in the activation of the various active sites onto the solid surface towards the adsorption process. The effects of surface heterogeneity on adsorption equilibrium have extensively studied, both experimentally and theoretically. The RF-IGC methodology constantly answers for the role of surface heterogeneity on the adsorption phenomena, through the local physicochemical quantities determined for various adsorption systems. The results obtained are in a good agreement with literature data as it has already been shown (Metaxa et al., 2009a).

Study of The Action of Sulfur Dioxide on Penteli Marble in the Presence or the Absence of Protective Materials-Evaluation of the Effectiveness of the Protective Materials for Monuments Against Sulfur Dioxide Corrosion

Calcareous stones, such as marble, suffer from the attack of sulfur dioxide in polluted atmospheres because of the transformation of calcite-content ($CaCO_3$) into gypsum ($CaSO_4 \cdot 2H_2O$) as final product [Elgohary, 2008]. The choice of the repair materials is a crucial point in stone conservation. Such operations, in fact, are much critical, as they can alter the structure of the original material and create new textural heterogeneity at the natural stone-composite interface. In order to avoid unsuccessfully results, the characteristics of materials used in conservation works and their compatibility, including long-term effects, with existing materials should be fully established. Thus, new impregnation products, as well as those which already exist, must be viewed with caution and be subjected to laboratory research before they applied on historic buildings. The choice between "traditional" and "innovative" materials and techniques should be determined case-by-case with preference given to those that are least invasive and most compatible with heritage values, consistent with the need for safety and durability. Protective materials, such as acrylic copolymers and siloxanes, have been largely used in conservation practice as coatings, consolidants and adhesives, because of their good adhesion, film forming properties and their environmental stability. These materials alter the physicostructural properties of the porous materials and change the physicochemical behavior of the interface between the work of art and the environment [Carreti & Dei, 2004]. Therefore, the characterization of the solid surfaces before and after the application of these materials is important for the evaluation of their ability to protect the historic monuments and buildings.

In order to study the action of SO_2 on Penteli marble, experiments were carried out by using the RF-IGC instrumentation and various physicochemical parameters (rate constants as well as equilibrium constants) were calculating by using a non-linear regression analysis PCprograms in GW-BASIC [Katsanos et al., 2003] for the experimental data. Afterwards, kinetic parameters such as adsorption rate constants k_1, adsorption/desorption kR and surface reaction rate constants k_2, as well as surface diffusion coefficients D_y, deposition velocities V_d and reaction probabilities y of SO_2 "on marble surfaces, at various temperatures," in the presence or in the absence of protective materials (an acrylic copolymer, Paraloid B-72 or a siloxane, CTS Silo 111) were calculated and discussed [Bakaoukas et al., 2005]. The results showed that both materials are good enough at low temperatures (303.2-323.2K), while at high temperatures (333.2-353.2K) siloxane acts better as protective material than acrylic copolymer. More specifically, the values of surface reaction rate constant k_2 in all cases where the marble was coated by the acrylic copolymer were bigger than those for pure marble, while they were smaller in all cases where the siloxane was used as protective material. Probably that happens because SO_2 interacts with the acrylic copolymer and not with the siloxane and a mechanism was proposed for this interaction. On the other hand, the values of the adsorption rate constant k_1 for the system (SO_2 + marble) were bigger than those for the systems (SO_2 + coated marble), thus indicating that the adsorption of SO_2 is more difficult in the case of coated marble than in the pure marble. In addition, the values of V_d and y in most cases where the marble was coated with anyone of the protective materials were smaller than those for pure marble, except from the higher temperature of 353.2_K where the differences before and after coating with the acrylic copolymer were negligible; probably, because in this high temperature SO_2 interacts more rapidly with the acrylic copolymer. Concerning the effect of coatings on Dy-values, it was found that in the absence of protective materials, at extreme temperatures ($>333.2_K$), the porous of the marble is being destroyed producing surface diffusion coefficients equivalent to diffusion coefficients in the gas phase. On the contrary, in the presence of protective materials, the later does not happen due to the shrink of the porous size of the marble[Bird et al., 2002; Carreti & Dei, 2004; Bakaoukas et al., 2005].

CONCLUSIONS

Among the various causes which are responsible for the destruction of cultural property, namely of historic artifacts and monuments, air pollution could be considered as an important one. Cultural goods are chiefly significant as a kind of evidence of past human activity. Conservation of cultural heritage allows this evidence to be consulted whenever new questions about the past are emerged. Thus it is well-understood that successful conservation has to be underpinned by a comprehensive understanding of the causes of decay and the factors controlling them. In order to estimate the impacts of air pollution on the various solid surfaces, including them of cultural heritage, in a real scientific basis, theory and experiment needs to cooperate in a way close to real systems. From this aspect of view, the new dynamic version of classic inverse gas chromatography, the so-called Reversed-Flow Inverse Chromatography (RFIGC), combining a powerful mathematical background with a very simple and smart experimental arrangement, attains this purpose by using time-resolved analysis in the way this method has developed it. Thus, by means of a simple PC-program, important local physicochemical quantities are determined, which characterize the main rate processes taking place at the phase boundaries, such as adsorption and desorption, aiming at their spreading in the bulk. Lately, this method has been successfully applied to the study of the impact of air pollutants on many cultural heritage objects, such as marbles, pigments of works of art, etc., by studying the topography of active sites in heterogeneous solid surfaces and their availability for adsorption, without using retention volume data, as it abandons the traditional role of the carrier gas in conventional chromatography and substitutes it with gaseous diffusion currents. The results of this method are based on a non-linear adsorption isotherm model and rate measurements over an extended period of time. All the chromatographic methods known up to now offer approximate functions for the probability density function for the adsorption energy, without any determination of local adsorbed parameters, as this technique does. All findings supplied from the application of this methodology in various systems gas/solid provide valuable information about the susceptibility of the examined artifact or statue to environmental gaseous pollutants.

REFERENCES

1. Adriaens, A. (2005). Review: Non-destructive Analysis and Testing of Museum Objects: An Overview of 5 Years of Research. Spectrochimica Acta Part B, Vol. 60, pp. 1503-1516.

2. Agelakopoulou, T.; Metaxa, E.; Karagianni, Ch.-S. & Kalantzopoulou-Roubani, F. (2009). Air Pollution Effect of SO2 and/or Aliphatic Hydrocarbons on Marble Statues in Archaeological Museums. Journal of Hazardous Materials, Vol. 169, pp. 182-189.

3. Arvanitopoulou, E.; Katsanos, N.A.; Metaxa, H. & Roubani-Kalantzopoulou, F. (1994). I. Simple Measurement of Deposition Velocities and Wall Reaction Probabilities in Denuder Tubes-II. High Deposition Velocities. Atmospheric Environment, Vol. 28, No. 15, pp. 2407-2412.

4. Bakaev, V.A. & Steele, W.A. (1992). Computer Simulation of the Adsorption of Argon on the Surface of Titanium Dioxide. 1. Crystalline

Rutile. Langmuir, Vol. 8, No. 5, pp. 1372-1378.

5. Bakaoukas, N.; Kapolos, J.; Koliadima, A. & Karaiskakis, G. (2005). New Gas Chromatographic Instrumentation for Studying the Action of Sulfur Dioxide on Marbles. Journal of Chromatography A, Vol. 1087, pp. 1693-176.

6. Bhargava, R. & Levin, I.W. (2003). Time-Resolved Fourier Transform Infrared Spectroscopic Imaging. Applied Spectroscopy, Vol. 57, No. 4, pp. 357-366.

7. Bird, R.B.; Stewart, W.E. & Lightfoot, E.N. (2002). Transport Phenomena (2nd ed.), Wiley, ISBN:0471410772, New York.

8. Carreti, E. & Dei., L. (2004). Physicochemical Characterization of Acrylic Polymeric Resins Coating Porous Materials of Artistic Interest. Progress in Organic Coatings, Vol. 49,No. 3, pp. 282-289.

9. Cazes, J. (2009-10-12). Encyclopedia of Chromatography (3rd ed.), CRC Press, ISBN: 1420084593, U.S.A.

10. Christmann, K. (1995). Some General Aspects of Hydrogen Chemisorption on Metal Surfaces. Progress in Surface Science, Vol. 48, No. 1-4, pp. 14-26.

11. Doménech-Carbó, A.; Doménech-Carbó, M.T. & Costa, V. (2009). Chapter 1: Application of Instrumental Methods in the Analysis of Historic, Artistic and Archaeological Objects, In: Electrochemical methods in archaeometry, conservation and restoration (1sted.), Scholtz Fritz (ed.), pp. 1-32, Springer-Verlag, ISBN: 978-3-540-92867-6, Berlin.

12. Elgohary, M.A. (2008). Air Pollution and Aspects of Stone Degradation "Umayyed LiwānAmman Citadel as a Case Study". Journal of Applied Sciences Research, Vol. 4, No. 6,pp. 669-682.

13. Giakoumaki, A.; Melessanaki, K. & Anglos, D. (2007). Review: Laser-Induced Breakdown Spectroscopy (LIBS) in Archaeological Science-Applications and Prospects. Analytical & Bioanalytical Chemistry, Vol. 387, No. 3, pp. 749-760.

14. Isnard, O. (2006). In Situ and/or Time Resolved Powder Neutron Scattering for MaterialsScience. Journal of Optoelectronics and Advanced Materials, Vol. 8, No. 2, pp. 411-417.

15. Jansen, A.P.J. (2008). Island Formation without Attractive Interactions. Physical Reviews B, Vol. 77, No. 7, article: 0732408, 4 pages.

16. Jenkins, R. (2000). X-Ray Techniques: Overview, In: Encyclopedia of analytical chemistry, Meyers, R.A. (eds.), pp. 1-20, John Wiley & Sons Ltd, ISBN: 9780470027318, U.S.A.

17. Katsanos, N.A., (1988). Flow Perturbation Gas Chromatography (1st ed.), Marcel Dekker, New York-Basel.

18. Katsanos, N.A.; Thede, R. & Roubani-Kalantzopoulou, F. (1998). Review: Diffusion, Adsorption and Catalytic Studies by Gas Chromatography. Journal of Chromatography A, Vol. 795, pp. 133-184.

19. Katsanos, N.A.; Gavril, D. & Karaiskakis, G. (2003). Time-Resolved Determination of Surface Diffusion Coefficients for Physically Adsorbed or Chemisorbed Species on Heterogeneous Surfaces, by Inverse Gas Chromatography. Journal of Chromatography A, Vol. 983, No. 1, pp. 177-193.

20. Katsanos, N.A. & Karaiskakis, G. (2004). Time-Resolved Inverse Gas Chromatography and its Practical Applications (1st ed.), HNB Publishing, ISBN: 0-9728061-0-5, New York.

21. La Russa, M.F.; Ruffolo, S.A.; Barone, G.; Crisci, G.M.; Mazzoleni, P. & Pezzino, A. (2009). The Use of FTIR and micro-FTIR Spectroscopy: an Example of Application to Cultural Heritage. International Journal of Spectroscopy, Vol. 2009, pp. 1-5.

22. Metaxa, E.; Kolliopoulos, A.; Agelakopoulou, T. & Kalantzopoulou-Roubani, F. (2009a). TheRole of Surface Heterogeneity and Lateral Interactions in the Adsorption of Volatile Organic Compounds on Rutile Surface. Applied Surface Science, Vol. 255, pp. 6468-6478.

23. Metaxa, E.; Agelakopoulou, T.; Bassiotis, I.; Karagianni, Ch. & Kalantzopoulou-Roubani, F. (2009b). Gas Chromatographic Study of Degradation Phenomena ConcerningBuilding and Cultural Heritage Materials. Journal of Hazardous Materials, Vol. 164,pp. 592-599.

24. Metaxa, E.; Agelakopoulou, T.; Karagianni, Ch.-S. & Kalantzopoulou-Roubani, F. (2009c).Study of the Adsorption of Ozone on the Surface of Ferric Oxide by Revered-FlowInverse Gas Chromatography. Instrumentation Science & Technology, Vol. 37, No. 5,pp. 584-606.

25. Miliani, C.; Rosi, F.; Brunetti, B.G. & Sgamelloti, A. (2010). In Situ Noninvasive Study of Artworks: the MOLAB Multitechnique Approach. Accounts of Chemical Research,Vol. 43, No. 6, pp. 728-738.

26. Osticioli, I.; Mendes, N.F.C.; Porcinai, S.; Cagnini, A. & Castellucci, E. (2009). Spectroscopic Analysis of Works of Art Using a Single LIBS and Pulsed Raman Setup. Analytical & Bioanalytical Chemistry, Vol. 394, No. 4, pp. 1033-1041.

27. Putzig, C.L.; Leugers, M.A.; McKelvy, M.L.; Mitchell, G.E.; Nyquist, R.A.; Papenfuss, R.R. &Yurga, L. (1994). Infrared spectroscopy. Analytical Chemistry, Vol. 66, No. 12, pp.26R-66R.

28. Quellette, J. (2004). Time-Resolved Spectroscopy Comes of Age. The Industrial Physicist, 2, pp. 16-19.

29. Roubani-Kalantzopoulou, F.; Metaxa, H.; Kalantzopoulos, A.; Kalogirou, E.; Sotiropoulou, V. & Katsanos, N.A. (1996). Contribution to the Mechanism of Marble Deterioration by Gas Chromatographic Studies, Koutsoukos, P. & Kontoyiannis, Ch. (Eds.), Eurocare-Euromarble EV496 Workshop 7, Patras, Greece, October 1996, pp. 33-38.

30. Roubani-Kalantzopoulou, F. (2004). Review: Determination of Isotherms by Gas-Solid Chromatography Applications. Journal of Chromatography A, Vol. 1037, No. 1-2, pp.191-221.

31. Roubani-Kalantzopoulou, F. (2009). Review: Time-Resolved Chromatographic Analysis and Mechanisms in Adsorption and Catalysis. Journal of Chromatography A, Vol. 1216, No. 10, pp. 1567-1606.

32. Sotiropoulou, V.; Vassilev, G.P.; Katsanos, N.A.; Metaxa, H. & Roubani-Kalantzopoulou, F. (1995). Simple Determination of Experimental Isotherms Using Diffusion Denuder Tubes. Journal of the Chemical Society Faraday Transactions, Vol. 91, pp. 485-492.

33. Thielmann, F. (2004). Review: Introduction into the Characterization of Porous Materials by Inverse Gas Chromatography. Journal of Chromatography A, Vol. 1037, No. 1-2, pp.115-123.

Chapter 9

EFFECT OF AIR POLLUTION ON ARCHAEOLOGICAL BUILDINGS IN CAIRO

Mohamed Kamal Khallaf

Restoration Department, Faculty of Archaeology, Fayoum Universiy, Egypt

INTRODUCTION

Islamic Cairo is a part of central Cairo noted for its historically important mosques and other Islamic monuments. It is overlooked by the Cairo Citadel. Islamic Cairo was founded in 969 AD as the royal enclosure for the Fatimid caliphs, while the actual economic and administrative capital was in nearby Fustat. Fustat was established by Arab military commander 'Amr ibn al-'As following the conquest of Egypt in 641 AD, and took over as the capital which previously was located in Alexandria. Al-Askar, located in what is now Old Cairo, was the capital of Egypt from 750 AD to 868 AD. [1] Ahmad Ibn Tulun established Al-Qatta'i as the new capital of Egypt, and remained the capital until 905 AD, when the Fustat once again became the capital. After Fustat was destroyed in 1168 AD /1169 AD to prevent its capture by the Crusaders, the administrative capital of Egypt moved to Cairo, where it has remained ever since. [2]It took four years for the General Jawhar Al Sikilli (the Sicilian) to build Cairo and for the Fatimid Calif Al Muizz to leave his old Mahdia in Tunisia and settle in the new Capital of Fatimids in Egypt. Fustat became a regional center of Islam during the Umayyad period. Later, during the Fatimid era, Al-Qahira (Cairo) was officially founded in 969 AD as an imperial capital just to the north of Fustat. [3] Over the centuries, Cairo grew to absorb other local cities such as Fustat, but the year 969 AD is considered the "founding year" of the modern city. In 1250 AD, the slave soldiers or Mamluks seized Egypt and ruled from their capital at Cairo until 1517 AD, when they were defeated by the Ottomans. [4] By the 16th century, Cairo had high-rise apartment buildings where the two lower floors were for commercial and storage purposes and the multiple stories above them were rented out to tenants. Napoleon's French army briefly occupied Egypt from 1798 AD to 1801 AD, after which an Albanian officer in the Ottoman army named Muhammad

Ali Pasha made Cairo the capital of an independent empire that lasted from 1805 AD to 1882 AD. [5] The city then came under British control until Egypt was granted its independence in 1922 AD. Cairo is a world heritage city. It contains possibly the finest collection of monuments in the Islamic world. It contains some of the best surviving monuments of the medieval period in the Islamic world. [6]. The wealth, prosperity, and power of Cairo are reflected in the grand architecture of the monuments that are crowded together into the Fatimid city and just beyond, Fig. (1). [7] Cairo's Islamic monuments are part of an uninterrupted tradition that spans over a thousand years of building activity. No other Islamic city can equal Cairo's spectacular heritage, nor trace its historical and architectural development with such clarity. [8] Cairo contains the greatest concentration of Islamic monuments in the world, and its mosques, mausoleums, religious schools, baths, and caravanserais, built by prominent patrons between the seventh and nineteenth centuries, are

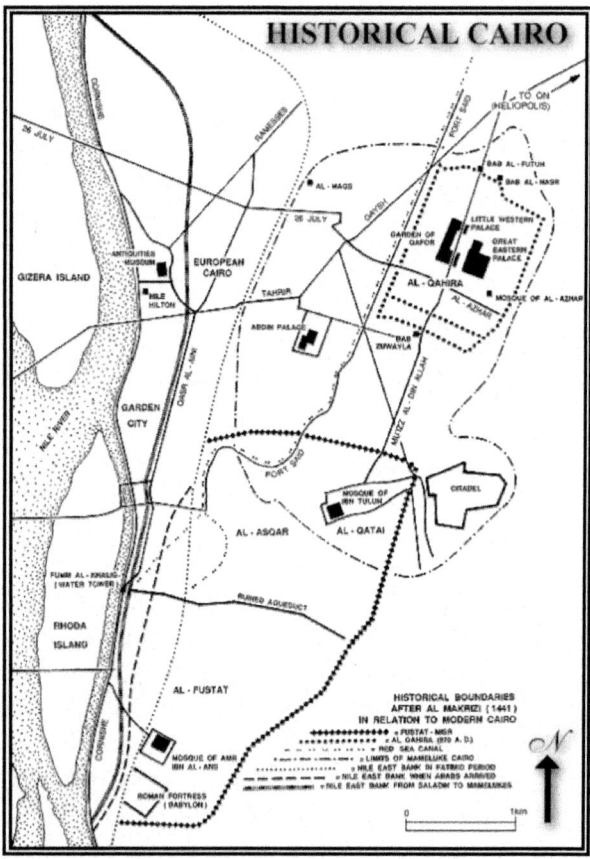

Figure. 1: Shows a map of historical Cairo. http://www.touregypt.net/Map08.htm.

among the finest in existence [9] , fig. (2) Shows some of Islamic archaeological buildings in Cairo. The air pollution in Cairo is a matter of serious concern. Greater Cairo's volatile aromatic hydrocarbon levels are higher than many other similar cities. Air quality measurements in Cairo have also been recording dangerous levels of lead, carbon dioxide, sulphur dioxide, and suspended particulate matter concentrations due to decades of unregulated vehicle emissions, urban industrial operations, and chaff and trash burning. There are over 4,500,000 cars on the streets of Cairo, 60% of which are over 10 years old, and therefore lack modern emission cutting features like catalytic converters. Cairo has a very poor dispersion factor because of lack of rain and its layout of tall buildings and narrow streets, which create a bowl effect. Cairo also has many unregistered lead and copper smelters which heavily pollute the city. The results of this have been a permanent haze over the city with particulate matter in the air reaching over three times normal levels. [10] Pollutants are deposited on the surface of stone from the air. Where the surface of the stone is totally dry, the stone is discolored as the deposits increase. Where the surface of the stone is moist, the pollutants are converted to acids that eat away the surface of the stone by dissolving the binder in the stone causing the stone particles or grains to separate and erode away easily. [11] Carbon dioxides, Nitric oxides, and Sulphur oxides product mineral acids in humid conditions. They dissolve the calcium and magnesium carbonates in limestone, marble, lime mortars, and plasters in archaeological buildings. Archaeological buildings in Cairo suffer from different deterioration phenomena for example, black crust formation, chemical alterations, disintegration between surface mineral grains, pitting, cracks, missing parts, erosion, and white stains. [12]. This chapter aims to study deterioration and decay of building materials in archaeological buildings in Cairo because of air pollution, Discussion and explanation of deterioration phenomena which forming in archaeological building in Cairo according to air pollution and Discussion of different methods and materials of treatment, restoration and conservation of building material in archaeological buildings from deterioration phenomena related to air pollution.

SOURCES OF AIR POLLUTION IN CAIRO

Air pollution plays a major role in the deterioration of building materials used in historic buildings. Industrial facilities such as factories and plants emit toxic gases into the atmosphere. Another major source of toxic emissions in Egypt is the widespread open-air burning of trash and waste. Waste landfills also give off methane, which, although not toxic, is highly flammable and can react in the air with other pollutants to become explosive. [13] There are numerous sources to air pollution in Egypt, as in other countries. However, the formation and levels of dust, small particles and soot are more characteristic in Egypt than presently found in industrialized countries. Some of the sources for these pollutants, such as industries, open-air waste burning and transportation, were also well known problems in most countries only 10 to 20 years ago. Another important source for particulate matter is the wind blown dust from the arid areas. Suspended dust (measured as PM10 and TSP) can be seen to be a major air pollution problem in Egypt. PM10 concentrations can exceed daily average concentrations during 98% of the measurement period. On the other hand it seems that the natural background of PM10 in Egypt may be close to or around the Air Quality Limit value. These levels can be found also in areas where local anthropogenic sources do not impact the measurements. Further measurements may be used in the future to quantify the relative importance of the different sources relative to a background level that varies dependent upon the area characteristics. In addition to particles, also SO_2 in urban areas and in industrial areas, as well as NO_2 and CO in the streets may exceed the Air Quality Limit value. Major industrial pollutants include sulphur oxides, nitrogen oxides, carbon monoxide and carbon dioxide. [14] For instance, Cairo is surrounded by various industrial sites. Thirty Kilometers to the south of Cairo is Helwan, where different factories produce iron, steel, coke, chemicals, automobiles, and cement. To the north of Cairo are Shubra Al – Khayma, Musturud, and Abu Zabal. In this area factories produce dyes, textiles, glass, ceramics, and chemical products. All of these factories emit different pollutants (gaseous, liquid, solid), which are carried by the dominant winds (north and northeast, and west or south, southwest) down to Cairo, many of the historic buildings are located. Every day Cairo receives a high dose of pollutants composed of 52 percent monocarbon oxide (CO), 14 percent sulphur dioxide (SO_2), 21 percent hydrocarbons, 10 percent dust, solid materials, and 2 percent (NOx) nitrogen oxides, The dust particles from the Muqattam hills to the east of Cairo was 27 gm / m2 / month in 1962. This increased to more than 60 gm / m^2 / month in 1988, with a particularly high a mount in the summer when the aerosols of dust in the air were more than 500 gm / m^2 / month [15]. Many Egyptians rely upon extremely old vehicles for transportation. These inefficient vehicles cause the carbon present in fuel to ineffectively react with oxygen during combustion, producing carbon monoxide or condensing to form particles of soot. The hydrocarbons do not combust completely and are released as gaseous hydrocarbons or absorbed by particles, increasing the particulate mass in the air. The speed at which pollutants disperse in the air is determined by meteorological conditions such as wind, air temperature and rain. Egypt and Cairo, particularly, have a very poor dispersion factor due to lack of rain and the layout of streets and buildings, which are not conducive to air flow. [16] Emissions that arise from the combustion of solid fossil fuels are of prime concern. Coal and oil both contain sulphur in varying amounts, and both therefore produce sulphur dioxide when burnt.

Figure. 2: Shows some of Islamic Archaeological buildings in Cairo: (A) El- Sultan Hassan Madrassa (1362 AD / 764 AH) and El – Refae Mosque. (B) El-Mosabeh Mosque(1792 AD / 1192 AH). (C) El- Ghouri Mosque in El – Sayeda Aisha Square (1504 AD / 909 AH). (D) El – Mahmoudya Mosque (1568 AD / 975 AH). (E) Qaitbay Sabil (1479 AD / 884 AH). (F) Taghri Bardi Mosque (1439 AD / 843 AH). (G) Azbak El- Yusufi Mosque. (H) Singer and Slar Mosque (1303 AD / 703 AH). (I) Lagen El – Sayfi Mosque (1296 AD / 696 AH).

There are a number of nitrogen oxides (NOx), but the one of principal interest as an air pollutant likely to have adverse effects on human health and soiling properties is nitrogen dioxide (NO_2). Nitrogen compounds are also contributors to the wet and dry deposition of acidic compounds on vegetation and buildings. Particulate matter is a term that represents a wide range of chemically and physically diverse substances that can be described by size, formation mechanism, origin, chemical composition, atmospheric behavior and method of measurement. The concentration of particles in the atmosphere varies across space and time and as a function of the source of the particles and the transformations that occur to them as they age and travel. Particles less than 10 mm in diameter (PM10) are often measured that include both fine and coarse dust particles. [17].

MATERIALS AND METHODS

Limestone and marble samples of original stones and crusts were collected from different deteriorated parts of Archaeological buildings, according to the decay and the damage levels fig. (3) as follows: - Limestone samples from El-Ghouri Mosque, El – Mahmoudya Mosque, Taghri Bardi Mosque and Lagen El – Sayfi Mosque. - Marble samples from Qaitbay Sabil, Taghri Bardi Mosque and Azbak El- Yusufi Mosque. Analytical study have been carried to selected samples by Polarizing Microscope [PLM], Scanning Electron Microscope [SEM], Energy dispersive X-ray analysis [EDX], X-ray diffraction (XRD) and FTIR analysis.

Figure. 3: Shows some details of Islamic Archaeological buildings, (A) , (B) , (G) El – Mahmoudya Mosque (1568 AD / 975 AH). (C) , (D) Qanibay Al – Ramah Mosque (1503 AD / 908 AH). (E) Qaitbay Sabil (1479 AD / 884 AH) (F) , (H) , (I) , (L) Taghri Bardi Mosque (1439 AD / 843 AH). (J) Lagen El – Sayfi Mosque (1296 AD / 696 AH). (K) Azbak El- Yusufi Mosque (1494 AD / 900 AH).

The x-ray diffraction analysis of samples was carried out using philips x-ray diffractometer. The operating conditions were: Generator applied on a Cu kα radiation (1.5418 A°) with Ni filter, 40 KV, 20 mA° target tube. Scavenging velocity 2° per minute and chart velocity 5 mm per minute were applied in Bulk sample powder. Fragments of crusts collected were prepared for observation using scanning electron microscope (SEM), operated at accelerating voltage of 30 kV. Infrared spectra were recorded employing a Nicolet Nexus 870 FTIR spectrometer. A small amount of samples were mixed with KBr and pressed into pellets, then scanned from 4000 to 400 cm−1.

RESULTS AND DISCUSSION

Limestone samples were examined by Polarizing Microscope (PM) and, it is found that: Samples consist mainly of fine-grained calcite besides presence of iron oxides, quartz, clay minerals and fossils include nummulite fossils these components increase the rate of stone decay [18], fig. (4 , A-F). On the other hand the thin section of fragments taken from marble objects shows that the major mineral is calcite, The crystals appeared in mosaic texture, The crystals have irregular faces and highly variable grain size, the cleavage planes of the calcite crystals and the presence of rare and very little amount of opaque minerals [19] fig. (4, G-I). When the limestone samples were examined by [SEM], it is found disintegration between calcite crystals and the stone lost the binding materials between grains by the effect of salts crystallization fig. (5, A-D). Examination of marble samples by [SEM] shows that, erosion of calcite crystals, presence of salts because of chemical reaction with climatic conditions, alteration of calcite into gypsum because of air pollution effect, voids and disintegration between grains by crystallization of salts stresses and lose of binding material [18] fig. (5, E-F). XRD data fig. (6) (A-D) shows that, the examined limestone samples consist of Calcite $CaCO_3$, Card No. (5-0586) in addition to Gypsum $CaSO_4.2H_2O$ Card No. (6-0046), Quartz SiO_2 ,Card No. (5-0490), Halite Card No.(5-0628) and Dolomite Card No.(11-078) . XRD data of the marble samples shows that, they consist of Calcite CaCO3, Card No. (5- 0586) in addition to Dolomite Ca,Mg(CO3)$_2$, Card No. (11-078), and Halite, NaCl Card No. (5-0628) fig (5-a). and Anhydrite, Card No (6-0226) fig (5-c). The surface of the marble is covered by a crust of Hydrated Calcium Sulphate (Gypsum) related to reaction with air pollution in presence of moisture. Gypsum crusts are the most common type of growth found on building surfaces. Gypsum is Calcium Sulphate Dihydrate, with the chemical formula $CaSO_4.2H_2O$. Gypsum crusts are formed on calcareous stones following SO_2 deposition to the surface in the presence of moisture, followed by the dissolution of Calcite and the precipitation of Gypsum. The black color of gypsum crusts is the result of the accumulation of particulate matter within the crust [20]. When the water evaporates from soluble salts as chlorides, it leaves behind concentrations of salt solutions which crystallize on the stone surfaces and between mineral grains of stone, this process cause disintegration and deterioration of stone [21]. Energy Dispersive X-ray analysis (EDX) of Limestone samples shows that it consists of calcium element (Ca), sulphur element (S) , Silicon element (Si) and Sodium element (Na) in addition to traces from other elements. The relative enrichment of Si, Al and Fe might be derived from the deposition of wind-borne articles since the archeological stone buildings in Cairo are near a road with much traffic [22] fig. (2). Rich-S is originated from SO_2 emitted by anthropogenic sources like combustion of fuels, automobile emissions, foundries and smelters. EDX data shows high content of calcium related to calcite mineral, silicon and aluminum due to clay minerals, silicon due to quartz mineral, iron related to iron oxides, sodium and chlorine due to presence of halite salt fig (7)(A-D). Rich-Ti-Mn is associated with industrial and urban emissions. The fly-ash particles play an active role in the damage processes affecting stone, since the content of transition metal oxides contribute to the catalytic oxidation of atmospheric gaseous SO_2 and to the sulphation of calcium carbonate. XRD, SEM, EDX results show that black crusts are essentially composed by gypsum crystals, fly ashes and soot, including some limestone and marble materials. Fly ashes usually are rich in Si and Al with higher or lower amounts of K, Fe, Ca, Ti and Cl. Combustion of fuel and natural gas in car engines and house heating originates carbon rich soot about one hundred times smaller than fly ashes. [23]. The term "atmospheric particulate material" refers to all airborne particles, so it is by definition nonspecific. It includes material from such diverse sources as, for example, vehicle emissions, the resuspension of surface dusts and soils and chemical reactions between vapours and gases in the atmosphere, which result in the formation of secondary particles [24]. Therefore emission inventories of PM relate to primary sources of PM only (not secondary sources) [25]. The principal types of primary particulate material are Petrol and diesel vehicles, the latter being the source of most black smoke [26]. Controlled emissions from chimney stacks. Fugitive emissions. These are diverse and mostly uncontrolled and include The resuspension of soil by wind and mechanical disturbance [27]. The resuspension of surface dust from roads and urban surfaces by wind, vehicle movements and other local air disturbance [28]. Emissions from activities such as quarrying, road and building construction, and the loading and unloading of dusty materials,[29]. Secondary particles are those arising when two gases or vapours react to form a substance that condenses onto a nucleation particle, [30]. The main sources of secondary particles are the atmospheric oxidation of sulphur dioxide to sulphuric acid and the oxidation of nitrogen dioxide to nitric acid; the sulphuric acid is present in air as droplets, the nitric acid as a vapour, [31]. Hydrochloric acid vapour (arising mainly from refuse incineration and coal combustion) is also present in the atmosphere, and both this and nitric acid vapour react reversibly with ammonia to form ammonium salts,[32]. Sulphuric acid reacts irreversibly in two stages to form either ammonium hydrogen sulphate or ammonium sulphate. These ammonium salts are formed continuously as sulphur dioxide and nitrogen dioxide are oxidised, and ammonia becomes available for neutralization, [33]. FTIR spectra of a limestone sample fig. (8) shows that the characteristic absorption peaks of $CaCO_3$ is at 1798, 1424, 874, 711 cm−1, the characteristic absorption peaks of, CO_3-apatite [$(Ca_5(PO_4)3)_2CO_3$] is at 565, 604, 1040 cm−1 and the characteristic absorption peaks of gypsum is at 672, 1623, 3408 cm−1 .The results of infrared spectroscopy are also confirmed by XRD analysis which provides information on the crystalline components. The limestone contains Calcite and Quartz, and Gypsum. In consideration of the high average relative humidity and rainwater in the environmental conditions in Cairo, the most probable process of crust formation on stone substrate is the absorption of sulphur dioxide in rainwater, liquid atmospheric aerosols or moist film supported on a stone surface, [34]where it is oxidized to form a sulphuric acid solution that dissolves the Calcium Carbonate by Gypsum formation. Kaolinite has been identified on a stone flake collected from a washed-out surface [35]. Its presence can be related to Calcite dissolution, which is strongly enhanced by its exposition to rainwater and winds[36]. The deposition of wind-born soil dust on the surface may also be a source of kaolinite. [37] The mineralogical, textural and physicochemical differences of the examined crusts suggest that it is unlikely that they have the same origin or the same pattern of development [38]. In Cairo, high relative humidity, frequent fogs, sulphur, nitrogen pollutants, carbonaceous and deposition of airborne particles either on exposed or sheltered areas of Cairo archaeological buildings[39]. In consequence of these processes, these deterioration products grow on sheltered areas leading to thick encrustations, which are washed-out on surfaces exposed to rainwater[40]. On the unsheltered surfaces, newly formed soluble salts, washed-out by water and percolated through the bedding planes of the stone substrate, create a network of parallel and deep fissures, which increase the stone susceptibility to further deterioration [41]. On the other hand archaeological buildings in Cairo suffer from soiling, fig. (3). Soiling is a visual effect resulting from the darkening of exposed surfaces following the deposition and accumulation of atmospheric particles [42]. Deposition, removal and accumulation processes are numerous and complex, [43] depending on the physical and chemical properties of the particles, the nature of the surface, the local meteorology and the pathways followed by rainwater after it hits the building surface[44]. As a result of these complex interactions, there can be substantial variations in the level of soiling observed on building surfaces. It is one of the effects of air [45].

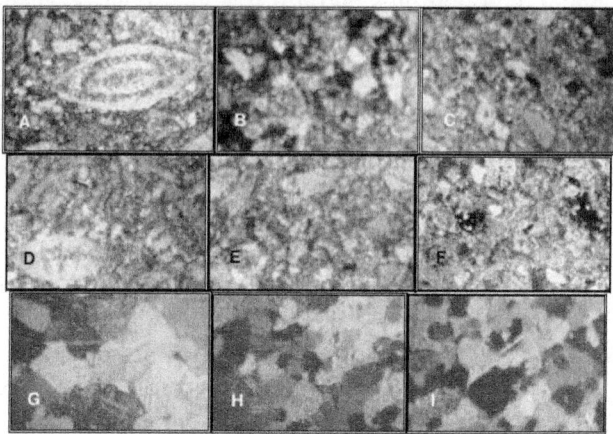

Figure. 4: (A-F) Thin section photomicrographs showing iron oxides, clay minerals, fossil and grains of quartz in a mass ground of fine- grained calcite. 60X (C.N). (G-I) shows that it is a mosaic texture; the calcite crystals have irregular faces and cleavage planes ,120X (C.N).

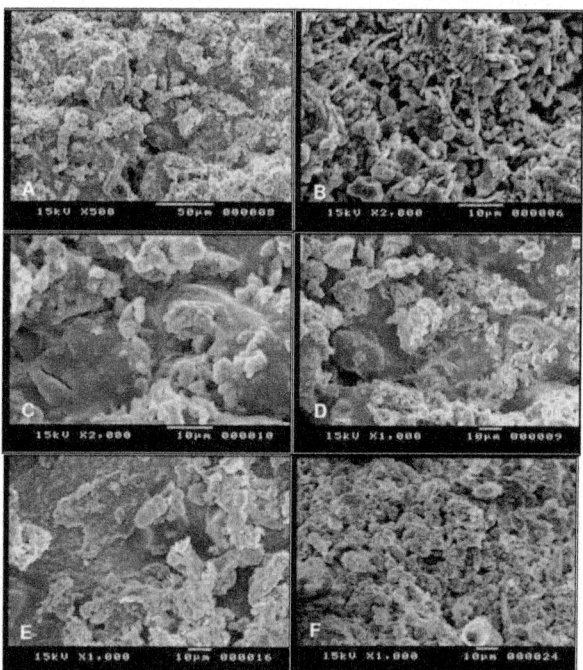

Figure. 5: (A-D) SEM photomicrographs of limestone samples showing the collapse of internal structure, salts crystallization between grains of limestone ornaments. (E-F)

photomicrographs of Marble samples showing voids due to lose of binding material erosion, discoloration, a coat of Carbon (C-D), chipping, fly ashes in a black crust and particles from the combustion of fuel oil and coal, containing a quantity of Carbon, Iron, Manganese and Sulphur.

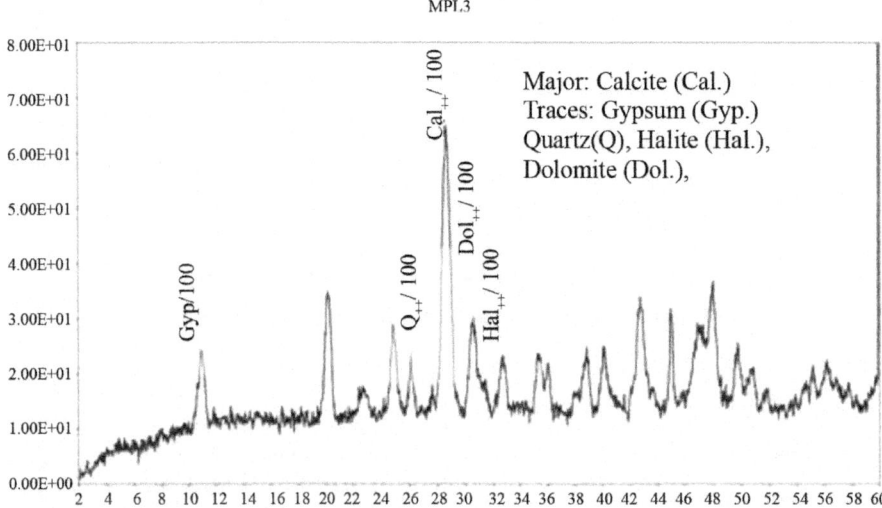

Figure. 6: (A) Shows XRD patterns of Limestone sample from El- Ghouri Mosque.

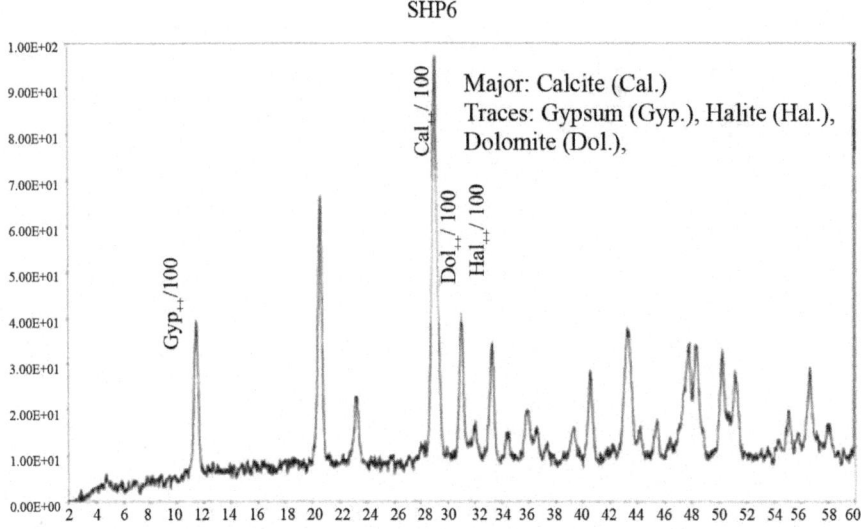

Figure. 6: (B) Shows XRD patterns of Limestone sample from El – Mahmoudya Mosque.

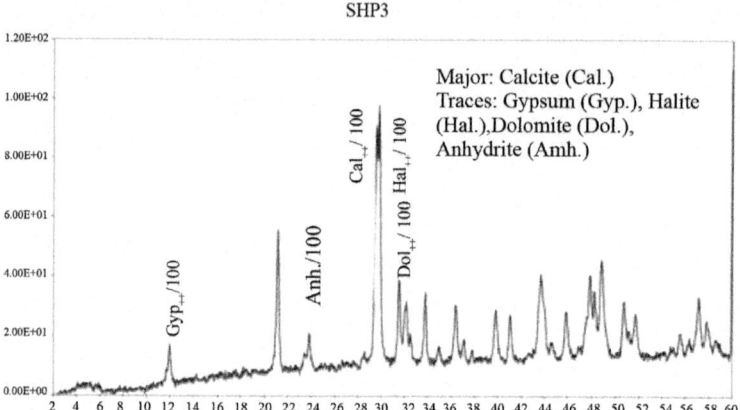

Fig. 6. (C) Shows XRD patterns of Marble sample from Taghri Bardi Mosque.

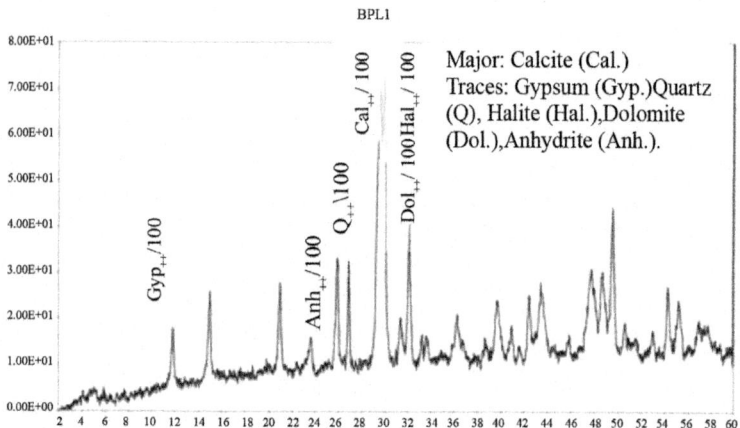

Figure. 6: (D) Shows XRD patterns of Marble sample from Qaitbay Sabil.

Atomic%	Weight%	Element
69.16	45.02	O K
1.98	2.17	Al K
5.40	6.17	Si K
2.27	2.96	S K
0.62	0.99	K K
1.69	2.75	Ca K
9.20	17.92	Ti K
0.45	1.00	Mn K
9.25	21.01	Fe K
	100.0	Total

Figure. 7: (A) shows EDX patterns of limestone sample from El- Ghouri Mosque.

Atomic%	Weight%	Element
49.75	36.59	C K
38.35	37.58	O K
0.46	0.65	Na K
0.21	0.41	P K
5.26	10.32	S K
0.52	1.14	Cl K
0.63	1.50	K K
4.81	11.81	Ca K
	100.0	Total

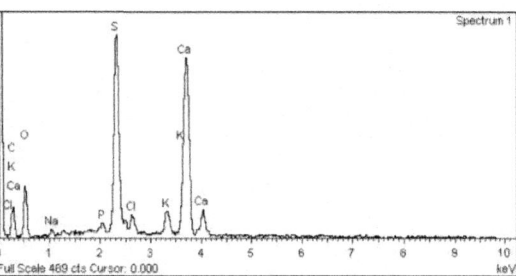

Figure. 7: (B) shows EDX patterns of limestone sample from El – Mahmoudya Mosque.

Atomic%	Weight%	Element
75.34	58.45	O K
1.35	1.51	Na K
0.72	0.85	Mg K
0.63	0.83	Al K
2.10	2.85	Si K
0.14	0.21	P K
7.27	11.30	S K
1.32	2.27	Cl K
1.35	2.57	K K
9.57	18.60	Ca K
0.21	0.56	Fe K
	100.0	Total

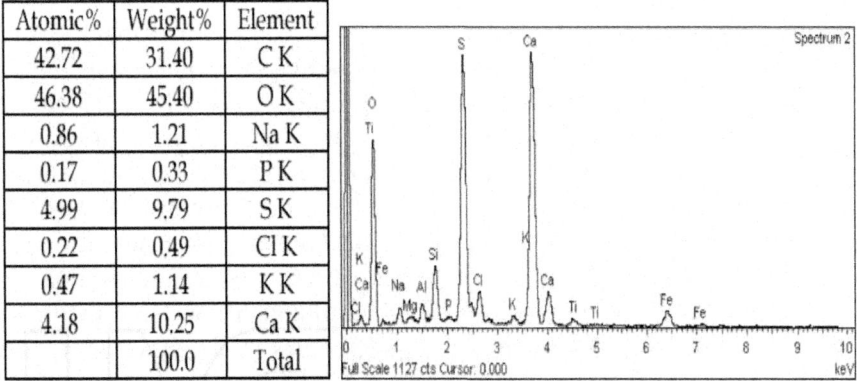

Figure. 7: (C) shows EDX patterns of Marble sample from Taghri Bardi Mosque.

Atomic%	Weight%	Element
42.72	31.40	C K
46.38	45.40	O K
0.86	1.21	Na K
0.17	0.33	P K
4.99	9.79	S K
0.22	0.49	Cl K
0.47	1.14	K K
4.18	10.25	Ca K
	100.0	Total

Figure. 7: (D) shows EDX patterns of Marble sample from Qaitbay Sabil.

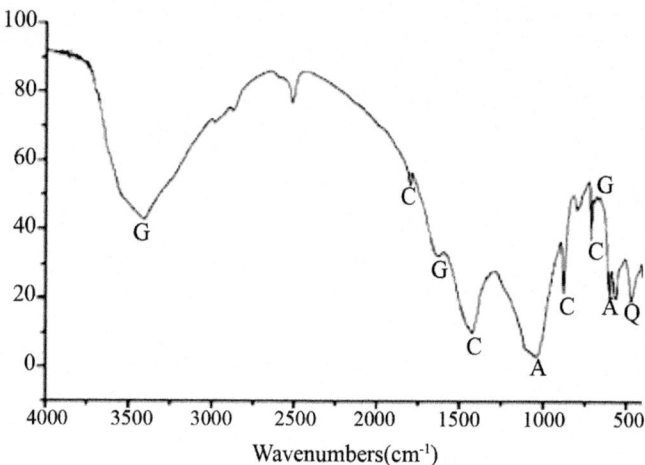

Figure. 8: FTIR spectra of limestone sample from El – Mahmoudya Mosque C, calcite (1798, 1424, 874, 711 cm−1); G, gypsum (672, 1623, 3408 cm−1); A, apatite (565, 604, 1040 cm−1); Q, quartz (469 cm−1).

TREATMENT AND CONSERVATION PROCESSES

There are many methods and materials of treatment, restoration and conservation of building material in archaeological buildings from deterioration phenomena related to air pollution Cleaning Methods and Materials. These methods include cleaning, extraction of salts, consolidation and Water-Repellent Coatings.

Cleaning

Masonry cleaning methods generally are divided into three major groups: water, chemical, and abrasive. Water methods soften the dirt or soiling material and rinse the deposits from the masonry surface [46]. Chemical cleaners react with dirt, soiling material or paint to effect their removal, after which the cleaning effluent is rinsed off the masonry surface with water. Abrasive methods include blasting with grit, and the use of grinders and sanding discs, all of which mechanically remove the dirt, soiling material or paint (and, usually, some of the masonry surface). Abrasive cleaning is also often followed with a water rinse. Laser cleaning, although not discussed here in detail, is another technique that is used sometimes by conservators to clean small areas of historic masonry. It can be quite effective for cleaning limited areas, but it is expensive and generally not practical for most historic masonry cleaning projects. Although it may seem contrary to common sense, masonry cleaning projects should be carried out starting at the bottom and proceeding to the top of the building always keeping all surfaces wet below the area being cleaned, [47]. The rationale for this approach is based on the principle that dirty water or cleaning effluent dripping from cleaning in progress above will leave streaks on a dirty surface but will not streak a clean surface as long as it is kept wet and rinsed frequently.

Removal and Extraction of Salts

The notion of the poultice has been adapted for the cleaning of historic buildings and a true poultice is intended to draw out deep-seated contaminants and staining from the surface of masonry and sculpture. In current practice the word poultice is extended to a wide range of cleaning materials and techniques, not all of which achieve a true poultice effect on the substrate. What might be termed the true or plain poultice contains water and the poultice medium only, relying on these ingredients to achieve the mobilisation and removal of the contaminant. The most common poultice medium is clay, although paper and cotton fabrics are also used, and talc, chalk and even flour are traditional poultice materials. A mixture of clay and paper fabric produces an absorbent and plastic mixture that is often favoured by conservators of stone sculpture.This plain or true poultice is normally used for desalination, to draw out soluble salts, or as a cleaning method on substrates such as limestone that respond to water cleaning. In these cases the poultice is allowed to dry out and the soiling and/or salts are drawn into the poultice by capillary action with the moisture. Multiple applications may be necessary to draw the salts from within the surface pores, [48]. Whatever the medium, the poultice is mixed with water to form a material that will adhere to the substrate. Clay forms a sticky mass that adheres well to stone and other surfaces. These plain poultices can be conveniently mixed by hand as required on site with the addition of water to the poultice medium. Alkaline poultice cleaners and strippers are commonly used for cleaning or degreasing masonry surfaces and for paint removal. Sodium hydroxide is the most common alkaline cleaning agent in proprietary cleaners for a range of masonry substrates, including limestone, sandstone, brick and terracotta and is the most common ingredient in proprietary paint removers. Care must be taken in the use of sodium hydroxide based cleaners to minimise risks to the building and the user. Sodium hydroxide based cleaners and strippers must be neutralised with acid afterwash. Adjacent, dissimilar building surfaces must be protected and personal protective equipment worn by the cleaning operative. In the field of stone conservation ammonium carbonate is added to clay and clay/ paper poultices to remove soiling from limestone. Ammonium carbonate is a less alkaline cleaner than sodium hydroxide. It works by reacting with calcium sulphate on the soiled surface to form calcium carbonate and soluble ammonium sulphate that can be rinsed off with water. These 'active' or 'chemical' poultices are all applied to a pre-wetted surface to minimize penetration of the chemical into the masonry surface and covered with plastic film to prevent the poultice drying out. The cleaning additives in these mixtures chemically dissolve the soiling or staining which is held to the surface of the poultice, and then both the cleaning agent and the contaminant are removed with the clay. Rinsing with water and, where necessary neutralization, follows to remove any soiling that remains on the surface and also to remove residues of the chemical cleaners. Strictly speaking these materials are clay-based cleaning packs rather than true poultices, but the word poultice is now widely used in the building cleaning industry [49].

Consolidation

Stone strengtheners based on ethyl silicates are generally applied plied by spraying or flooding. It is usually also possible to treat moveable parts by immersing them in a bath. Compresses can serve as an alternative to immersion .they ensure maximum length of con – tact between the stone strengthener and the stone[48]. Equipment employed for flooding includes electrical pumps airless sprays and simple hoses. The pressure has to be kept as low as possible since the aim is to apply the material to the surface so that it will be absorbed naturally by the stones capillary system the excess will run off and be absorbed immediately by untreated areas below[49]. Several wet – on – wet treatments are generally needed applied at intervals of 20 to 30 minutes .the exact number of treatments quantity of material and desired minimum penetration depth have to be ascertained by preliminary tests and trials .the construction materials must be dry since the active in- gradient in the stone strengthener, ie, the ethyl silicates reacts with moisture. The moisture required by the stone strengthener for chemical deposition of the silica gel is sup- plied by the construction material which always has a certain sorption moisture content varying in equilibrium with the atmospheric humidity [50]. The best working conditions are a relative humidity of 40 to 70 % and a surface temperature on the construction material of 10 to 25 c each coating operation be so arranged that the entire surface can be covered in one working day . otherwise there is the danger that gel which gas been deposited in the pore system will prevent the strengthener from penetrating further this in turn might cause gel to be deposited in the surface regions of the stone and to gloss or crust formation . Very often instead of the whole object only small sections are treated such as a precious or- nametag detail or areas that are severely damaged in such cases it is advisable to follow up the last treatment with a solvent wash suitable solvents are hydrocarbons methyl ethyl ketene and ethyl alcohol [47]. Freshly treated surfaces must be covered for 2 to 3 days against the rain. Considerable loss of the active ingredient by evaporation may occur at temperatures exceeding 25 c at such temperatures the freshly consolidated surfaces have to be protected against direct sunlight. Temperatures below 5 c cause the stone strengthener to react very slowly this may result very slowly this may result in discoloration or glaze on the surface [51]. The total time needed for the stone strengthener to deposit the silica gel depends on the relative humidity and the temperature. it varies form one to at most three weeks therefore before any further restoration work is carried out on period of roughly one week should elapse this will allow 90 to 95 % of the silica gel to be de –posited . On no account should water be added to the ethyl silicate preparation in an attempt to speed up the reaction this can result in extensive glazing of the surface that is extremely difficult to remove if indeed this is at all possible.

Water-Repellent Coatings

Water-repellent coatings are formulated to be vapor permeable, or "breathable". They do not seal the surface completely to water vapor so it can enter the masonry wall as well as leave the wall. While the first water-repellent coatings to be developed were primarily acrylic or silicone resins in organic solvents, now most water-repellent coatings are water-based and formulated from modified siloxanes, silanes and other alkoxysilanes, or metallic stearates [49]. While some of these products are shipped from the factory ready to use, other waterborne water repellents must be diluted at the job site. Unlike earlier water-repellent coatings which tended to form a "film" on the masonry surface, modern water-repellent coatings actually penetrate into the masonry substrate slightly and, generally, are almost invisible if properly applied to the masonry. They are also more vapor permeable than the old coatings, yet they still reduce the vapor permeability of the masonry [48]. Once inside the wall, water vapor can condense at cold spots producing liquid water which, unlike water vapor, cannot escape through a water-repellent coating. The liquid water within the wall, whether from condensation, leaking gutters, or other sources, can cause considerable damage. Waterrepellent coatings are not consolidants. Although modern water-repellents may penetrate slightly beneath the masonry surface, instead of just "sitting" on top of it, they do not perform the same function as a consolidant which is to "consolidate" and replace lost binder to strengthen deteriorating masonry. Even after many years of laboratory study and testing, few consolidants have proven very effective. The composition of fired products such as brick and architectural terra cotta, as well as many types of building stone, does not lend itself to consolidation. Some modern water-repellent coatings which contain a binder intended to replace the natural binders in stone that have been lost through weathering and natural erosion are described in product literature as both a water repellent and a consolidant The fact that the newer water-repellent coatings penetrate beneath the masonry surface instead of just forming a layer on top of the surface may indeed convey at least some consolidating properties to certain stones. However, a water-repellent coating cannot be considered a consolidant. In some instances, a water-repellent or "preservative" coating, if applied to already damaged or spalling stone, may form a surface crust which, if it fails, may exacerbate the deterioration by pulling off even more of the stone [52].

ACHIEVEMENTS AND PLANNED ACTIVITIES FOR IMPROVEMENT OF AIR QUALITY IN CAIRO

Air quality represents a major priority for the Egyptian Ministry of state for Environmental Affairs, Egyptian Environmental Affairs Agency as it has dangerous impacts on the public health and it is effect on archaeological buildings. This concern encompasses a number of trends [53]:

Alleviating the Vehicles' Emissions

Through the coordination and effective cooperation between the Ministry of Environment and the Ministry of Interior, the decree of the Minister of Interior was issued:

- To link between the issuance of the licenses of the Vehicles and its emissions testing, and the start of the implementation of this decree in the Qaluibia and Giza governorates. Such decree provides a new hope of the improvement of air quality and the first step of overcoming the problem of the vehicles' emissions, to be applied in many other governorates. This decree is essential for the reinforcement of Law No. 4 for 1994 on the protection of Environment. [53].

- The Ministry of State has already, in collaboration with USAID through the Cairo Air Improvement Project, delivered the traffic departments in Giza and Qaluibia governorates 38 devices for vehicles' emission testing, in addition to training those who are designated to the technical inspection of vehicles using diesel and benzene. It is worth mentioning that the application of the issuance of the vehicles' licenses in both governorates has started from June 1, 2003 on vehicles' emissions testing to combat the emissions of Carbon monoxide and Hydrocarbons.

- The cooperation between the Ministry of Environment and the Ministry of Interior has resulted in the establishment of the environment police: the first police stations to be inaugurated will be in the Regional Branch of the Egyptian Environmental Affairs Agency in Greater Cairo and El-Fayoum as well as in Beni Siweif.

Relocation of the Heavily Polluting Activities Outside the Populated Areas

Due to the variety of pollution sources especially within Greater Cairo, the Ministry of Environment has formulated a plan of the relocation of the polluting activities outside the populated areas, among them the smelters, quarries, potteries, crackers, brick factories and coal and lime facilities as well as 1206 mining factories and 6000 textiles factories. This funding plan is based on the contribution of the owners of these activities, applying the principle "Polluter pays". The estimated budget of this plan is L.E. 1745 million; the share of the government is about 15% of its total. In addition to that, the government provides soft loans for the relocation of these polluting activities to the desert. The owners of these activities contribute to the remainder of cost for 4 years starting from July 1, 2003 to June 30, 2007. The Ministry has, in cooperation with the competent governorates, identified the places of relocation of these polluting activities in El-Amal region in the Ain El Sokhna Road for all Cairo smelters and in Akrasha region for Qaluibia smelters, in addition to relocation of coal facilities to the industrial zone in Belbeis as well as the brick factories to Arab Abu Saad region. [53].

Combating the Industrial Pollution

As for the plan of the Ministry of Environment for pollution sources control in the big factories, it has prepared a plan in two phases as follows: The first phase: factories in need of limited funds for approximately L.E. 23.13 million to combat pollution discharged from them. The second phase: factories in need to huge funds for about L.E. 545.9 million. In this respect, the EEAA is implementing some of the projects that make available the funding and technical support for the industrial establishments, such as the Industrial Pollution Abatement Projects providing grants and soft loans offered by the World Bank as well as technical assistance as a grant from the Finnish Government. In addition, there is the Environment Protection Fund for the Public Sector Industries funded by the German Construction Bank that provides Euro 25.56 million as a grant from the German Government, representing a partial funding of 50 % of the necessary investments for the implementation of industrial waste treatment projects as well as soft loans presented from the Egyptian banks participating in this project[54].

The Environmental Inspection on the Establishments

Since the start of the practical application of Law No. 4 for 1994 and after the termination of the grace period provided by the Law and its Executive

Regulation, the Ministry has established the Environmental Inspection Unit at the central level and prepared a manual of the policies and procedures for the inspection unit, which is considered the first manual in this field. This manual has specified the role and authorities of the environmental inspection in comparison to the other supervisory agencies concerned with the inspection on the establishments. This manual reemphasizes that the periodical follow-up and inspection are the effective means of the non-replication of violations.

The Safe use of the Treated Sewage Water in the Irrigation of Forests

For further improvement of air quality and the reduction of dust and sand rates, arising from Al-Khamaseen wind, the Ministry of State for Environmental Affairs is implementing the Green Belt Project around Greater Cairo (Cairo-Giza-Qaluibia) along 100 km on the sides of the circular road with a width of 10-25 m, [55].cultivating it with Acacia and Cypress trees. This project aims at protecting the citizens of the Greater Cairo from dust and sands and conserving their health. In addition, it provides job opportunities to the graduated youth whether in the implementation of the project or its maintenance, besides using the treated sewage water for the irrigation of these trees to be economically made use of. This project is implemented in four phases in the three governorates: starting from Cairo Governorate in the region from El-Moneib Bridge to Misr Ismailia Desert Road, in Qaluibia Governorate to El-Kanater establishment, and in Giza Governorate to El-Moneib Bridge. The total cost of the project is L.E. 13.7 million. [53]. The Green Belt is not the last project implemented by the Ministry for the improvement of air quality but there is also the National Programme for the Safe Use of Treated Sewage Water, in collaboration with the Agriculture, Irrigation, Housing, Local Development and Environment Ministries as well as the different governorates. The concept of this project depends on the investment of treated sewage water since Egypt produces about 3 billions m3 annually at the cost of 14 Piastres/meter with a total of approximately L.E 14 million, and turns this problem into a social, environmental and economical value. Instead of disposing this treated water into water channels and contaminating it, it can be used in afforestation. This project achieves several social, economic and environmental benefits as it basically improves air quality through the plantation of trees that are the source of Oxygen for they intake Carbondioxide and produce Oxygen[53]. In addition, it helps in combating desertification, protecting water resources and soil from pollution, building green belts and wind obstructers, to be used in producing woods instead of importing them. It also helps in providing job opportunities for the youth and establishing the new urban communities side by side with these forests. [54]. There are successful attepts for this project in Serabuim area in Ismailia, Sadat City, Asuit, Sohag, Luxor, Qena, New Valley, Tour Sinai, El-Saaf, and Aswan. This national project is carried out at several phases. The first phase is executed in an area of 82940 thousand Fedan around 72 Sewage stations in the different governorates all over the Republic at the cost of L.E. 5 thousand/ Fedan, providing a collective revenue during the lifetime of the forest, i.e. 12 years. The implementation is carried out for 8 thousands Fedan annually.

Manufacturing the Construction Materials from Rice Straw Using Unconventional Technology

There is no doubt that success that will come out of the real partnership between the Government and the Private Sector in the relocation of polluting activities outside the residential regions, based on environmental principles and standards supported by environment friendly technologies will directly assist in combating the Black Cloud phenomenon that we suffer from in October annually. Scientists from the Scientific Research Academy, the National Research Center, the Environment Research Council, the Meteorology Organization and the Specialized National Councils have a consensus that the real reasons for the Cloud are confined to a climate phenomenon, namely, the existence of high pressure that appears every year at the same time, accompanied by a thermal change and stability of wind, which all lead to the accumulation of pollution in Cairo air. [53].

CONCLUSION

The danger to archaeological buildings from air pollution comes from two main sources – gases that increase the corrosivity of the atmosphere and black particles that dirty lightcolored surfaces. Acid rain comes from oxides of sulphur and nitrogen, largely products of domestic and industrial fuel burning and related to two strong acids: sulphuric acid and nitric acid. Sulphur dioxide (SO_2) and nitrogen oxides (NOx) released from power stations and other sources form acids where the weather is wet, which fall to the Earth as precipitation and damage both heritage materials and human health. In dry areas, the acid chemicals may become incorporated into dust or smoke, which can deposit on buildings and also cause corrosion when later wetted. Atmospheric chemistry is, of course, far more complex than this and a variety of reactions occur that may form secondary pollutants that also attack materials. Particulate matter is much more complicated because it is a mixture rather than a single substance – it includes dust, soot and other tiny bits of solid materials produced by many sources, including burning of diesel fuel by trucks and buses, incineration of garbage, construction, industrial processes and domestic use of fireplaces and woodstoves. Particulate pollution can cause increased corrosion by involvement in a number of chemical reactions and, often more importantly, it is the source of the black matter that makes buildings dirty. The influence of heavily polluted atmosphere in the urban environment results in different weathering patterns, mainly in the form of crusts. It might be assumed that the analytical results of Polarizing Microscope, XRD, SEM, EDX and IR. alone are not sufficient to clarify and interpret the growth mechanisms of crusts. However, they do provide valuable information about changes in compositions of crusts and original rock, and the relationship between crusts composition and air pollution. The compositions of the crusts collected from areas on the archaeological stone buildings with different decay patterns show that the deterioration is mainly due to the atmospheric pollutants and its extent is strongly dependent on the surface exposition to the environment. According to the obtained results, an appropriate conservation plan will be developed, that includes the steps of cleaning and consolidation, in order to identify the most suitable materials and methodologies to remove the deterioration crusts avoiding the loss of original substrate and ensuring an increased cohesion to deteriorated stone.

REFERENCES

1. Creswell, (1959), The Muslim Architecture of Egypt, Oxford, P.112.

2. Beattie, Andrew (2005). Cairo: A Cultural History (illustrated ed.). New York: Oxford University Press.

3. 3.Butler, Alfred J. (2008). The Arab Conquest of Egypt - And the Last Thirty Years of the Roman Dominion. Portland, Ore: Butler Press.

4. Behrens-Abouseif, Doris (1992). Islamic Architecture in Cairo (2nd ed.). Brill. Daly, M. W.; Petry, Carl F. (1998). The Cambridge History of Egypt: Islamic Egypt, 640-1517. Cambridge, UK: Cambridge University Press.

5. Glassé, Cyril; Smith, Huston (2003). The New Encyclopedia of Islam (2nd revised ed.). Singapore: Tien Wah Press.

6. Rose, Christopher; Linda Boxberger (1995). "Ottoman Cairo". Cairo: Living Past, Living Future. The University of Texas Center for Middle Eastern Studies.

7. Mortada, Hisham (2003). Traditional Islamic principles of built environment. Routledge. p. viii.

8. Williams,C., Islamic Monuments in Cairo: The Practical Guide, American University in Cairo Press,2004.

9.   Anoniou, J. "Historic Cairo - A Walk through the Islamic City", American University in Cairo Press,1999.

10. Watt, J. et al.Creighton NP et al., (1990) Soiling by atmospheric aerosols in an urban industrial area. J Air Waste Manag Assoc. 40, 1285–1289.

11. Aksu R, Horvath H, Kaller W, Lahounik S, Pesava P and Toprak S (1996) Measurements of the deposition velocity of particulate matter to building surfaces in the atmosphere. J Aerosol Sci. 27, S675–676

12. Davidson CI, Tang F, Finger S, Etyemezian V and Sherwood SI (2000) Soiling patterns on a tall limestone building: changes over 60 years. Environ Sci Technol. 34, 560– 565.

13. Air Pollution Levels Measured in Egypt Exceeds Air Quality Limit Values.(2002), EEAA/EIMP, Ministry of state for Environmental Affairs, Egyptian Environmental Affairs Agency.

14. Hopkins; N. 2003"The Environmentalist: Living with Pollution in Egypt".

15. Abo El-Ela, A., The Impact of Environmental pollution on the Mosque of Al-Azhar and the complex of Al-Ghuri, In :

16. The Restoration and Conservation of Islamic Monuments in Egypt, The American University in Cairo Press,1995.Pp. 99-114.

17. Del Monte M, Sabbioni C and Vittori O (1981) Airborne carbon particles and marble. deterioration, Atmos Env. 16, 2253–2257.

18. Mohamed ,K. Khallaf, (2006), Role of Investigation and Analytical Methods in study of Archaeological Stone Ornaments's Deterioration and Its Treatment,The Seventh International Symposium on New Trends in Chemistry " Analytical Chemistry for a better Life " Egyptian Journal of Analytical Chemistry – Volume (15), January .

19. Mohamed ,K. Khallaf , (2006) Analysis and Preservation of Marble in Archaeological Buildings, The Seventh International Symposium on New Trends in Chemistry " Analytical Chemistry for a better Life " Egyptian Journal of Analytical Chemistry – Volume (15), January.

20. Mohamed ,K. Khallaf, (2008), Degradation and Conservation of Marble Floors in Archaeological Buildings, 5th Symposium of the Hellenic Society for Archaeometry, 8-12 October, 2008, Athens, Greece. October.

21. Mohamed ,K. Khallaf ,(2006), Environmental Detrioration and Conservation Studies of Building Materials of Qaitbay Citadel, Rosetta City, Egypt, Civil Engineering Research Magazine, Faculty of Engineering, Al - Azhar University, Volume(28) No.(1), January.

22. Saiz-Jimenez C. (editor) 2004, Air pollution and cultural heritage. A.A. Balkema Publishers, Taylor & Francis Group plc, London.

23. Parker A. (1955) The destructive effect of air pollution on materials. National Smoke Abatement Society. London. pp 3–15.

24. Watt J. (1998) Automated Characterisation of Individual Carbonaceous Fly Ash Particles by Computer Controlled Scanning Electron Microscopy -Analytical Methods and Critical Review of Alternative Techniques. Water AirSoil Pollut. 106, 309–327.

25. Pesava P, Aksu R, Toprak S, Horvath H and Seidl S (1999) Dry deposition of particles to building surfaces and soiling. Sci Total Env. 235, 25–35.

26. Pio CA et al., (1998) Atmospheric aerosol and soiling of external surfaces in an urban environment. Atmos Env. 32, 1979–1989.

27. Brimblecombe P 2003, The effects of air pollution on the built environment. Imperial College Press, London.

28. Hamilton RS and Mansfield TA 1991, Airborne particulate elemental carbon: its sources, transport and contribution to dark smoke and soiling, Atmos. Env. 25, 715–723.

29. Hinds WC (1999), Aerosol Technology: properties, behaviour and measurements of airborne particles, 2nd edition. Wiley ISBN 978-0-471-19410-1.

30. Tidblad, J., Mikhailov, A.,& Kucera, V. (2000). Acid deposition effects on materials in subtropical and tropical climates. Data compilation and temperate climate comparison. SCI Report 2000:8E, Swedish Corrosion Institute, Stockholm, Sweden.

31. Kucera, V., Tidblad, J. (2005). Comparison of environmental parameters and their effects on atmospheric corrosion in Europe and in South Asia and Africa. Proc. 16th Int. Corrosion Congress, Beijing.

32. Cole, I. S. (2000). Mechanisms of atmospheric corrosion in tropical environments. ASTM STP 1399. In S. W. Dean, G. Hernandez-Duque Delgadillo & J. B. Bushman (Eds), American Society of Testing and Materials. West Conshohocken, PA.

33. Maeda, Y., Moriocka, J., et al., (2001). Materials damage caused by acidic air pollution in East Asia. Water, Air and Soil Pollution, 130, 141–150.

34. Beloin NJ and Haynie FH (1975) Soiling of building materials. J Air Pollut Control Ass. 25, 393–403.

35. Hamilton RS and Mansfield TA (1992) The soiling of materials in the ambient atmosphere. Atmos Env. Part A – Gen Topics. 26, 3291–3296.

36. Lanting RW (1986). Black smoke and soiling in aerosols: research, risk assessment and control strategies. In Proceedings of the Second US-Dutch Symposium, Ed. Lee SD, Lewis Publishers, Williamsburg, VA.

37. Mansfield TA and Hamilton RS (1989). The soiling of materials: models and measurements.

38. Parker A. (1955) The destructive effect of air pollution on materials. National Smoke Abatement Society. London. pp 3–15.

39. K.L. Gauri, G.C. Holdren, (1981) Pollutant effects on stone monuments, Environ. Sci. Technol. 15 (4), 386–390.

40. F. Delalieux, C.P. Cardell, V. Todorov, (2001) Environmental conditions controlling the chemical weathering of the Madara

41. Horseman monument, NE, J. Cult. Herit. 2, 43–54.

42. A. Moropoulou, K. Bisbikou, K. Torfs, (1998) Origin and growth of weathering crusts on ancient marbles in industrial atmosphere, Atmos. Environ. 32 (6), 967–982.

43. P. Maravelaki-Kalaitzaki, (2005) Black crusts and patinas on Pentelic marble from the Parthenon and Erechtheum (Acropolis, Athens): characterization and origin, Anal. Chim. Acta 532, 187–198.

44. C. Vazquez-Calvo, M. Buergo, R. Fort, (2007),Characterization of patinas by means of microscopic techniques, Mater. Charact. 58, 1119–1132.

45. Brimblecombe P and Grossi CM (2005) Aesthetic thresholds and blackening of stone buildings. Sci Total Env. 349, 175–198. Fig. 4.15 Variation of soiling with PM10 concentration (white painted steel) 124

46. Mack, R.C. and Grimmer, F.A.: Assessing Cleaning and Water-Repellent Treatments for Historic Masonry Buildings, Washington DC, (2003).

47. Mohamed ,K. Khallaf, (2008), Interfacial Characteristics of Polymeric Coatings for Archaeological Stones Conservation., Sixth International Conference: Science and Technology in Archaeology and Conservation, Rome, Italy, December 9th – 13th.

48. Ana Luque, Giuseppe Cultrone and Eduardo Sebastián: (2010) The Use of Lime Mortars in Restoration,Work on Architectural Heritage, In book: Materials, Technologies and Practice in HistoricHeritage Structures, Edited by, Maria Bostenaru Dan, Springer , New York .

49. Clifton, J.R.: (2005) Stone Consolidating Materials, A Status Report, Cool, Documents, August.

50. Gansicke, S., and J. Hirx. (1997) Mortars as A filling materials for the compensation of losses in objects. Journal of the American Institute for Conservation 36:17-29.

51. Wheeler, G.: (2005) Alkoxysilanes and the Consolidation of Stone, Columbia University Press, U.S.A.

52. Noll ,W., Chemistry and technology of silicones. Academic Press, New York, (1986).

53. The Cairo air improvement project (2004), Final Report | March, Prepared by: Chemonics International Inc.1133 20t h Street, NW, Washington, DC 20036 / USA, Prepared for: USAID/Egypt, Office of Environment , Contract 263-C-00-97-00090- 00.

54. Khoder, M.I. (2007). " levels of volatile organic compounds in the atmosphere of Greater Cairo". Atmospheric Environment (Air Pollution Research Department, National Research Centre, Dokki, Giza) 41 (3): 554–566.

55. Hopkins N., (2003) "The Environmentalist: Living with Pollution in Egypt" A.A. Balkema Publishers, Taylor & Francis Group plc, London.

Chapter 10

INFLUENCE OF AIR POLLUTION ON DEGRADATION OF HISTORIC BUILDINGS AT THE URBAN TROPICAL ATMOSPHERE OF SAN FRANCISCO DE CAMPECHE CITY, MÉXICO

Javier Reyes

Autonomous University of Campeche,mexico

INTRODUCTION

The role of atmospheric pollution in degradation of historic building has been studied for long time along the world because it increases stone decay and the lost of historic materials (Massey, 1999; Graedel, 2000; Monna, 2008). The preservation of Cultural Heritage is considered a strategic factor in countries integration because of their economical, social and cultural implications (Cassar et al., 2004; Sessa 2004, Moropoulou and Konstanti, 2004). Latin-american countries have an important building legacy from prehispanic, colonial and modern periods. This is the case of México which currently count with 15 sites included in UNESCO´s cultural heritage list. Most of them are located in urban areas like Mexico City, Morelia, Guanajuato, San Miguel de Allende and San Francisco de Campeche, between others. San Francisco de Campeche is a small City located at the south east of Mexican Republic, just in the occidental coast the Peninsula of Yucatan, inside the Gulf of Mexico Basin (Fig. 1). The City was founded in 1527, by Spanish colonizers leaded by Francisco de Montejo, "el Mozo". During the XVII century, it was the only point for exportation of goodness from Yucatan to Europe. Because of these conditions, French, Netherlanders and British pirates considered the city a legitimate target. At that time, authorities designed an impressive military defensive system to protect the City and their inhabitants. Forts, batteries and a rampart surrounding San Francisco de Campeche urban core were built by using calcareous materials based on masonry structures made with limestone quarry blocks joined and covered with mortars made with slike lime and sahacab, a typical calcareous clay material used since prehispanic period for building construction. Nowadays, about 1500 buildings are located into the historic and architectonic complex included in 1999 in the UNESCO`s Cultural Heritage List.

ince their construction, these buildings have been exposed to the action of environmental agents that induce their deterioration. For long time, natural parameters like high relative humidity, extended rainfall periods and the effects of marine aerosols were the principal factors related with buildings degradation (Zendri, 2001, Cardell et al., 2003). Karstification, crust formation, lost of components and biodegradation are typical pathologies of degradation observed in the buildings. Nevertheless, in the last decade, the City has been under a dynamic development. As a consequence, a sensible increase in automobile units has been registered in specific areas of the city, including the historical centre.

Figure. 1: The State of Campeche located at the South East of Mexican Republic. Red dots indicate the location of San Francisco de Campeche City and Iturbide town, current environmental monitoring sites operated by the Corrosion Research Center (CICORR).

Automotive emissions generate atmospheric particles and corrosive gases like sulphur dioxide (SO_2) and nitrogen oxides (NO_x) that, in contact with environmental humidity produce acid precipitation that dissolve calcareous materials, or induce black crust formation (Lipfert, 1989; Gobi, et al., 1998, Kucera, 2007). Systematic studies related to atmospheric pollution and their effects in historic building degradation at San Francisco de Campeche City are scarce.

STONE DECAY

Stone materials have a natural tendency to degradation as a consequence of change in their chemical stability when they are extracted from the quarry and submitted to the building fabric, atmospheric action and change in air quality. Before industrial revolution, natural agents were the main cause of stone buildings degradation, sometimes through suddenly destructive actions as earthquakes, volcanic eruptions or hurricanes. Most of the times, acting in slow weathering process. Nevertheless, with the appearance of the industrial society, atmospheric pollution got a major role in building deterioration. In natural conditions atmospheric water is the main agent associated to stone degradation. Its influence is especially important in tropical climates, where high relative humidity and large rain forest period along the year guarantee water availability to lead chemical reaction over stone substrata or to produce secondary pollutant's potentially harmful for stone materials.

In San Francisco de Campeche City, historic buildings were constructed using calcareous stone materials including quarry blocks and mortars. Calcareous stone and traditional mortars used during buildings construction or restitution works usually show a wide interval of porosity (Reyes et al., 2010; Torres, 2009). It is well known that water circulation in porous stones and their exchange with atmosphere or ground, affect their behavior and durability. The flux of water across porous structure of stones and mortars is consequence of wet-to dry- cycles, that induce chemical reactions and salts crystallization leading materials lost and decreasing their mechanical capabilities. Furthermore, direct impact of rainfall is cause of erosion on stone surface and the appearance of run-off inside of masonry structures. On the other hand, when water table level is high, a capillary effect could appear. Then, a continuous flux of soluble salts inside and outside materials stone structure is established. Water presence also facilitates the development of microorganism colonies and the growth of superior plants. In both cases, their consequences on stone materials are chemical and mechanical damage. In urban environments, decay of historic buildings is strongly influenced by the presence of atmospheric pollutants like SO_2, NO_x, atmospheric particles and acid rain. In the atmosphere water drops incorporates carbon dioxide (CO_2) to produce the weak carbonic acid (HCO_3^-), which is partially dissociated according to the next reaction:

$$CO_2 + H_2O \leftrightarrow H^+ + HCO_3^-$$ (1)

As a consequence, water acquires a pH of 5.65. It means that in unpolluted atmosphere water tends toward acid. Under this condition, dissolution of calcareous materials is possible. Dissolution of carbonates inside walls as their migration and deposition to the evaporation front lead the formation of crusts, as is demonstrated in the next reaction:

$$CaCO_3 + H_2O + CO_2 \rightarrow Ca(HCO_3)_2$$
(2)

Soluble calcium bicarbonate ($Ca(HCO_3)_2$) is transported by water to the surface of stone and mortars across porous system of built and decorative elements. When water evaporates CO_2 drags (equation 3).

$$CaCO_3 + H_2O + CO_2 \rightarrow Ca(HCO_3)_2$$
(3)

Formation rate of $Ca(HCO_3)_2$ depends on CO_2 levels, that is the reason why in urban environments with high levels of this gas, carbonation of calcareous materials is most important than in rural environment. On the other hand, recrystallized calcite is bigger in size and more porous than the microcrystalline original calcite. The increase in size is extreme harmful for stone materials, because it creates conditions for a deep penetration of acidic solutions (like acid rain), insoluble salts and gases like SO_2 and NO_x. Acid rain is produced when gases like SO_2 or NO_x reacts with water drops, increasing their acidity under pH value of 5.65 to form the so called acid rain. Acid rain is a global phenomena and its effect can be observed at long distances from their precursor sources (Bravo et al., 2000).

The presence of a minimum water amount is enough to oxidize SO_2 to sulphuric acid (H_2SO_4) according to the next reactions:

$$2SO_{2(g)} + O_{2(g)} \rightarrow 2SO_3$$
(4)

$$SO_{3(g)} + H_2O_{(l)} \rightarrow H_2SO_4$$
(5)

H_2SO_4 can easily react with calcareous materials to form gypsum ($CaSO_4$.$2H_2O$) as is indicated in equation (6)

$$CaCO_3 + H_2SO_4 \rightarrow CaSO_4 \cdot 2H_2O + CO_2 .$$
(6)

Gypsum formation is a serious problem because when it crystallizes gradually expands up to 30 % of their original size (Feddema, et al., 1987). $CaSO_4.2H_2O$. It is highly soluble at predominant temperatures in tropical regions, so it requires a minimum water amount to dissolve and lead a fast migration to evaporation front by capilar mechanisms. When gypsum lost humidity, it can recrystallize into porous, where induce the formation of microcracks and fatigue of materials. In urban environments, gypsum incorporates into their mineral structure atmospheric particles, dust and biomass to form the so called black crust. NO_x formation depends on environmental conditions and the kind of pollutant present in the atmosphere. It is expressed in the next reactions:

$$2NO + O_2 \rightarrow 2NO_2 \tag{7}$$

$$2NO_2 + H_2O \rightarrow HNO_3 + HNO_2 \tag{8}$$

Ozone (O_3), also can also react with nitrogen oxide (N_o) and nitrogen dioxide (NO_2):

$$NO_2 + O_3 \rightarrow NO_3 + O_2 \tag{9}$$

$$NO + O_3 \rightarrow NO_2 + O_2 \tag{10}$$

The products of these reactions establishes an equilibrium with dinitrogen pentoxide, which react with water to form nitric acid (HNO_3):

$$NO_2 + NO_3 \leftrightarrow N_2O_5 \tag{11}$$

$$N_2O_5 + H_2O \rightarrow 2HNO_3 \tag{12}$$

In urban zones, O_3 and NO also react with water to form HNO_3

$$2NO + O_3 + H_2O \rightarrow 2HNO_3 \tag{13}$$

Nitric acid dissolves calcareous stone to produce calcium nitrate ($CaNO_3)_2$):

$$CaCO_3 + 2HNO_3 \rightarrow Ca(NO_3)_2 + H_2O + CO_2 \tag{14}$$

$Ca(NO_3)_2$ is more water soluble than $CaCO_3$. If it is present, is transported across porous capillars to finally crystallize on monuments surface to be washed during rainy events (Allen et al, 2000). Deposition mechanisms also play an active role in historic building deterioration. Atmospheric particles and aerosols are transported by wind toward monumental structures.

Here, they are incorporated into neo-mineral matrix of degradation products or participate in oxidation reactions induced by carbonaceous particles or metals like iron (Fe), vanadium (V), and nickel (Ni) content in dust. In coastal zones, marine aerosols also contribute to deterioration of stone. It is primary composed by sea water along with particles naturally generated by the action of the wind on the seawater surface to introduce ionic species into the atmosphere, principally chlorides and sulfates (Stefanis et al., 2009). Chlorides are a destructive agent of porous materials. Because of its high solubility, it penetrates into porous network, and crystallizes inside the material. Its crystallization produces disruptive pressure forces that lead to microcracks formation (Cardell et al., 2003). On the other hand, suspended particles are also natural substrata for oxidation reactions (Primerano et al.,

2000). New products eventually reach stone surface were they originates physical, chemical and aesthetic changes (Fig. 2). Once stone materials have been sensitized by physical or chemical factors it is more sensible to the action of biological agent causing biodegradation. Biodegradation is an undesirable change in materials properties caused by the action of microorganisms, animals and plants. The presence of microorganisms causes the formation of biofilms. Biofilms are sessile communities adhered to substrate enclosed in a polymeric matrix producing metabolites with capabilities to initiate, promote or magnify stone degradation through modification in pH levels, ionic concentrations, and redox conditions at the interfase between substrate and surrounding media to produce chemical and physical alterations (Gorbushina et al., 2002; Ortega-Morales, 2003; Guiamet et al., 2005; Little y Ray, 2005).

a   b

Figure. 2: General aspect of degradation at Forts San Pedro (a) and San Carlos (b), historic buildings of San Francisco de Campeche City.

DEGRADATION OF HISTORIC BUILDINGS: THE CASE OF SAN FRANCISCO DE CAMPECHE

Meteorological Conditions

San Francisco de Campeche City is located under a gently slooping flood plain. The City is limited at the Norwest by the Gulf of México and at the South, Southeast and Southwest by a group of softened hills. Under these conditions, the natural expansion of the city follows to South and Southeast direction. The City presents a tropical summer rain forest climate (Aw) (Castro Mora, 2002). Table 1 concentrate the annual average value of meteorological parameters registered during 1992 to 2002 period at National Meteorological Service station (SMN), located into the installation of the aeronaval airport of the City.

Table 1: Annual average value of meteorological parameters registered during 1992 to 2002 period at National Meteorological Service station (SMN), at San Francisco de Campeche City.

Year	Meteorological parameter					
	Precipitation (mm)	Temperature (° C)	Relative humidity (%)	Atmospheric pressure (mb)	Predominant Wind direction	Wind velocity (m.s⁻¹)
1992	1224.30	27.20	73	12198.50	SE	3.40
1993	1294.30	27.60	71	12168.70	E	3.60
1994	1084.80	27.30	73	12166.10	E	2.90
1995	1688.40	26.90	74	12149.10	SE	3.10
1996	938.50	26.40	74	12159.20	ESE	3.10
1997	1115.60	27.30	74	12153.40	SE	2.60
1998	815.40	27.80	71	12143.50	ESE	2.70
1999	1227.70	26.70	73	12164.20	ESE	2.80
2000	927.00	26.70	72	12168.20	E-SE	2.60
2001	1004.70	27.00	73	12159.20	E	3.10
2002	1297.20	27.10	71	12157.50	E	2.80

The existence of high relative humidity values along the year can be observed and the persistent sum of rainfall covering an extended period from June to November. Those conditions guarantee water availability for occurrence of chemical and physic processes able to deteriorate calcareous stone materials through binder dissolution mechanisms, including also the penetration of soluble salts and atmospheric pollutants (Corvo et al., 2010). Two characteristics regional meteorological phenomena affecting coastal zones are related to inland humidity penetration from the sea: along the autumn season, tropical storms (hurricanes) carry on humidity from Caribbean Sea raising up rainfall precipitation levels. It is especially worthily in September and October. During winter, cool dry fronts come in from North America, drag humidity when they cross the Gulf of Mexico warm water, increasing haze episodes in the coast and eventually the rainfall events. In this period rainfall events tend to minimum, and an extended dry season from November to May begin. In spite of those situations, during this period, San Francisco de Campeche City temperature rise up to its maximum levels, while relative humidity falls to the lowest value.

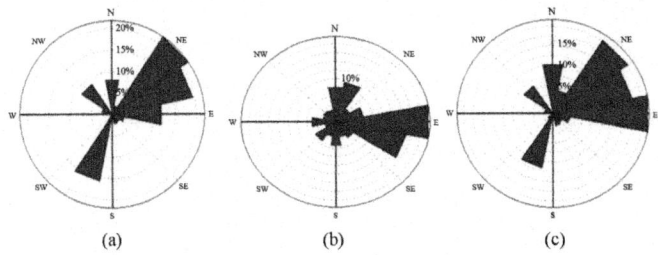

(a)   (b)   (c)

Figure. 3: Characteristics wind pattern observed at San Francisco de Campeche City during 2007. (a). Dry season, (b) rainy season, (c) polar front season. (Miss, 2008).

Along the year, three wind patterns can be observed in dependence of meteorological conditions at San Francisco de Campeche City (Fig. 3). Dry season is characterized by winds from E-ENE and SW directions that increase dust level at the atmosphere. At the rainy season, the wind pattern is dominated by E-ESE and a small contribution from N-ENE, due to, eventually, strong tropical storms hitting the city. In winter, when polar front reaches the coast of Campeche, winds from E-NE, N NW and SW are more frequent.

Environmental Conditions

San Francisco de Campeche City is an emerging place located at the occidental coast of the Peninsula of Yucatan. At the present time, it has about 235,000 inhabitants. Until the last decade of the XX century the City was considered a place of scarce economical and industrial development, since the main productive activities were administration, fishing, and processing of food. There are no installed heavy industries were installed, except by a power plant located at Lerma town, about 6 km SE from downtown. Nevertheless in 1999, the historic and architectonic complex of the City was included into UNESCO´s Cultural Heritage List. It considered the city as a historic and cultural reference in Mexico and other countries. As a consequence, an intense urban and economical development occurs, mainly due to the raising of cultural sector. Also an increasing of infrastructure needs because of the parallel demographic expansion that is occurring in the city. Environmental problematic like water supply, solid residues management, residual water disposition and atmospheric pollution are associated with urban development. Studies about the number of existing automotive units ordered in 2003 by the Government of Campeche State demonstrated that during 1996 to 2003 period, San Francisco de Campeche City suffered an increasing of 8% the vehicular units, while between 2002 to 2003 the increasing was of 13.13 %, this situation cause serious vial problems. According to this study, projections for 2010 indicates an increase of 69,130 units (Government of the State of Campeche, 2004). Under these conditions, it is expected an increase in atmospheric pollution level.

Atmospheric pollution is mainly related with industrial and vehicle exhaust emissions. Gases like ozone (O_3), carbon oxides (CO, CO_2), nitrogen oxides (NO_2, NO_3), sulfur dioxide (SO_2), and atmospheric particles (PST, PM10, PM2.5), have been used to indicate air quality in urban areas. Those pollutants can be the origin of health diseases, changes in environmental conditions and degradation of materials. In this sense, they are precursors of acid rain and the blackening of stone materials in historic buildings and monuments (Reyes et al., 2009; Corvo et al., 2009). It is interesting to report that since 1992, atmospheric corrosion under structural materials aluminium (A_l), carbon steel (F_e), copper (C_u) and zinc (Z_n), was monitored in five sites distributed across urban city area (Fig. 4) (Reyes, 1998; Cook, et al., 2000; Corvo et al., 2008). These studies were performed according to criteria established by the program ISO CORRAG (Tidblad et al., 2000). During the study, an estimation of corrosion rates was carried out considering deposition rate of corrosive parameters like chloride ions (C_l-), SO_2, and their correlation with the temperature-humidity complex represented as time of wetness (TOW) (Tables 2 and 3). It was established that the corrosion rate at exposures sites depends strongly of their distance to the coast, since C_l- levels decrease when distance increases (Corvo et al., 2008). Here, SO_2 deposition rate was very low, except for Technological Institue of Campeche (ITC) site located at Lerma town, closer to the Power Station (Table 3).

Figure. 4: Map of San Francisco de Campeche City. Red dots indicate the location of selected monitoring sites for atmospheric corrosion research. ITC, Technological Institute of Campeche, CRIP, Regional Center for Fisheries Research; CICORR, Corrosion Research Center; INAH, National Institute for Anthropology and History; SMN, National Meteorological Service.

It has the highest SO_2 content between all exposure sites. Nevertheless, it does not appear as decisive as chloride and TOW in prediction of corrosion rate as usually occurs in Regional Center for Fisheries Research (CRIP) and CICORR stations. The results of the study also shows that in San Francisco de Campeche atmospheric corrosion rates are lower than those located in Mexican and Cuban coastal stations, where industrial and marine influence was more important. The only exception to this rule was the station located at the CRIP, located at 4 meters from the coastline. It is an interesting data in order to consider the possible effects of atmospheric condition on stone materials decay.

Table 2: Estimation of atmospheric corrosivity at selected monitoring sites in San Francisco de Campeche City

Station	Atmospheric corrosivity			
	Al	Cu	Fe	Zn
SMN	Medium to high	Medium	Medium	Medium
CICORR	Medium to high	High	High	High
INAH	Medium to high	Medium	Medium	Medium
CRIP	Medium to high	High	High	High
ITC	Medium to high	Medium	Medium	Medium

Table 3: Corrosive parameters registered at San Francisco de Campeche City selected monitoring sites.

Station	Distance to the coast (m)	Corrosive parameters		
		SO_2 mg.m^{-2}.day^{-1}	Cl⁻ mg.m^{-2}.dia^{-1}	TOW annual hours
SMN	4.000	2.42	19.98	4576
CICORR	0.300	2.61	70.50	4894
INAH	0.615	1.47	18.08	3271
CRIP	0.004	2.64	76.20	4572
ITC	0.300	15.83	29.50	3380

Atmospheric Pollution

During the dry season of 1998, exceptional natural fires were declared along the south east of Mexican Republic. It was especially worthily at Campeche State, where several health problems like skin, respiratory and ocular diseases were observed in people. For the first time, atmospheric particles considering the fraction of atmospheric particles with diameter below 10 μm (PM_{10}) fraction was measured at San Francisco de Campeche following procedures established by Official Mexican Standards. The study (carried out in May 21th, 1998), yielded average value of 40 μg.m-3,that was considered below health risk levels for inhabitants. This study was the only reference of atmospheric pollution measured at San Francisco de Campeche City until 2005, when an initiative to study the effects of the environment on degradation of Cultural Heritage was driven by Autonomous University of Campeche (Reyes, 2005a, 2005b). Atmospheric parameters like SO_2, atmospheric particles (TSP and PM10 fractions) and acid rain were measured in different periods during 2005 to 2009 following Mexican and International Standards (NOM, US-EPA, ISO and UNE). Until the beginning of this project, there was no additional information on air pollutants measured using standard methods in the City of San Francisco de Campeche.

Present Atmospheric Pollution Levels at San Francisco De Campeche City

We proceed to determine the levels of air pollutants in the city of San Francisco de Campeche, considering two important aspects: its effect on materials and the possibility of using standardized methods to generate a database that could be used as a reference on air quality in the city (Reyes 2005a, 2005b; Miss 2008, Villaseñor, 2008., Dzul 2010, Góngora 2010, Quirarte, 2010). Two atmospheric pollution stations were placed at "home of Lieutenant of the King" and San Pablo Buildings, historic buildings belonging to Centro INAH-Campeche (INAHNational Institute of Anthropology and History). Another station was installed on the Corrosion Research Center (CICORR), main Campus of the Autonomous University of Campeche (Fig.4). Passive (SO_2, NO_X, Cl-), active (Total Suspended Particles –TSP and PM10 fraction) and automatic (SO_2) samplers were employed. Also, wet precipitation was sampled by using a wet/dry rain sampler. The results of the sampling are condensed on Table 4. Table 4, shows medium, maximum and minimum deposition rates and concentrations of the different types of pollutants determined using standardized methods at selected atmospheric monitoring stations.

Table 4: Average deposition rate and concentration of the different types of pollutants determined by standard methods at monitoring stations in San Francisco de Campeche City during 2006 to 2009

Pollutant	Method	Standard	Medium value	Maximun value	Minimun value	Station
SO_2 (mg.m^{-2}) Feb. 2007 to Feb. 2009	Sulphation plate (passive)	ISO 9225:1992	1.31	3.52	0.58	INAH
Cl^{-1} (mg.m^{-2}) Feb. 2007 to Feb. 2009	Wet Candle (passive)	ISO 9225:1992	20.28	31.90	3.53	INAH
NO_X (mg.m^{-2}) Feb. 2007 to Feb. 2009	Diffusion tubes (passive)	UNE EN 13528	9.82	13.48	4.71	INAH
TSP (mg.m^{-2}) Nov. 2006 to Dec. 2008	High-volume sampler (Active)	NOM-035-SEMARNAT-1993	47.23 48.71	101.30 106.67	15.26 23.89	INAH CICORR
PM_{10} (mg.m^{-2}) May to August 2007	Low volume sampler (Active)	US-EPA standard	3.54 3.30	8.69 9.72	1.49 1.34	INAH CICORR
SO_2 (mg.m^{-3}) Jan. 2007 to Jan. 2008	Fluorescence (Automatic)	NOM-038-SEMARNAT-1993	6.95	74.70	1.30	INAH

Passive Methods

Passive methods consist of an absorbent substrate that reacts with a specific chemical compound in the atmosphere. Afterwards, the samplers are removed and analyzed quantitatively in the laboratory. These devices work by principles of deposition or diffusion, but they are not considered appropriated for air quality studies; however, they provide trends on the spatial-temporal distribution. Fig. 5 shows the values of SO_2, NO_X and Cl- determined by passive methods in the historic center of the city of San Francisco de Campeche (urban-marine atmosphere). These data can be compared with results obtained in the rural monitoring station installed at Iturbide town (Fig. 6), about 100 km E far away from the City (Fig. 1). As can be seen, the values of all pollutants are higher in the city of San Francisco de Campeche in comparison with Iturbide, due to the urban nature of the city. Pollutants such as SO_2 and NO_X are usually produced during combustion of fossil fuels and emitted into the atmosphere by motor vehicles. These gaseous pollutants are considered acid contaminants because they corrode metals and stone materials due to its ability to form acid solutions in contact with environmental humidity on the surface of materials (Tercer 1998; Massey, 1999; Zappia et al., 1998; Allen et al., 2000). The levels of airborne salinity in a particular site depend upon the geographical position and the existence of orographic accidents. Its marine origin causes a preferential distribution in coastal areas.

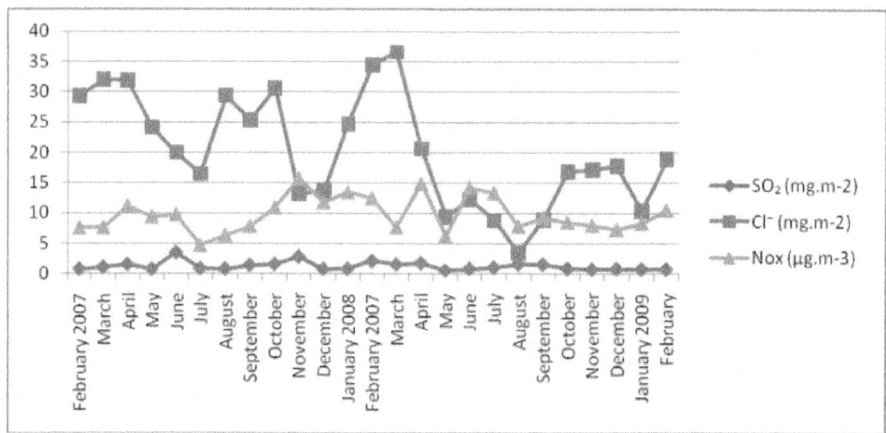

Figure. 5: SO_2, NO_X and Cl-, levels determined by passive methods in the urban marine atmosphere of San Francisco de Campeche City (INAH station).

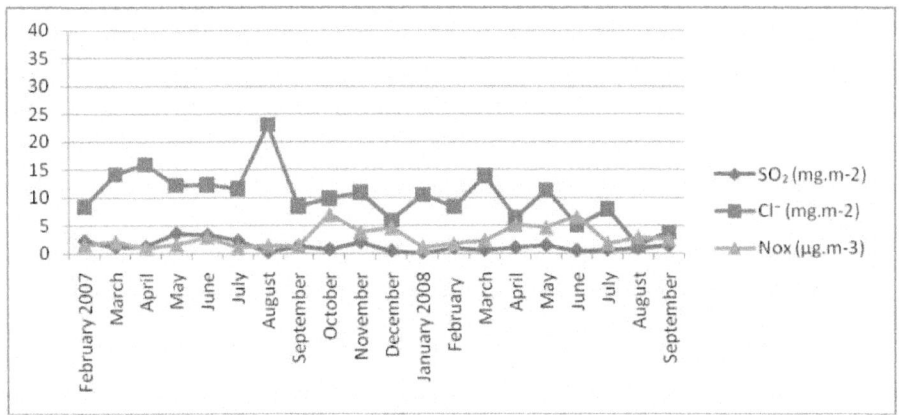

Figure. 6: SO_2, NO_X and Cl- levels determined by passive methods in the rural monitoring station installed at Iturbide Town.

Its concentration decreases when the distance to the coastline increases. This distribution also depends on the speed and wind direction. Higher levels of airborne salinity are expected near the coastline It is appropriate to mention that despite the proximity to the coast of INAH station (600 m), marine aerosol levels are relatively lower than those observed in Boca del Río coastal stations (600 m from shore line) or Coatzacoalcos petrochemical complex (1000 m from shore line) (Carpio et al., 1996; Reyes 1998; Cook et al., 2000). It occurs because the prevailing wind patterns in Campeche is most of the year from the E (they are called offshore winds) (Fig. 3), opposing the entry of masses of moisture from the Gulf of Mexico (Reyes, 1998; Cook et al 2000). This wind regime, suffers slight modifications relatively constant during winter, since the winds from the N increases in intensity and frequency, so that marine aerosol levels tend to rise, increasing the potential corrosivity of the atmosphere.

Atmospheric Particles

On the other hand, active methods involve a flow of air through an absorbent medium or a physical collecting medium. A suction pump is used. Samples thus obtained are quantitatively analyzed in the laboratory. Two types of samplers are used: high volume and low volume. Two sampling sites were selected: the "Home of Lieutenant King", central building of INAH-Campeche, located in the historic center of the city of San Francisco de Campeche, and the building of the CICORR in the main campus of Autonomous University of Campeche. The level of total suspended particles (TSP) was determined at both sites during the period August 2006-October 2008. PM10 fraction of airborne particles was recorded during the period May to August 2007. Table 4 display the average, maximum and minimum values determined for the corresponding sampling periods. Table 4 shows statistics for data sets obtained for PST in both sampling stations. In all cases the maximum, minimum and average values were higher for CICORR related to INAH station although a "t" test performed showed no significant differences between the average obtained in both sampling sites (t = 1.57225 p> 0.05). Moreover, during the sampling period, none of the stations exceeded the maximum permissible limit for Mexican Standard (210 g.m-3), as shown in Fig. 7. Higher average values of TSP were monitored during the month of July coinciding with the end of the dry season and beginning of summer rainfall season. Average TSP values were found to be 47.23 and 48.71 g.m-3 for INAH and CICORR monitoring stations respectively. Several authors suggest that in drought periods, atmospheric particles concentration is higher than in rain periods, those because of the lack of washing of the atmosphere caused by rainfall (Muñoz et al. 2001; Miss 2008).

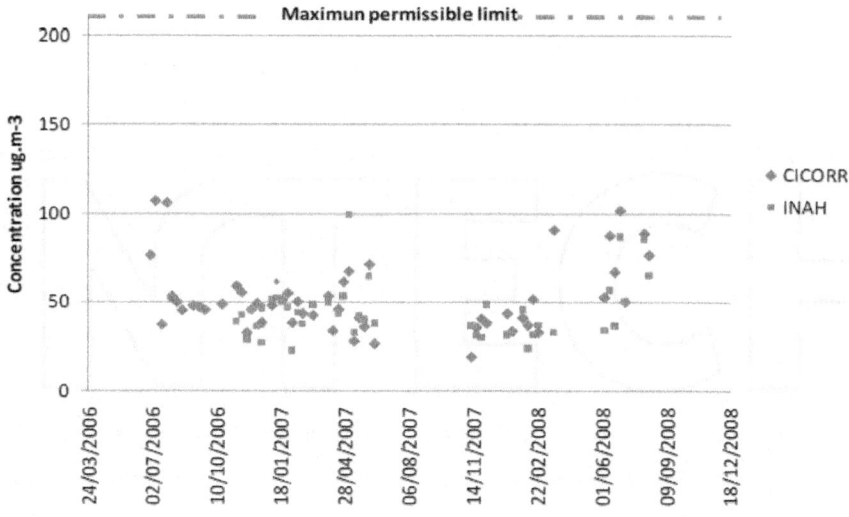

Figure. 7: TPS at San Francisco de Campeche monitoring sites during the period August 2006- September 2008. Red dotted line represents the maximum permissible limit of 240 µg.m-3 According to Mexican Legislation

The city of San Francisco de Campeche is located in the middle of a small valley, surrounded at N, S and E by hills, with elevations not higher than 150 meters. The Bay of Campeche is located in the W. Many of these hills are suffering continuous erosion and clearing of land for the construction of living houses or are employed by construction companies as sources of construction materials. These activities give rise to soil erosion and constant dust storms, which in times of drought contribute to increased levels of local

TSP. In a regional scale, the prevailing winds in the dry season (April to July), converge towards the sea ground by the E-NE quadrant and an important component S-SW (Fig. 3a). It contributes to the transport of atmospheric particles, originated in farming areas, eroded land and cattle ranches in the state, which add to the locally originated TSP. The role of rainfall in the levels of TSP is evident in urban and industrialized areas, since water acts as a purifier of particles in the atmosphere (Muñoz et al., 200; Sosa et al., 2006; Miss, 2008), also the wind disperse atmospheric particles and reduce their content at the atmosphere. It is confirmed by the minimum average value of 15.26 ug.m^{-3} recorded during the month of March 2007 at INAH, when cool fronts introduce strong wind velocities and eventually rain episodes. During the period from August to November there is a significant decrease in the levels of TSP on both stations as a result of purifying effect of seasonal rains which masses are originated in the Caribbean Sea (Fig. 3b). The presence of polar fronts in the Gulf of Mexico during the period from December to March becomes a factor of atmospheric instability that contributes to the dispersion of pollutants and the introduction of humidity from the ocean in coastal areas (Reyes 1998). It coincides with the monthly average minimum of 35.25 g.m^{-3} registered at CICORR during December 2006, precisely at the end of the rainy season and early winter seasonal fronts when the wind increases in strength and components N-NE direction (Fig. 3c). Atmospheric particulate matter PM10 fraction was determined during the end of the dry season and the beginning of the rainy season (May-August 2007). A Student "t" test to compare the arithmetic means of data sets collected at stations CICORR and INAH was used. The test results indicated no significant difference between values observed in the testing sites (t = 0.612, p> 0.5). At both monitoring sites, the concentration of PM10 follows the same tendency being the maximum concentration of 9.72 mg.m^{-3} and minimum concentration of 1.34 mg.m-3 for CICORR, while for INAH, maximum and minimum concentrations were 8.69 and 1.49 mg.m^{-3}, respectively (Table 4). Regarding the maximum concentrations obtained during evaluation, values of 8.69 and 9.72 mg.m^{-3} for CICORR and INAH were determined, respectively. These values represent no health risk to people and the environment because do not exceed the average value of 120 ug.m^{-3} in 24 hours established by the Mexican Standard (Dzul, 2010). Respecting the average values, a concentration of 3.54 mg.m^{-3} and 3.30 mg.m^{-3} was determined for INAH and CICORR, respectively, indicating a slight difference in concentration between both sites which follow the same behavior. According to the results, a higher concentration of PM10 particles in the CICORR station was found with respect to INAH. This behavior coincides with that observed previously for TSP in both seasons, given the prevalence of similar environmental conditions (Miss, 2008). CICORR station is surrounded by trees and by the athletic field of the Autonomous University of Campeche. In the West side of CICORR is located Juan de la Barrera Street, showing steady traffic during the morning and tends to diminish in the evening during the class activities, a period which coincided with the sampling. INAH station is located in the center of the city of San Francisco de Campeche in an urban area with heavy traffic flow during most of the day.

Sulfur dioxide

SO_2 is considered as an indicator of atmospheric pollution in urban sites. It has been included in air quality indexes in several cities along the world (Valeroso, et al., 1992, Shifer et al., 2000; Raavindra et al., 2003). Industrial emissions and vehicle exhaust are the mains source of this pollutant which is precursor of acid rain and black crust formation (Mala, 1999; Primerano et al., 2000; Reyes 2004; Reyes et al., 2004). This parameter was monitored during January 2007 to January 2008 in the historic center of San Francisco de Campeche City (INAH station), by using a visible fluorescence automatic equipment (NOM-038-SEMARNAT-1993). Fig. 8 shows the behavior of SO2 during the sampling. Maximum, minimum and medium values are reported in Table 4. According to the results, both 24 hours maxima and annual arithmetic average were reported below maximum limits established by Mexican Standard. It means that its effects in health are limited. Nevertheless, the behavior of SO2 during sampling period indicates a continue increase in their atmospheric concentration.

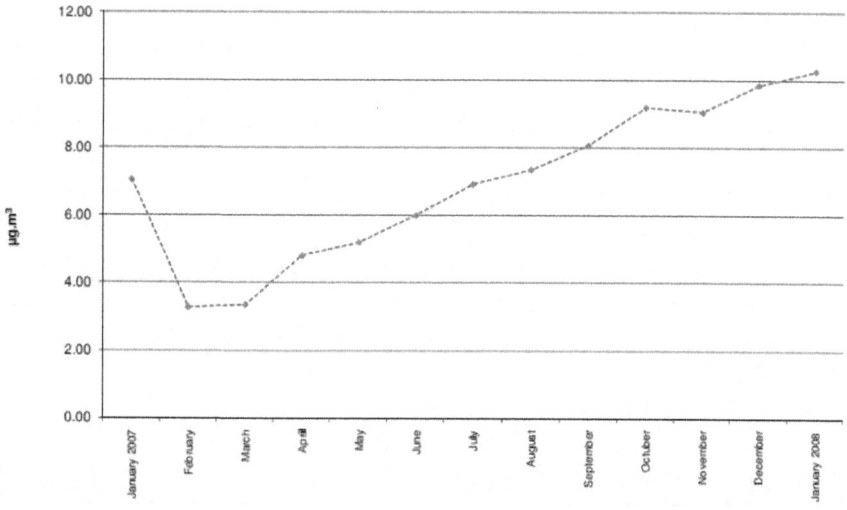

Figure. 8: Monthly average value of SO_2 registered in San Francisco de Campeche City Historic Center (INAH station), during January 2007 to January 2008.

The last one is critical for environmental air quality because this situation may be consequence of an increase in the number of automobiles in the city. That is a critical situation because it could generate traffic jam conditions in the historic center of the city. Vehicle exhausts create adverse conditions that allow the initiation of degradation mechanisms in stone materials, as have been observed in several historic cities along the world (Primerano, et al., 2000).

Acid Rain

During the years of 2006 and 2007, a wet sample collecting campaign was carried on by using an automatic wet/dry sampler (US-EPA, 1994) installed at the INAH station (Quirarte, 2010). A total of 147 samples were obtained. Table 5 shows the maximum, minimum and average weighted pH registered during the campaign. Fig. 9, represents the tendency in change of pH value along the rainy period. It is important to note that in both years, a natural tendency to alkalinity exists in rain water pH. During the months corresponding to dry season (from December to June) rain water pH are usually higher than 6. This general tendency changes from July to November, period in which the atmosphere has been washed of dust particles by the rainy season. Then, the minimum values of pH are reached and eventually, sporadic acid rain events can be observed, probably as a consequence of atmospheric transport (Quirarte, 2010).

Table 5: Maximum, minimum and average ponderated pH registered at San Francisco de Campeche City.

Year	Number of samples	PH			% of acid samples
		maximum	minimum	average	
2006	83	7.54	5.19	6.04	12
2007	73	7.80	4.97	6.39	5

Torres (2009), studied the ionic enrichment in rain samples collected at INAH station during 2007. The study indicate an enrichment on sulphates (SO_4^-), nitrates (NO_3^-), calcium (Ca^{2+}) and Cl^- ions. SO_4^- and NO_3^- are acidic compounds present as a consequence of human activity, while $Ca2+$ is dragged from alkaline soils of Peninsula of Yucatan, because it is transported by the wind and incorporated to the rain drops in the atmosphere, contributing to the neutralization of acidic compounds.

Under this condition, rain acidity is not a determinant factor in recession rates of calcareous materials, since volume and intensity of precipitation seems like key factor in deterioration of the historic building at San Francisco de Campeche City.

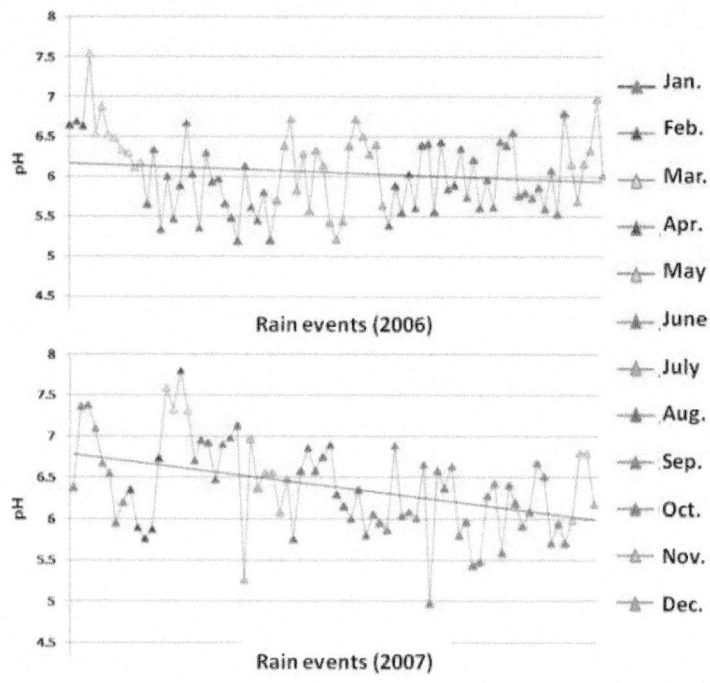

Figure. 9: Tendencies of rain pH during 2006 and 2007 at San Francisco de Campeche City.

Degradation of Historic Buildings in San Francico De Campeche City

Two representative building from the old military complex of the City were studied in order to analyze the influence of environmental condition on degradation of their mansory structure: Forts San Carlos and San Pedro (Fig. 10). Both buildings were constructed in masonry base structure made by calcareous stone quarry blocks and mortars, made with slike lime and stone dust named sahacab. Fort of San Carlos is a pentagonal-shaped structure located at the city´s bastions-andrampart system´s northwestern corner, in front of the south of Gulf of Mexico shoreline. Until mid of the XX century, when, state government, public works reclaimed some portion of land from the original previous shorefront, three walls suffered direct wave impact and tidal movements. At present, the State and Municipal Government office buildings as well as the State Congressional offices and Legislature auditorium are located adjacent to Fort San Carlos. Continuous vehicular movements flow through this immediate area, which houses peripheral urban core lanes and formal entrance into the 8th Street downtown historic district. Fort San Pedro crowns the city´s bastion- and- rampart system´s southeaster sector located at the Southeastern sector. While functioning as a bastion again possible inland attacks and "watchtower" for surrounding neighborhoods located to the south, southeast and southwest, this structure does not receive direct marine aerosols and tidal movements as noted in the case Fort San Carlos located in the northern parapet perimeter. These factors suggest that deterioration followed a slow natural process over a long time period. However, at present this Colonial construction is surrounded by traffic jammed streets, municipal bus terminals and intense anthropogenic activity in the immediate area.

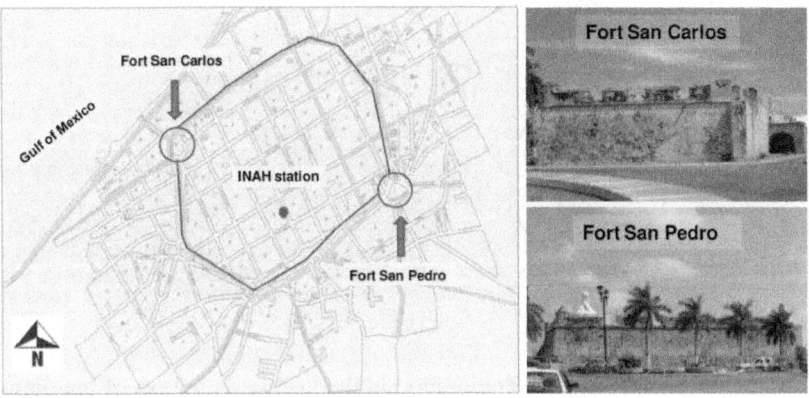

Figure. 10: Location of Forts San Carlos and San Pedro at the historic center of San Francisco de Campeche City. Also location of INAH station is showed.

In spite of consider the effects of environment in degradation of historic buildings, samples were collected from their walls and mineral alteration was investigated by XRD analysis in a Bragg–Brentano geometry X-ray diffractometer (Siemens D5000), and analyzed under the following conditions: Cu Kα radiation ($\models$=1.5416 Å) and operational conditions of 25 mA and 35 kV at a step size of 2°/2θ/min in the 2–60° range 2θ. Table 6, shows the mineral phases identified during the analysis in crusts samples from both Forts. Calcite ($CaCO_3$), a rhombohedric form of calcium carbonate, seems like the major compounds in all the samples. As have been described before, tropical climate guarantee the water availability to lead dissolution of calcium carbonate content in calcareous materials and their later recrystallization to form crusts. Also minerals like, aragonite ($CaCO_3$), sodium silicate ($Na2Si_4O_9$), quartz (SiO_2), dolomite ($CaMg(CO_3)_2$) and portlandite ($CaOH_2$) were present.

There are mineral components of limestone and traditional mortars employed during the construction of the Forts or the utilization of cements to make modern mortars during recent preservation works. Aragonite ($CaCO_3$), is a polymorphous of calcium carbonate and is present in bioclastic limestones. The identification of neomineral phases like whewellite ($C_2CaO_4.H_2O$), and wheddellite ($C_2CaO_4.2H_2O$) keep relation with bio-deterioration phenomena. Calcium oxalates are formed during oxalic acid dissolution of calcareous materials (Arocena et al., 2007). Oxalic acid is produced by metabolic activity of microorganisms like cyanobacteria and lichens (Del monte y Sabbioni, 1985; Rampazi et al., 2004). In the walls of Forts San Carlos and San Pedro, was evident the colonization by abundant microbial communities (Fig. 11).

a   b

Figure. 11: Aspect of the biodeterioration in the historic buildings of San Francisco de Campeche City. (a) Fort San Carlos. (b). Microbial community at West wall of Fort San Pedro.

On the other hand, it is important to note the presence of gypsum in Fort San Pedro samples while it was absent in Fort San Carlos ones. Gypsum is a neomineral product formed as a consequence of SO_2 reaction with $CaCO_3$ in urban environments (Graedel et al., 2000; Reyes et al., 2010b). It is an indicator of the certain pollution level in specific areas submitted to the pressure of vehicular and industrial emissions. San Pedro Fort is localized in the east area of the historic centre of the city. All their walls (except the west), are bordered by heavy traffic jams avenues, while south and southwest walls are very close to a bus station from Municipal Urban System.

Table 6: Mineral phases identified in samples from Forts San Carlos and San Pedro by XRD Samples CNC and PCC correspond to the Convent of San Francisco de Asís (Havana City). (+) present, (-) not present [1] ICD card number 29-0713 [2] ICD card number 33-650

Mineral phase	Sample code												CNC	PCC
	1	2	3	4	5	6	7	8	9	10	11	12		
Calcite	+	+	+	+	+	+	+	+	+	+	+	+	+	+
Aragonite	+	-	+	+	-	+	+	+	-	+	+	+	-	-
Sodium fedespar	+	-	+	+	+	+	+	+	+	+	+	+	-	-
Quartz	+	+	+	+	+	-	-	-	-	-	-	+	+	+
Orthoclase	-	-	-	-	-	-	-	-	-	-	+	-	-	-
Dolomite	-	-	-	+	+	+	-	-	-	+	+	-	-	-
[1]Goethite	-	-	-	-	-	-	-	-	-	+	-	-	-	-
[2]Iron hydroxide carbonate	-	-	-	-	-	-	-	-	-	+	-	-	-	-
Clay minerals	-	-	-	-	-	+	-	-	-	+	-	-	-	-
Whewellite	+	+	+	-	-	-	+	+	-	-	+	-	-	-
Bassanite	-	-	-	-	+	-	-	-	-	-	-	-	-	-
Weddellite	-	+	-	-	-	-	-	-	+	-	-		-	-
Portlandite	-	-	-	-	-	-	-	+	-	-	+	+	-	-
Hidroxiapathite	-	-	-	-	-	-	-	-	+	-	-	-	-	-
Gypsum	-	-	-	-	-	-	-	-	+	-	+	+	+	+

(Columns 1–8: Fort of San Carlos; Columns 9–12: Fort of San Pedro; CNC, PCC: Convent of San Francisco de Asís)

City of Havana: a Comparison of Air Pollution and Stone Degradation

The City of Havana

The City of Havana was founded on November 16, 1519 by Spanish conquest Diego Velázquez de Cuellar. Its historical center was declared a World Heritage Site by UNESCO in 1982. Havana was strengthened in the XVII century by order of the Spanish kings who signed as "Key to the New World and bulwark of the West Indies". In 1763 construction began on the fortress of San Carlos de la Cabaña, the largest built by Spain in the New World, which shored up the defensive system of Havana after the British occupation. The port of Havana was considered one of the most important of the region during the colonial era and one of the strategic points for Spain, which is why the bay was protected with a very important network of fortifications, including the Tower of San Lazarus, El Morro de La Habana, the Fortress of San Carlos de la Cabaña, the Castle of "La Fuerza" and other fortress dedicated to protecting the harbor and the city. During the colony, Havana was also the major transshipment point between the New World and Europe. As a result Havana was the most fortified City in the Americas. Most examples of early architecture can be seen in military fortifications such as Fortress San Carlos de la Cabaña (1558 - 1577) and the Morro Castle (1589 -1630). The Convent of San Francisco de Asis, is a religious building of Baroque architecture located in the plaza of the same name in the Old Havana (Figure 12). Construction began in 1548 until 1591, although it opened in 1575, fully completed nearly 200 years later, with a series of structural reforms that occurred from 1731 to 1738. It has a tower of 48 yards high, which in colonial times was the tallest structure in the city for several centuries.

Figure. 12: The Convent of San Francisco de Asis (a). Location of the Convent into the Historic Center of Havana City.

Degradation of Historic Buildings: A Comparative Havana Vs San Francisco De Campeche

Nowadays Havana is a City having about 2.2 million inhabitants and different types of industries, particularly around the Bay, a different situation respecting the Mexican City of San Francisco de Campeche. At Havana, air pollution levels are higher than those observed in the Mexican City (Corvo et al., 2010). In this order, a comparison of the influence of air pollution on stone buildings degradation can be made between both cities located in tropical climate. San Francisco de Campeche City shows a tendency to alkaline rain water with percent of acid rain event of 12% and 5% during 2006 and 2007 respectively (Quirarte, 2010); however, in Havana City, during the period 1981-1994, rain having a pH lower than 5,6 oscillated between 25% and 75% of the samples. It indicates a general tendency to acid rain in Havana. On the other hand, Table 7 shows the results of atmospheric contamination measured in San Francisco Convent and the Basilica. It can be noted that there is an evident difference in the deposition level of sulfur compounds between Havana and San Francisco de Campeche sites (Table 4). Havana sites show a significant higher deposition of sulfur compounds respecting San Francisco de Campeche. The two selected monitoring sites were located inside San Francisco de Asis Convent and Basilica Minor. This building is located at less than 200 m from Havana Bay shoreline. Under indoor conditions, deposition rate is usually lower than outdoors. One of the monitoring sites was located inside the Basilica building, in the concert hall at about 3 m from the floor. The second monitoring site was located in the Chorus, in the same Basilica Building, at about 10 m from the floor. Evaluation was carried out beginning September 2006 up to March 2007. Chloride deposition rate was negligible because it was determined in indoor conditions, it is very well known that chloride aerosol significantly decreases in indoor conditions; however, in outdoor conditions, in sites near Havana Bay, an average chloride deposition around 10- 20 mg.m^{-2}d^{-1} has been measured. It is important to note that even under indoor conditions, values of sulphur and nitrogen compounds inside the Convent are higher than those reported for San Francisco de Campeche outdoors. It confirms that air pollution in Havana City is significantly higher (Corvo et al., 2010; Reyes et al., 2010).

Table 7: Air pollution levels inside San Francisco de Asis Convent and Basilica Minor in Havana, Cuba. Neg: Negligible.

City	Site	Sulphur compounds deposition rate (mg.m^{-2}d^{-1})			Chloride deposition rate (mg.m^{-2}d^{-1})			NO$_2$ concentration (µg.m^{-3})		
Havana		Ave.	Max	Min	Ave.	Max	Min	Ave.	Max.	Min.
Indoor	Basilica	10.50	12.50	6.51	Neg.	Neg.	Neg.	16.35	26.08	6.23
Indoor	Chorus	11.60	14.65	7.60	Neg.	Neg.	Neg.	16.29	24.49	11.50

a   b

Figure. 13: Main Façade of San Francisco de Asis Convent and Basilica Minor in Havana, Cuba (a). Black crust deposits (b).

Crust representative samples were taken from the façade of the Convent of San Francisco de Asis and analyzed according the same procedure previously described by Forts San Carlos and San Pedro samples (Fig. 13). Mineral composition of Cuban samples is included in Table 6 (CNC and PCC samples). Black crusts formed at the Basilica façade (outdoors) in Obispo Street show gypsum as a predominant phase with small amounts of calcite and quartz. It means that black crust composition is almost completely gypsum due to contamination by atmospheric SO_2. No presence of nitrogen degradation product was detected. Crust formed at Forts San Carlos and San Pedro (San Francisco de Campeche) is mainly formed by calcite, the original main content of the stone. Different minor phases are: aragonite, dolomite, and quartz. The presence of whewellite and weddelite in the samples is an index of the influence of biological activity in stone deterioration, although the presence of bassanite in sample 5 from Fort San Carlos, shows the influence of environmental SO_2. It is important to note that gypsum was identified in samples 9, 11 and 13 corresponding to Fort San Pedro, but not at Fort San Carlos. Gypsum is produced by the action of SO_2 over calcareous materials. The comparison between crust composition in Campeche and Havana is a demonstration of the role of air pollution in deterioration of stone buildings. According to the present results, the influence of sulphur contamination is higher than nitrogen contamination, because degradation products do not show nitrogen compounds in its composition. Sulphur dioxide is highly soluble in water; however, nitrogen dioxide is not significantly soluble, it could be a cause for a higher influence of sulphur compounds in stone degradation. In addition, nitrogen degradation compounds are more soluble than sulphur degradation compounds, so the first are easily eliminated by rain.

CONCLUSIONS

The present contribution, showed a general description of the current air quality conditions at San Francisco de Campeche City. From the health point of view, SO_2, TSP and PM10 fraction are below the limits of risk considered by Mexican Legislation. The creation of a local air monitoring program in order to prevent an increase of atmospheric pollution levels is necessary as a consequence of the recent economical, demographic and urban expansion suffered by the City. In this order, although SO_2 concentration was always below critical risk levels, it suffered a continuous increase during the monitoring period. From the materials point of view, the tropical climate and the presence of natural and anthropogenic pollutants create conditions for degradation of both, metals and stony materials. In this order, the degradation of historic building in San Francisco de Campeche City shows a closer relationship with the effect of natural environmental factors, led by water actions that induce mechanisms of salt dissolution and recrystallization across wet todry cycles. The majority presence of calcium carbonates in crust formed on walls of Forts of San Carlos and San Pedro seems to confirm this fact. On the other hand, in spite of the low levels of atmospheric pollutants observed in the City, the presence of gypsum (Fort San Pedro) and bassanite (Fort San Carlos), is an indicator of a growing influence that the anthropogenic pollution could have on deterioration mechanisms. The last one result clear in the case of Fort San Pedro, which actually is under high environmental pressure. In contraposition, samples from San Francisco de Asis Convent (Havana), show gypsum as a majority neomineral phase. Gypsum is produced in urban environments with high content of SO_2, which agrees with the higher levels of atmospheric pollution detected at Havana City. In case of increase of pollution levels at San Francisco de Campeche City a similar situation will be found.

ACKNOWLEDGEMENTS

The realization of this contribution was possible thanks to the support of FOMIX CAMP2005-C01-025 Project (Urban Environmental Influence on degradation of colonial military and religious buildings at Campeche City) financed by the Government of State of Campeche and the Council of Science and Technology of México. Also thanks to Centro INAH-Campeche for their giving facilities to the development of the project.

REFERENCES

1. Allen, G., El-Turki, A., Hallam, K. R., McLLaughlin, D., Stacey, M. (2000). The role of NO2 and SO2 in degradation of limestone. British Corrosion Journal. 35, 35-38.

2. Arocena J., Siddique T., Thring R., Kapur S. (2007) Investigation of lichens using molecular techniques and associated mineral accumulations on a basaltic flow in a Mediterranean environment. Catena 70, 356-365.

3. Castro-Mora, J. (2002). Geological and mineral monography of the State of Campeche. Council for Mineral Resources. Secretary of Economy. Mexico. ISBN 968-6710-95- 7.

4. Bravo, H., Soto, R., Sanchez, P., Torres, R., Granada, L. (2000) Chemical composition ofprecipitation in a Mexican Maya región. Atmospheric Environment. 34, 1197- 1204.

5. Cardell, C., Delalieoux, F., Roumpopoulos, K., Moropoulou, A., Auger, F., Van Grieken, R. (2003). Salt-induced decay in calcareous stone monuments and buildings in a marine environment in SW France.

Construction and Buildings Materials. 17, 165-179.

6.   Carpio, J., Reyes, J., Salazar, J., Parra, A., Martínez, L. (1995) Influence of industrial pollutantson the atmospheric corrosivity of petrochemical installation. Paper LA9635.

7.   Proceedings 2nd Latinoamerican Corrosion Congress. Maracaibo, Venezuela. Cassar, M., Brimblecombe, P., Nixon, T., Price, C., Sabbioni, C., Saiz-Jimenez, C., Van Balen, K. (2004). Sustainable solutions in the conservation and protection of historic monuments and archeological remains: a critical assessment of European research needs. In: Air Pollution and Cultural Heritage. Sáiz-Jiménez Ed. Balkema TheNetherlands. ISBN. 905809622.

8.   Cook, D., Van Orden, A., Reyes, j., Oh, S., Balasubramanian R., Carpio, J., Towsend, H. (2000). Atmospheric corrosion in marine environments along the Gulf of Mexico. In: Marine corrosion in tropical environments. Sheldon, W. Dean, Guillermo, Hernandez-Duque Delgadillo, James B, Bushman Eds. American Society for Testing and Materials. 75-97. STP1399.

9.   Corvo, F., Pérez, T., Martin, Y., Reyes, J., Dzib, L., Gonzalez, J., Catañeda, A. (2008). Corrosion Research Frontiers. Atmospheric Corrosion in Tropical Climates". On the concept of time of wetness and its interaction with contaminants deposition. In electroanalytical Chemistry: New research". G. M. Smithe Ed. Nova Science Publishers Inc.62-9.ISBN 978-1-60456-347-4.

10.   Corvo, F., Pérez, T., Reyes, J., Dzib, L., Gonzales-Sanchez, J.,Catañeda, A. (2009). Atmospheric corrosion in tropical humid climate. In: Environmental degradation of infrastructure and cultural heritage in coastal tropical climate.J. Gonzalez-Sanchez, F. Corvo, N. Acuña-González. 1-34. ISBN: 978:81-7895-426-4.

11.   Corvo, F., Reyes, J., Valdes, C., Villaseñor, F., Cuesta, O., Aguilar, D., Quintana, P. (2010). Water, Air and Soil Pollution. 205, 359-375.

12.   Del Monte, M., Sabbioni, C. (1986). Chemical and biological weathering of an historical building: Reggio Emilia Cathedral. Science of the Total Environment. 50, 65-182.

13.   Dzul, B. (2010). Elemental analysis of atmospheric particles, PM10 fraction at San Francisco de Campeche City by using PIXE spectroscopy. Undergrade thesis. Autonomous University of Campeche, San Francisco de Campeche, México.

14.   Gongora, H. (2010). Influence of corrosive environmental parameters on degradation of historical and industrial materials in three microclimates

of the State of Campeche.Undergrade thesis. Autonomous University of Campeche, San Francisco deCampeche, México

15. Gobbi, G., Zappia, G., Sabbioni, C. (1998). Sulphite quantification on damaged stones and mortars. Atmospheric Environment. 32(4), 783-789.

16. Gorbushina, A. (2007). Minireview: Life on the rocks. Environmental Microbiology 9(7): 1613–163.

17. Government of the State of Campeche (2004). Diagnostic of vial and transport system at San Francisco de Campeche City. Campeche, México

18. Graedel, T., (2000). Mechanisms for the atmospheric corrosion of carbonate stone. Journal of the Electrochemical Society, 147 (3), 106-109.

19. Guiamet, P., Gómez de Saravia, S., Nuñez G. (2005). Biodeterioration of buildings stone bycyanobacteria, bacteria and fungi. In Ortega-Morales, B. O; Gaylarde, C. C.;Narváez-Zapata y Gaylarde P.M. (Eds). LABS 5 Latín América Biodegradation and Biodeterioration Symposium. Autonomous University of Campeche and National Polytechnic Institute. 51-53. ISBN 968-5722-34-X.

20. Kuchera, V., Tidblad, J., Kreislova, K., Knotkova, D., Faller, M. Reiss, D. (2007). UN/ECE ICP materials dose-response functions for the multi-pollutant situation. Water Air and Soil Pollution: Focus. 7, 249-258.

21. Little, B.J., Ray, R. (2005). The role of fungi microbiologícally influenced corrosion. En: OrtegaMorales, B. O; Gaylarde, C. C.; Narvaez-Zapata y Gaylarde P.M. (Eds). LABS 5

22. Fifth Latín América Biodegradation and Biodeterioration Symposium. Autonomous University of Campeche and National Polytechnic Institute. 51-53. ISBN 968-5722-34-X.

23. Lpifert, F. (1989).Atmospheric damage to calcareous stones. Atmospheric Environment. 23,415-429.

24. Massey, S. (1999). The effects of ozone and NOX on the deterioration of calcareous stone. The Science of the Total Environment. 227, 109-121.

25. Mala, B. (1999). Global transport of anthropogenic contaminants and the consequences for the Arctic marine ecosystem. Marine Pollution Bulletin. 38 (5), 356-379

26. Miss M. (2008). Characterization of aliphatic hydrocarbons and analysis of spacial and temporal variation of atmospheric particles at San Francisco de Campeche City.

27. ndergrade thesis. Autonomous University of Campeche, San Francisco deCampeche, México.

28. Monna, F., Puertas, A., Leveque, F., Losno, R., Fronteau, G., Marin, B. (2008). Geochemical records of limestone facades exposed to urban atmospheric contamination as monitoring tools?. Atmospheric Environment. 227, 109-121.

29. Moropoulou, A., Konstanti, A. (2004). European strategies for preservation and managements of Historic Cities: knowledge based decision-making approach. In: Air Pollution and Cultural Heritage. Sáiz-Jiménez Ed. Balkema The Netherlands.ISBN. 905809622.

30. Muñoz, A., Rodriguez, A., Villalobos, P., MUnive, Z., Marttelo, O., Diaz, G., Bravo, C. Gomez, A. (2001). Suspended particles, policyclic aromatic hydrocarbons and mutagenesis at Southwest of Mexico City. Revista Internacional de Contaminación Ambiental. 17, 193-204.

31. Ortega-Morales, B.O. 2004. Biofilms fouling ancient limestone Maya monuments in Uxmal, México. Current Microbiology 40, 81-85.

32. Primerano, P., Marino, G., Dipascale, s., Mavilia, L., Corigliano, F. (2000). Possible alteration of monuments caused by particles emitted onto the atmosphere carrying strongprimary acidity. Atmospheric Environment. 34, 385-401.

33. Quirarte, O. (2010). Variation of rain acidity in San Francisco de Campeche City and Calakmul Biosphere Reserve (archeological zone), during 2006 and 2007.

34. Undergrade thesis. Autonomous University of Campeche, San Francisco de Campeche, México.

35. Rampazzi, l., Andreotti, A., Bonaduce, I., Colombini, M, P.; Colombo, C., Toniolo, L. (2004).

36. Analytical investigation of calcium oxalate films on marble monuments. Talanta. 63, 967-977.

37. Ravindra, K., Mor, S., Kamgotra, J., Kaushin, C. (2003). Variations in spatial patterns of criteria air pollutants before and during initial rain in Monsoon. Environmental Monitoring Assessment. 87, 145-153.

38. Reyes, J. (1998). Influence of the main atmospheric and air quality factors on metals atmospheric corrosion along the southeast cost of the Gulf of México M. Phil. Thesis. Universidad Veracruzana. Boca del Rio, Veracruz. Mexico.

39. Reyes, J. (2004). Diagnostic criteria for the identification of particular matter organic compounds from vehicular exhaust and their application to study the deterioration of the Cathedral of Seville. PhD. Thesis. University of Seville. Spain.

40. Reyes, J. Hermosin, B., Sáiz-Jimenez, C. (2004). Organic analysis of aerosols in Seville Atmosphere (2004). In: Air Pollution and Cultural Heritage. Sáiz-Jiménez Ed. Balkema The Netherlands (ISBN. 905809622).

41. Reyes, J. (2005a). Research Project CONACYT-46434-Y. National Council for Science and Technology. Autonomous University of Campeche. Mexico.

42. Reyes, J. (2005b). Research Project FOMIX-CAMP-2005-C01-028. National Council for Science and Technology –Goverment of the State of Campeche. Autonomous University of Campeche. México.

43. Reyes, J., Hermosín, B., Sáiz-Jimenez, C. (2006). Organic Composition of Seville Aerosols. Organic Geochemistry, 37, 2019-2025,

44. Reyes, J., Torres, F., Ché, I., Corvo1, F., Pérez, T., Bravo, H., Sánchez, P., Aguilar, D., Quintana, P. (2010a), Dissolution of Traditional Mortars under Artificial Rain Conditions: A Laboratory Test. In: José Luis Ruvalcaba Sil, Javier Reyes Trujeque, Jesús A. Arenas Alatorre, Adrián Velázquez Castro Eds.

45. Proceedings of 2nd Latin-American Symposium on Physical and Chemical Methods in Archaeology, Art and Cultural Heritage Conservation & Archaeological and Arts Issues in Materials Science - IMRC 2009, Mexico.195-200. ISBN: 978-607-02-2017-3.

46. Reyes, J., Corvo, F., Espinosa-Morales, Y. Valdes, C., Bartolo, P., Aguilar, D., Quintana, P., Hermosín, B., Saiz-Jimenez, C. (2010b). Analysis of Black Crust from a Cuban Historic Building . In: José Luis Ruvalcaba Sil, Javier Reyes Trujeque, Jesús A.

47. Arenas Alatorre, Adrián Velázquez Castro Eds. Proceedings of 2nd LatinAmerican Symposium on Physical and Chemical Methods in Archaeology, Art and Cultural Heritage Conservation & Archaeological and Arts Issues in Materials Science - IMRC 2009, Mexico. 195-200. ISBN: 978-607-02-2017-3.

48. Sessa, C. (2004). Cultural Heritage protection with integrated policies. In: Air Pollution and Cultural Heritage, Sáiz-Jiménez Ed. Balkema The Netherlands. ISBN.905809622.

49. Schifter, I., Díaz, L., López-Salinas, E. (2006). "Assessment of new vehicles emissionscertification standars in the Metropolitan area of México city. Environmental Monitoring Assessment. 114: 419-432.

50. Sosa-Echeverria, R., Bravo-Alvarez, H., Sánchez-Alvarez, P., Soto-Ayala R., AlarcónJiménez,A., Khal, J. (2006). Determination of total suspended particles during five research cruises at Gulf of Mexico continental

platform. Ingeniería, Investigación y Tecnología. VII (2), 71,-83.

51. Stefanis, N., Theoaulakis, P., Pilinis, C. Dry deposition effect of marin eaerosol to thebuilding stone of the medieval city of Rhodes, Greece. Building and Environment. 44, 260– 270.

52. Tercer, L. (1998). Laboratory experiments on the investigation of the effects of sulfuric acid on the deterioration of carbonate stone and surface corrosion. Water, Air and Soil Pollution. 114, 1-12.

53. Tidblad, J., Mikhailov, A., Kuchera, V. (2000) Application of a model for prediction of atmospheric corrosion in tropical environments. In: Marine corrosion in tropical environments. Sheldon, W. Dean, Guillermo, Hernandez-Duque Delgadillo,James B, Bushman Eds. American Society for Testing and Materials. 18-32.STP1399.

54. Torres, F. (2009). Effects of rain in stony materials from the Campeche State culturalheritage. M.Phil. thesis. Autonomous University of Campeche, San Francisco de Campeche, México.

55. Valeroso, I. I., Monteverde, C. A., Estoque, M. A. (1992). Diurnal variations of air pollutionover metropolitan Manila. Atmosfera. 5: 241-257.

56. Villaseñor, F. (2008). The use of passive techniques for sampling atmospheric pollutants in the State of Campeche. Undergrade thesis. Autonomous University of Campeche,San Francisco de Campeche, México.

57. Zappia, G., Sabbioni, C., Riontino, C., Gobbi, G., Favoni, O. (1998). Exposure test of building materials in urban atmosphere. The Science of the Total Environment. 224, 235-244.

58. Zenddri, E., Biscontin, G., Kosmidis, P. (2001). Effects of condensed water on limestone surfaces in a marine environment. Journal of Cultural Heritage. 4, 283-289.

Chapter 11

PLASMA-BASED DEPOLLUTION OF EXHAUSTS: PRINCIPLES, STATE OF THE ART AND FUTURE PROSPECTS

Ronny Brandenburg

Leibniz Institute for Plasma Science and Technology, Germany

INTRODUCTION

Nowadays non-thermal plasma technologies are state of the art for the generation of ozone as an important oxidant for water cleaning or bleaching, the incineration of waste gases or for the removal of dust from flue gases in electrostatic precipitators. Furthermore their possibilities of gas depollution are well known. Plasmas contain reactive species, in particular ions, radicals or other oxidizing compounds, which can decompose pollutant molecules, organic particulate matter or soot. Electron beam flue gas treatment is another plasma-based technology which has been successfully demonstrated on industrial scale coal fired power plants. This chapter aims a comprehensive description of plasma-based air remediation technologies. The possibilities of exhaust air pollution control by means of non-thermal plasmas generated by gas discharges and electron beams will be summarized. Therefore plasma as the 4th state of matter, its role in technology and the principle of plasma-based depollution of gases the will be described. After an overview on plasma-based depollution technologies the main important techniques, namely electron beam flue gas treatment, gas discharge generated plasmas including plasma-enhanced catalysis and injection methods will be described in separate sections. In these sections selected examples of commercially available or nearly commercialised processes for flue gas treatment or the removal of volatile organic compounds and deodorization will be described, too. Current trends and concepts will be discussed.

PLASMAS AND PLASMA-BASED DEPOLLUTION TECHNOLOGIES

In physics and chemistry, plasma is an ionised gas containing free electrons, ions and neutral species (atoms and molecules) characterized by collective behaviour. Plasma is often referred as the "4th state of matter" since it has unique physical properties distinct from solids, liquids and gases. In particular, due to the presence of charge carriers plasmas are electrically conductive and respond strongly to electromagnetic fields. It contains chemically reactive media as well as excited species and emits electromagnetic radiation in various wavelength regions. The majority of matter in the visible universe (stars, interplanetary and interstellar medium) is in the plasma state. Lightnings, sparks, St'Elmos fires and the polar aurorae are examples for natural terrestrial plasmas. Furthermore, since more than 150 years plasmas are generated artificially by supplying energy to gases, liquids or solids. Such plasmas are used and under investigation for various applications, e.g. surface modification, chemical conversion, light generation or controlled nuclear fusion. Natural as well as artificial plasmas cover an extremely wide range of parameters like temperatures, particle densities and pressure. Broadly speaking, plasmas can be distinguished into thermal and non-thermal plasmas. In thermal plasmas all present species (electrons, ions and neutral species) are in the local thermal equilibrium, i.e. all species have the same mean free kinetic energy (temperature). Such plasmas are produced in fusion experiments with temperatures higher than 10^4 K. Contrary, in other situations most of the coupled energy is primarily released to the free electrons which exceed the temperatures of the heavy plasma components (ions, neutrals) by orders of magnitude. Such mixtures of energetic electrons in a relatively cold mass of ions and neutrals are called non-thermal or non-equilibrium plasmas. If the gas temperature stays nearly at or slightly above room temperature the plasma is termed "cold plasma". Even in non-equilibrium plasmas the gas temperature can increase to some 10^3 K. In such cases it is called "hot non-thermal plasma" or "translational plasma" since it marks the transition to the thermal regime. In fact cold as well as translational plasmas are used for gas depollution. The most common method for plasma generation for technological and technical application is by applying an electric field to a neutral gas. If the applied field exceeds a certain threshold (breakdown field strength) a gas discharge and thus plasma is formed. There are many different designs of plasma sources for depollution and the most important will be described in the next two sub-sections. Alternatively by the interaction of an electron beam with gaseous medium plasma can be generated. Such electron beam generated plasmas are used in the so-called electron beam flue gas treatment, which is further described in section 3 of this chapter.

Plasma-Based Depollution by Means of "Hot" Plasmas

Plasma pollution control can be done by an increase of the gas enthalpy by means of hot (i.e. thermal or translational) plasmas. Such plasmas are widely used for the incineration of gaseous but also liquid and solid waste. An overview is given in (Hammer, 1999). Typical examples are high-intensity arc or plasma torches. Electric arcs discharges are driven between two electrodes (see fig. 1 a) by high current (10 to 1000 A). Thus in arc plasmas high energy and current densities are reached (10^7–10^9 J m−3; 10^7–10^9 A m^{-2}). High-current arcs at atmospheric pressure can be characterized as thermal plasmas reaching temperatures in the range 5,000–50,000 K (Kogelschatz, 2004), which makes them very useful for material processing (welding, cutting, spraying) and waste treatment.

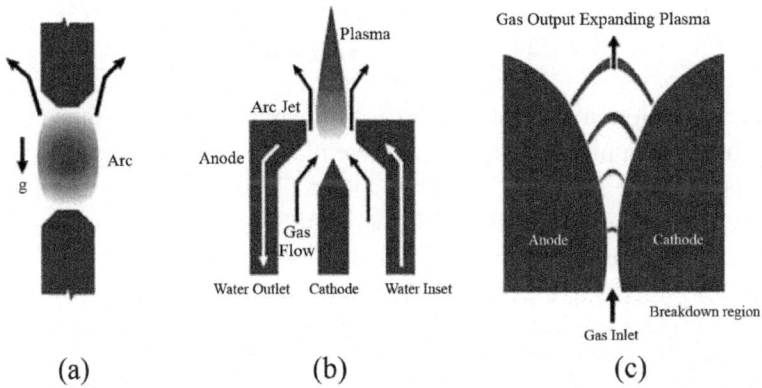

(a)   (b)   (c)

Figure. 1: General schemes of thermal and translational plasmas (a) free burning arc discharges in vertical and horizontal configurations; (b) plasma torch; (c) gliding arc

In plasma torches (also referred to as plasmatrons or plasma guns) the electrical energy is coupled into the working gas inside a nozzle and a high gas flow leads to the expansion outside the nozzle as a plasma jet (fig. 1 b). A large variety of plasma torches has been developed. The majority of commercial torches uses direct current arc, inductively coupled radio frequency discharges or microwave excited plasmas as the heat source and atmospheric-pressure air as working medium. The power consumption of plasma torches is in the range of several kW up to some MW. As a very rough estimation, the energy costs for conversion of noxious compounds is about 20 eV/molecule. This corresponds to 0.1 to 1 kg/kWh, a value which is comparable to that obtained in non-thermal plasmas (Hammer, 1999). Gliding arcs (fig. 1 c) are another example for translational plasmas studied for gas depollution and other applications. They consist of at least two diverging electrodes which are passed by a gas flow.

The discharge starts at nearest distance between the electrodes, is spreading by gliding along the electrodes in the direction of the gas flow which leads to cooling of the plasma. Microwave driven plasma torches at atmospheric pressure are typical examples for translational plasmas (non-thermal plasmas at elevated gas temperatures up to 4,000 K). However the gas temperature is high enough to decompose stable organic molecules. In particular nozzle-type microwave plasma source (MPS) (see e.g. Jasinski et al., 2002) has been used for the destruction of gaseous pollutants - mainly vapours of organic solvents - of relatively high concentration, up to tens of vol.%. The nozzle-type MPSs first appeared as structures based on microwave coaxial line components (see e.g. Cobine & Wilbur, 1951) where the microwave plasma was induced in the form of a plasma "flame" at the open end of a rigid coaxial line, at the tip of its inner conductor. The power-handling capability of coaxial-line-based microwave discharges is generally limited to much less than 1 kW due to the low thermal strength of the coaxial line components. Parallel with the coaxial-line-based nozzle-type MPSs so-called waveguide-based nozzle-type MPSs have been developed (e.g. Yamamoto & Murayama, 1967; Moisan et al., 1994, 2001). In these applicators the microwave plasma is also induced in the form of a plasma flame at the tip of a field-shaping structure that is similar to that of the coaxial-line based MPSs. However, the microwave power is fed into this structure from a waveguide, usually rectangular at 2.45 GHz. In advanced devices, the microwave power is delivered to the field-shaping structure in form of a conductor with a conical nozzle through a waveguide with a reduced-height section (fig. 2 a).

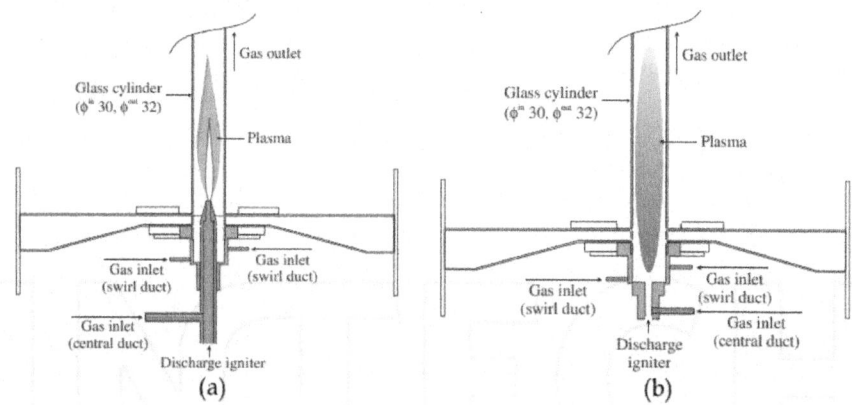

Figure. 2: Sketches of the waveguide-based cylinder-type MPS (a) and waveguide-based nozzle-type MPS (b). Dimensions are given in mm.

Since both microwave discharges, the coaxial-line-based and waveguide-based one, are gas flowing systems, they are particularly suitable for processing various gases or materials carried by gases. Recently, a new MPS was developed (e.g. Uhm et al., 2006) based on the rectangular waveguide with a reduced-height section, where the discharge is generated inside of a dielectric cylinder with a swirl flow of the working gas. There are no nozzles in the system (see fig. 2 b). It was successfully used for destruction of refrigerant HFC 134a (Jasinski et al., 2009) with destruction mass rate and corresponding energetic mass yield of up to 34.5 kg h^{-1} and 34.4 kg per kWh of microwave energy absorbed by the plasma, respectively.

Plasma-Based Depollution by Means of "Cold" Non-Thermal Plasmas

In cold non-thermal plasmas the free energetic electrons are able to produce radicals and other reactive species (e.g. ions) which react with the pollutant molecules or particles. Furthermore, if ions can be extracted from the discharge, fine particles can be charged and thus filtered electrically from the flue gas (Grundmann et al., 2007). Additional a biological decontamination of air due to plasma treatment has been reported (e.g. Müller & Zahn, 2007).

Cold Non-Thermal Plasma Sources for the Depollution of Gases

As already mentioned, non-thermal plasmas in gas streams at atmospheric pressure can be generated in two ways. Either with the injection of a high energetic electron beam (so-called electron beam flue gas treatment, EBFGT) or the generation of a gas discharge by means of a sufficient high voltage applied to two electrodes (gas discharges). In discharge generated plasmas the electrons have lower mean energies than in electron beam produced plasmas. Thus plasma chemical reactions can differ and usually in electron beam generated plasmas the energy efficiency is better. However, discharge generated plasmas give the chance to construct more compact after treatment systems for small and medium size gas streams. To generate plasmas with electron beams special electron accelerator units are needed. Electrons are produced via thermionic emission from a cathode and accelerated inside the vacuum tube. The electron beam transits from the beam generation environment at vacuum pressure (10^{-5} mbar) into the flue gas stream at atmospheric conditions via a beam window and than through a secondary window (Chmielewski et al., 1995). Due to a beam alignmentsteering system the beam will scan across or along the flue gas stream. Beam scanning and window cooling is necessary to avoid destruction of the titanium windows. The beam acceleration ranges from 0.7 to 1.2 MeV, allowing the beam to penetrate the windows without excessive energy loss. The maximum power per accelerator available nowadays is up to 400 kW, total beam power in installations exceed 1 MW (Department of Energy, 2010). Next generation electron beam techniques use radio frequency cavity systems instead of DC transformators (Edinger, 2008). This enables pulsed driven beams with optimized energy control. To generate plasmas by gaseous discharges several possibilities exists (Becker et al., 2005; Fridman, 2008; Kogelschatz, 2004). The most common discharge types are dielectric barrier discharges (DBDs) and corona discharges. For both types different configurations and geometries, namely cylindrical and planar, exist as shown in fig. 3. DBDs, also referred to as barrier discharges or silent discharges are characterized by the presence of at least one dielectric layer between the electrodes (Kogelschatz, 2004; Wagner et al., 2003). Typical materials for dielectric barriers are glass, quartz and ceramics. Fig. 3 a shows a so-called volume barrier discharge in cylindrical geometry. The discharge gap is usually in the range of 1 mm. Fig. 3 c is a planar surface barrier discharge, i.e. both electrodes (metal meshes) are in direct contact with the dielectric plates. Another type of DBD is the so-called coplanar discharge where both electrodes are embedded in the dielectric material. Due to the capacitive coupling of the insulating material to the gas gap DBDs can only be driven by alternating feeding voltage or pulsed DC voltages. When a sufficient voltage is applied to the electrodes, electrical breakdown occurs most commonly as number of individual discharge filaments or microdischarges (Kogelschatz, 2002). Microdischarges have a small duration (tens of nanoseconds in air), small size (diameter about 100 µm) (Brandenburg et al., 2005) and are distributed over the whole surface area. Due to the local charging of the dielectric surface after microdischarge inception the local electric field is weakened leading to the extinction of the microdischarge after several ten nanoseconds. Thus the barrier prevents the formation of a spark or arc discharge, keeping the plasma in the non-thermal regime. Despite the numerous applications of DBDs the knowledge on microdischarge development and thus plasma parameters and elementary processes within these microplasmas is not sufficient, although the multitude of subsequent microdischarges determines the efficiency and selectivity of the exhaust gas treatment. Special feature of DBDs are so-called packed bed reactors, where dielectric or ferroelectric pellets (e.g. alumina oxide Al_2O_3, titanium oxide, TiO_2 or barium titanate $BaTiO_3$) are packed between two electrodes (see fig. 4; Holzer et al., 2005; Yamamoto et al., 1992). Due to spontaneous polarization of the ferroelectric a high electric field at the contact points of the pellets is formed resulting in microdischarge inception. The use of pellets is disadvantageous in terms of pressure drop but lead to uniform distribution of gas flow and plasma in the reactor. Furthermore the pellets can be used as catalyst enabling direct interaction between plasma and catalyst. Corona discharges are characterized by a non-uniform configuration of the electric field, which is achieved by special electrode geometries, e.g. point-to-plane, wire-to-plane (see fig. 3 d) or coaxial wire-in-cylinder configurations (see fig. 3 b). The non-uniformity of the discharge gap enables breakdown at lower voltages allowing low current, non-thermal plasma channels based on the streamer mechanism. Thus coronas often show a filamentary character like DBDs. The electrode gap can be set to several centimetres, which is favourable for large scale applications and minimizes pressure drops. Corona discharges are usually DC-driven discharges, but for environmental applications they are often driven by high voltage pulses with rapid voltage rise (several kV per ns) and short duration (some tens of ns). This concept also referred to as pulsed corona discharges (PCD).

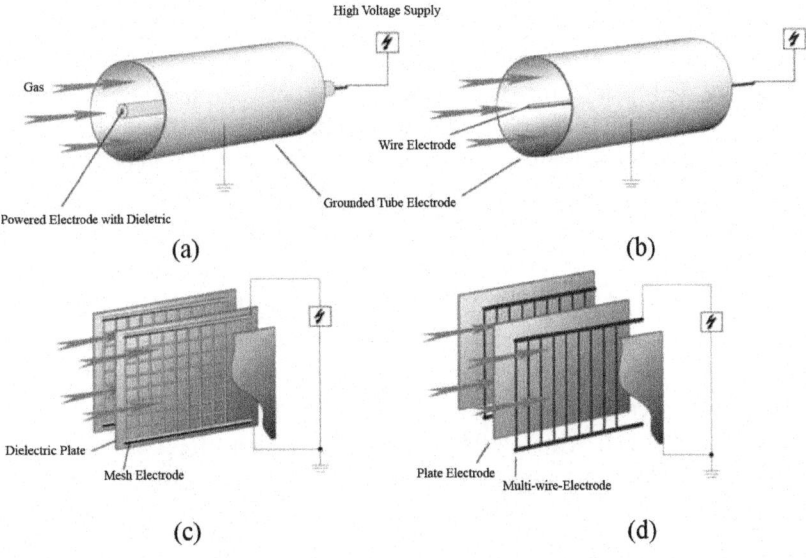

Figure. 3: Typical configurations of barrier (a, c) and corona discharges (b,d) for gas treatment (a) cylindrical asymmetric volume barrier discharge, (b) cylindrical wire-in-tube corona arrangement, (c) plate-like surface barrier discharge, (d) multineedle-plate-corona arrangement

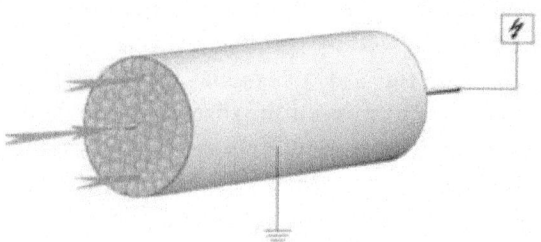

Figure. 4: Example of a packed bed reactor with special pellet filling

DC-driven corona discharges are established in pollution control as electrostatic precipitators (ESP) for dust removal of flue gases. In this application the active plasma is restricted to the region closed around the wire electrode. Between this so-called active zone and the opposite electrode (so-called collecting electrode made as plate or cylinder) a passive zone of low conductivity is formed. Ions generated in the active plasma zone enter the passive zone and drift to the collecting electrode. On their way they charge solid particles or droplets which migrate to the collecting electrode. The charged particles precipitate onto the collecting surfaces, are neutralized, dislodged and removed. Various types of dust, mist, droplet etc. down to submicron size can be removed under dry and wet conditions with high efficiency and low pressure drop (Kogelschatz, 2004). Thus ESP technology uses physical aspects of corona discharge and not the chemical processes, although the promotion of plasma chemistry is possible, too. To overcome the "back corona effect" or to decrease the power consumption pulsed operation was proposed (Mizuno, 2007; H.H. Kim, 2004). The back corona effect is obtained with high resistivity dust (e.g. cement particles), which leads to the formation of insulating dust layers on the collecting electrode which reduces the emissions of ions. Alternatively sulphur trioxide can be injected into the flue gas stream to lower the resistivity of the particles. An interesting concept of corona discharge is the (corona) radical shower discharge, which was developed in particular for NOx- and later for combined NOx- and SOx-removal (Ohkubo et al., 1996; J.P. Park et al., 1999). The discharge only treats a portion of the total contaminated exhaust flow. The treated gas with plasma generated active species is then injected in the total exhaust gas flow like a shower. Typically DBD and PCD reactors require different supply waveforms with efficiencies (i.e. overall consumed plug power vs. power dissipated into the plasma) as high as possible. DBD reactors are most often supplied using alternating, sinusoidal voltage while the corona discharge systems are pulsed supplied. In case of DBD in many cases classical 50 or 60 Hz supplies are used with highvoltage transformers (Sasoh et al., 2007; Kostov et al., 2009). Due to operating conditions higher operation frequency is often necessary in order to increase the discharge power. The average power control is critical for the yield of the chemical processes. Modern supply system designs include power amplifiers with highvoltage transformers (Francke et. al., 2003; Mok et al., 2008) or many solid-state switch based power electronic converter topologies, often resonant ones (Casanueva et al., 2004). Since resonant operation complicates fluent control of the output power, often a time-averaged burst (so-called pulse density modulation - PDM) technique is used (Fujita & Akagi, 1999). Basic configurations of non-thermal plasma supply systems are depicted in fig. 5.

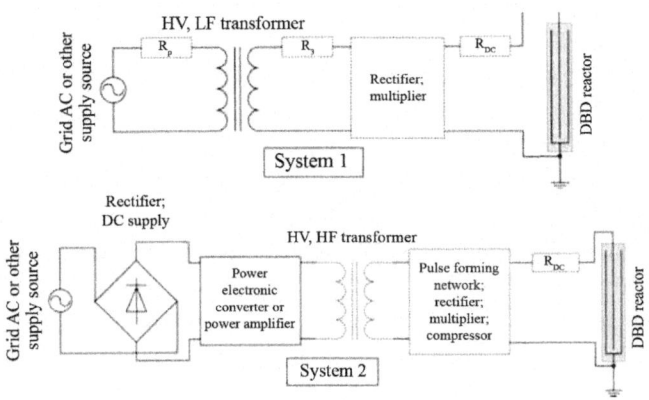

Figure. 5: Basic configurations of power supplies: low frequency systems (left) and high frequency systems (right).

Generally low frequency or high frequency systems are used. In the case of low frequency primary or secondary transformer side current limiting resistors are sometimes used (R_p or R_s), in case of pulsed D_C supplies sometimes a reactor current-limiting resistor is implemented (R_{DC}). These types of supplies usually have limited efficiency ratings (about 40% for low power systems) and due to low operating frequency large weight/volume consumption. In case of controllable systems an adjustable transformer is sometimes used. High frequency supplies usually use a rectifier as the first power electronic converter. Then different configurations and topologies are used, in many cases a high frequency – high voltage transformer (HF, HV). Sometimes additional pulse forming networks are implemented in order to shape the output voltage waveform. Considering the supply voltage waveform itself a set of different patterns can be defined. Most common is the use of high voltage, AC, sinusoidal supply. In order to influence the average reactor power pulse density modulation technique is sometimes used. Optimization of effectiveness as well as voltage potential distribution levelling sometimes results in a discontinuous, bipolar waveforms. Pulsed high voltage power supply systems are constructed in a variety as large as in the case of AC sources. In case of large installations, due to high peak values of voltage (up to several MV) and current (up to 0.5 MA), pulse modulators are constructed implementing pulsed thyristors, gas switches (thyratrons, krytrons) or spark gap switching apparatus. These technologies however, due to the principle of operation allow only a low frequency of operation and a limited lifetime. Classical constructions often implement the so called Marx generator topology (Marx, 1928) and Fitch generator topology (Fitch et al., 1968) in connection with magnetic pulse compression, which reaches efficiency rating of up to 76 %. Solid state technology enables much higher operating frequencies and very long lifetime but have a limitation of maximum allowable blocking voltage and maximal repeatable peak current per single power semiconductor. Typically high voltage MOSFET transistors and HV IGBT transistors are used for power electronic supply systems. In order to overcome single element limitations power switching stacks are produced. Nowadays typical efficiency values of up to 96 % are possible. New concepts of non-thermal plasma sources for the treatment of gases are fused hollow cathodes (FHC). The FHC cold atmospheric plasma source is based on the simultaneous generation of multiple hollow cathode discharges in an integrated open structure with flowing gas (Barankova & Bardos, 2002; 2003). The hollow cathode discharges are nonthermal because of the population of high energy electrons due to the pendulum motion of accelerated electrons between the repelling space charge sheaths at the opposite walls either in cylindrical or planar configurations. For operation at atmospheric pressure small hollow cathode inner diameters (about 200 to 400 μm) are required. The operational stability of the FHC systems is excellent; the plasma is uniform and does not exhibit streamers. The FHC systems allow generation of cold plasma in both monoatomic and molecular gases and the upstream FHC concept with aerodynamic stabilization was successfully tested for gas conversion. The power consumption of FHC has been reported to be about 1–3 orders lower than for other non-thermal atmospheric plasma sources. The FHC design for conversion experiments is based on experimental results obtained with a tuneable radial cathode slit system and different FHC structures (Barankova & Bardos, 2010). A minimum separation of the cathode walls depends both on the type of the gas (monoatomic or molecular) and on the type of generation (pulsed DC or radio frequency). Beside gas conversion the concept has been successfully used for surface treatment, activation and cleaning of temperaturesensitive materials.

Fundamentals

Chemical processes in non-thermal plasmas are based on non-thermal activation of particles via collisions. The quality and quantity of collisions is determined by the density and the kinetic parameters (e.g. mean velocity, collision frequency). In general three different phases has to be distinguished. The first phase is characterized by the electrical breakdown of the gas (e.g. in form of short-lived microdischarges as described above) where free electrons with high kinetic energies are produced via ionising collisions. These electrons undergo further electron molecule collisions, namely ionisation (1, 3), dissociation (2, 3), excitation (4) and electron attachment (7). Furthermore Penning-ionisation and dissociation (5, 6); charge transfer (8) and ion reactions are possible. All mechanisms have quite different reaction rates due to its different energy thresholds. For example for dissociation energies between 3 and 10 eV are sufficient, while ionisation requires energies more than 10 eV and electron attachment happens at energies of some eV or lower. Indeed, the exact values are determined by the electronic configuration of the molecule being considered. The reaction rate further depends on the gas temperature which depends on the vibrational excitation level of molecules. The second stage of non-thermal plasma chemistry is the radical formation and removal stage, where a multitude of anorganic reactions takes place. In particular radicals are generated through direct electron impact molecule dissociation and ionization as well as ion-molecule reactions (10), dissociate recombination of ions and electrons (11), attachment and detachment reactions (12) (Chang, 2008).

Ionisation: $$AB + e^- \rightarrow AB^+ + 2e^- \tag{1}$$

Dissociation: $$AB + e^- \rightarrow A + B + e^- \tag{2}$$

Dissociative ionisation: $$AB + e^- \rightarrow A^+ + B + 2e^- \tag{3}$$

Excitation: $$AB + e^- \rightarrow AB^* + e^- \tag{4}$$

Penning-Ionisation: $$M^* + A_2 \rightarrow A_2^+ + M \tag{5}$$

Penning-Dissociation: $$M^* + A_2 \rightarrow 2A + M \tag{6}$$

Attachment: $$AB + e^- \rightarrow AB^-$$
$$AB + e^- \rightarrow A^- + B \tag{7}$$

Charge transfer: $$AB^+ + C \rightarrow AB + C^+ \tag{8}$$

Recombination:	$AB^+ + e^- \rightarrow AB$	
	$A^+ + B^- \rightarrow AB$	(9)
Ion-Molecule reaction:	$I^+ + AB \rightarrow products$	(10)
Dissociate recombination:	$AB^+ + e^- \rightarrow products$	(11)

Detachment: $\qquad\qquad\qquad\qquad\qquad AB^- \rightarrow A + B + e^-$ (12)

In air plasmas reactive oxygen species are generated by direct electron collisions (13-16), via Penning-processes (17-19) and charge exchange (20) with subsequent ion-molecule reaction (21) from O_2 and H_2O. Furthermore in non-thermal plasmas generated in oxygen containing atmospheres at low gas temperatures ozone, and other a strong oxidizing agents like O, ïOH and HOï 2 will be formed.

$$e^- + O_2 \rightarrow 2\,O(^3P) + e^- \tag{13}$$

$$e^- + O_2 \rightarrow O(^3P) + O(^1D) + e^- \tag{14}$$

$$e^- + O_2 \rightarrow O_2(^1\Delta) + e^- \tag{15}$$

$$e^- + H_2O \rightarrow O^\bullet + {}^\bullet OH + e^- \tag{16}$$

$$N(^2D, {}^3P) + O_2 \rightarrow O(^3P) + NO$$
$$N(^2D) + H_2O \rightarrow {}^\bullet OH + NH \tag{17}$$

$$O(^1D) + H_2O \rightarrow 2\,{}^\bullet OH \tag{18}$$

$$N_2(A) + H_2O \rightarrow {}^\bullet OH + H + N_2 \tag{19}$$

$$M^+ + H_2O \rightarrow M + H_2O^+ \tag{20}$$

$$H_2O^+ + H_2O \rightarrow {}^\bullet OH + H_3O^+ \tag{21}$$

$$O_3 + {}^\bullet OH \rightarrow HO^\bullet_2 + O_2 \tag{22}$$

$$H + O_2 + M \rightarrow HO^\bullet_2 + M \tag{23}$$

Many molecules are readily attacked by free radicals. Decomposition of hazardous compounds is archived without heating of the flue or off-gas. Due to the presence of oxygen, water vapour and ozone, oxidizing reactions are dominant. The resulting chemistry is quite complex and depends on the gas mixture itself as well as the temperature. A complete description of all processes is outside the scope of this chapter and only the main important aspects will be discussed in the following. For more detailed and comprehensive information the reader is referred to several books and review papers, e.g. (Fridman, 2008; Penetrante & Schultheiss, 1993; H.H. Kim, 2004; Chang, 2008). Regarding the removal of saturated hydrocarbons (denoted as RH, e.g. alkane), the process start with dehydrogenization reactions (24, 25) followed by the oxidation of the remaining organic radical Rï (26). The latter reaction result in the formation of peroxy radicals ROï 2 (26) which are further oxidized down to CO_2 and H_2O (total oxidation) or trigger a radical chain reaction with alkyl hydroperoxide radicals R-OOH (27). In case of unsaturated hydrocarbons additionally radical addition following oxidation, radical chain reaction or polymerisation of hydrocarbons are taking place.

$$R\text{-}H + O^\bullet \rightarrow R^\bullet + {}^\bullet OH \tag{24}$$

$$R\text{-}H + {}^\bullet OH \rightarrow R^\bullet + H_2O \tag{25}$$

$$R^\bullet + O_2 \rightarrow R\text{-}O\text{-}O^\bullet \tag{26}$$

$$R_i\text{-}O\text{-}O^\bullet + R_j\text{-}H \rightarrow R_i\text{-}OOH + R_j{}^\bullet \tag{27}$$

In plasma-based flue gas treatment for NO and SO_2 removal desired reductive reaction paths are of minor importance. Oxidative processes (28 - 30) lead to the formation of NO_2. The oxidation up to N_2O_5 is possible (see section 5). If hydrocarbons are present (e.g. ethene, propene, propane) HOï 2 and peroxy radicals become the dominant oxidizers (30, 31) and the energy required to oxidize NO molecule can be reduced. However, to remove NOx from the gas a heterogeneous chemical process for NO_2 reduction must follow the plasma treatment. In a similar way SO_2 oxidation to SO_3 by means of plasma treatment is possible, while SO_3 needs to be removed chemically.

$$NO + O(^3P) + M \rightarrow NO_2 + M \tag{28}$$

$$NO + O_3 + M \rightarrow NO_2 + O_2 + M \tag{29}$$

$$NO + HO_2 + M \rightarrow NO_2 + {}^\bullet OH + M \tag{30}$$

$$NO + R\text{-}O\text{-}O^\bullet \rightarrow NO_2 + R\text{-}O^\bullet \tag{31}$$

Following the removal stage aerosol particles are formed through reaction of larger radicals with cluster ions and molecules. Aerosol formation is a quite important process since aerosol surface reaction rate is a few orders of magnitude higher then the electronic, ionic and radical reactions. The removal processes are promoted due to heterogeneous reactions. Regarding SO_2 the stimulation of chain oxidation mechanism by plasmas in liquid droplets or ionic clusters at humid gas conditions is known (see Fridman, 2008). In order to compare different concepts and technologies different aspects must be considered. The main focus is the efficiency evaluation, but costs for investment and operation (warranty intervals, consumption of additives) need to be taken into account, too. Several examples are described, see (Chang, 2008) and references therein. There is no universal parameter for the energy efficiency and the conditions of operation in research and application vary to a great extend. Most widely used parameters are the Specific Input Energy (SIE, or specific energy density SED) and the G-value. The SIE is the dissipated discharge power divided by the gas flow rate Q (32). In general the gas flow rate Q relates to standard or normal conditions (Temperature $T_N = 273.15$ K, pressure $p_N = 100$ kPa) and SIE is given in J/sl or kWh/Nm^3. The SIE is a reliable scaling parameter and together with the energy efficiency of pollutant removal η (also referred too as energy yield, i.e. mass of removed pollutant Δ_{mPol} divided by consumed energy of the plasma E_{PL}) a good economic evaluation can be done by $\eta(SIE)$ characteristics (Chang, 2008). It should be mentioned again, that a comprehensive evaluation must consider the efficiency of the power supply transformation, too (i.e. $P_{tot} > P_{PL}$).

$$SIE = P_{Plasma} / Q_N \tag{32}$$

$$\eta = \Delta m_{Pol} / E_{PL} \tag{33}$$

$$G(-A) = \beta_A \; p_A \; Q_N \; N_0 / (E \, R \, T) \tag{34}$$

The G-value is adapted from radiolysis and refers to the number of molecules of reactant consumed per 100 e_V of energy absorbed (Baird et al., 1990; Penetrante et al., 1996). It is defined as given in (34), where A is removed specie, β_A percentage of destroyed contaminants, p_A partial pressure of A, N_0 Avogadro constant, E used energy and R gas constant. In plasmas G-value gives the number of radicals generated per 100 e_V. Another value to be considered is the chemical selectivity S_A of one possible chemical product A. It is given by the ratio of its concentration (or number density of molecules etc.) and the sum of concentrations of all possible products of one reaction.

ELECTRON BEAM FLUE GAS TREATMENT (EBFGT)

Electron beam flue gas treatment technology is one among the most promising advanced air pollution control techniques. EBFGT is a dry-scrubbing process of simultaneous SO_2 and NO_x removal, where no waste (except by-products) is generated. The main components of flue gases are N_2, O_2, H_2O, and CO_2, with SO_x and NO_x in much lower concentrations. Ammonia NH_3 may be present as an additive to support the removal of SOx and NOx. The electron energy is transferred to the gas components present in the mixture in proportion to their mass fraction. The fast electrons slow down by collisions, secondary electrons are formed which plays an important role in overall energy transfer and the plasma is formed in the flue gas. Then, fast electrons interact with gas creating various ions and radicals, the primary species formed include N_2^+, N^+, O_2^+, O^+, H_2O^+, OH^+, H^+, CO_2^+, CO^+, N_2^*, O_2^*, N, O, H, OH, and CO. In case of high water vapor concentration the oxidizing radicals ïOH, HO_2 ï and $O(^3P)$ as well as excited ions are the most important products. These species take part in a variety of ionmolecule reactions, neutralization reactions, dimerization etc. SO_2, NO, NO_2, and NH3 cannot compete with the reactions because of very low concentrations, but react with N, O, ïOH, and HOï2 radicals. After humidification and lowering of the temperature, flue gases are guided to reaction chamber, where irradiation by electron beam takes place. NH_3 is injected upstream of the irradiation chamber. There are several pathways of NO oxidation known. In the case of EBFGT the most common are as follows (Tokunaga & Suzuki, 1984):

$$NO + O(^3P) + M \rightarrow NO_2 + M \tag{35}$$

$$O(^3P) + O_2 + M \rightarrow O_3 + M \tag{36}$$

$$NO + O_3 + M \rightarrow NO_2 + O_2 + M \tag{37}$$

$$NO + HO^{\bullet}_2 + M \rightarrow NO_2 + {}^{\bullet}OH + M \tag{38}$$

After the oxidation NO_2 is converted to nitric acid in the reaction with ïOH according to the reaction (39) and HNO_3 aerosol reacts with NH_3 giving ammonium nitrate. NO is partly reduced to atmospheric nitrogen.

$$NO_2 + {}^{\bullet}OH + M \rightarrow HNO_3 + M \tag{39}$$

$$HNO_3 + NH_3 \rightarrow NH_4NO_3 \tag{40}$$

There can be also several pathways of SO_2 oxidation depending on the conditions. In the EBFGT process the most important are radio-thermal and thermal reactions. Radio-thermal reactions proceed through radical oxidation of SO_2 in the reaction (41) and HSO_3 creates ammonium sulphate in the following steps (42) and (43).

$$SO_2 + \cdot OH + M \rightarrow HSO_3 + M \tag{41}$$

$$HSO_3 + O_2 \rightarrow SO_3 + HO\cdot_2 \tag{42}$$

$$SO_3 + H_2O \rightarrow H_2SO_4 \tag{43}$$

$$H_2SO_4 + 2NH_3 \rightarrow (NH_4)_2SO_4 \tag{44}$$

The thermal reaction is based on the following process:

$$SO_2 + 2NH_3 \rightarrow (NH_3)_2SO_2 \tag{45}$$

$$(NH_3)_2SO_2 \xrightarrow{O_2, H_2O} (NH_4)_2SO_4 \tag{46}$$

The total yield of SO_2 removal consists of the yield of thermal and radio-thermal reactions that can be written as follows (Chmielewski, 1995).

$$\eta_{SO2} = \eta_1(\phi,T) + \eta_2(D, \alpha_{NH3}, T) \tag{47}$$

Where η, φ, T, D and αNH_3 are process efficiency, gas humidity, gas temperature, dose deposited (amount of energy transferred to gas by means of irradiation) and ammonia stoichiometry (NH_3 concentration in relation to stoichiometric value) respectively. The yield of the thermal reaction depends on the temperature and humidity and decreases with the temperature increase. The yield of the radio-thermal reaction depends on the dose, temperature and ammonia stoichiometry. The main parameter in NOx removal is the dose. The rest of parameters play minor role in the process. Nevertheless in real, industrial process, dose distribution and gas flow conditions are important from the technological point of view. The technology was originally implemented in coal fired power plants but can be applied for the cleaning of off-gases from various combustion processes. A complete EBFGT installation is schematically shown in fig. 6. After the boiler fly ash is removed from the flue gas by an ESP and cooled down and humidified in spray towers. Cooled and humidified gases are than exposed to the electron beam radiation after the injection of ammonia. The high-energetic electrons are forming the plasma and initiate a series of the above listed reactions which lead to the removal of the SOx and NOx by forming ammonium sulphate $(NH_4)2SO_4$ and ammonium nitrate $NH4NO3$ respectively. The reacted gas then passes through a particulate removal device (e.g. ESP) to remove the ammonium sulphate and ammonium nitrate which are used as fertilizers. Pilot and industrial installations demonstrated the feasibility of this technology for effective flue gas purification. The process was implemented in industrial scale in Pomorzany Power Plant (Poland) for total capacity of 270,000 Nm₃/h of flue gas. SO₂ removal efficiency above 95 % and NOx removal above 75 % were reported for optimal treatment conditions. A dose of up to 10 kGy (1 kGy = 1 kJ/kg flue gas) is required for NOx removal, while SO₂ can be removed in proper conditions at lower energy consumption. Nowadays most technical problems occurred in the prototype installations has been solved (Chmielewski et al., 2004).

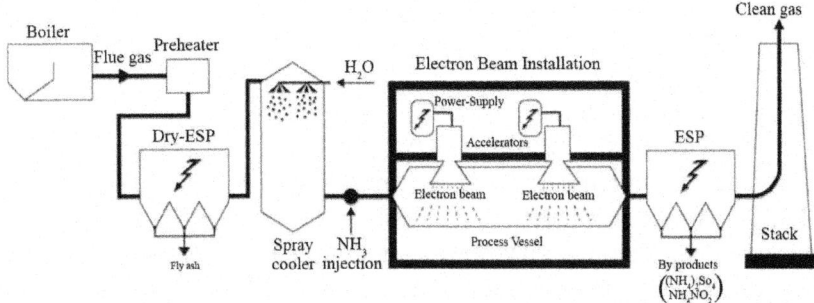

Figure. 6: Scheme of an EBFGT process

In recent investigations the electrical energy consumption could be decreased and the increase of system availability is in progress, too. The new applications concern application of electron beam for flue gases treatment from high sulphur oil fired boiler performed for Saudi Aramco Company (Basfar et al., 2008). In addition the removal of VOCs, dioxins, mercury and other pollutants from flue gases using EBFGT has been investigated. In the case of VOCs, decomposition the process itself is based on the similar principles as primary reactions concerning SO₂ and NOx removal i.e. free radicals attack on organic compounds chains or rings causing VOCs decomposition. For chlorinated aliphatic hydrocarbon' decomposition (e.g. chloroethylene), Cl- dissociated secondary electron attachment and Cl, OH radicals reaction with VOCs play very important roles (Sun et al., 2006). The most important development concerns application for the reduction of polychlorinated dibenzodioxins (PCDDs, so-called dioxins) and polychlorinated dibenzofurans (PCDFs) emission from municipal solid waste incinerators (Hirota et al., 2003). Electron beam irradiation demonstrated high levels of mercury oxidation at the bench scale, and the technology might help to improve mercury removal in wet scrubbers or wet ESPs when employed as a primary or secondary mercury oxidation technique (J.C. Kim et al., 2010).

AIR-DEPOLLUTION BY MEANS OF DISCHARGES GENERATED PLASMAS AND PLASMAENHANCED CATALYSIS

Several examples on the use of gas discharges for depollution of exhaust air will be discussed in the following. This will cover the removal of volatile organic compounds (VOCs) and deodorization, NOx- and SOx removal and removal of particulate matter (PM), e.g. soot.

VOC-Removal and Deodorization

Plasmas have been demonstrated to be able to decompose VOCs and thus odour molecules very efficiently for low decontamination levels (< 1 g_{Corg}/m^3) as like in deodorization issues. Odour emission is a significant problem in the production of food, for farms, in gastronomy and kitchens as well as in waste management. Typical odour molecules are in fact VOCs, namely aldehydes, fatty acids, alkanes, formic acid, amines or esters. VOC contamination is an issue of increasing importance for the depollution of exhaust air, i.e. offgases from industrial processes. For such problems plasma can be better suited than classical methods like wet scrubbing, adsorption or thermal processes because of lower energy consumption. In (Rafflenbeul, 2008) a plasma based process with an energy consumption of about 35 kWh for 70,000 m3/h exhaust air is described, while an odour reduction of up to 99 % is possible. Due to its compactness plasma devices can be easily integrated in existing systems and processes. Furthermore no waste or waste water is generated. However, the application of non-thermal plasma has to be reckoned for every specific exhaust problem and in fact, in industrial practice plasmas are combined with catalysts, absorbing agents and other methods of depollution. E.g. undesirable by-products can be formed since the plasmachemical conversion is not selective or influenced by the gas composition and properties, such as residual humidity or temperature. Energetic efficiency has been found to be best at low contamination levels and low gas flows. Special attention must be paid to the geometrical properties of the reactor (length, cross section) since it influences and determines the residence time as well as the back pressure. The residence needs be optimised for a successful treatment (1 to 3 sec are given in literature), while the back pressure should be as low as possible in order to ensure a proper integration in an exhaust air system (Rafflenbeul, 2008; Müller & Zahn, 2007). For deodorization applications NTP is often enhanced with catalyst or absorption methods. The plasmaNorm process (airtec competence GmbH) comprises a three-stage treatment unit (Müller et al., 2006). In the first stage the polluted gas is stripped of solids, aerosols and particulates by means of a pre-filter. Appropriate filter media such as bag filters for damp or oily air are used according to the air impurities to be removed. A surface DBD serves as the second stage, where pre-filtered air is subjected to reactive radicals and ions initiating oxidation reactions and the decomposition of VOCs and other contaminants. Finally compounds not yet oxidised are retained in an activated carbon bed, which is described as a storage reactor that, among other effects, revert residual ozone to atmospheric oxygen. The economical, long serviceable life of the activated carbon, as it regenerates itself during the process is promoted as one of the main special characteristic of this technology. It is successfully used in gastronomy and kitchens (large scale and private households) as well as food processing industry. E.g. the exhaust from 1.5 MW ovens for convenience products made of meat generating an exhaust stream of 8,000 Nm3/h can be deodorized (Langner, 2009). In (Rafflenbeul, 1998) a commercial plasma process combined with a biofilter as pre-filter and oxidation catalyst as after-filter is described. Biodegradable compounds in higher concentrations are decomposed in the biofilter, while the subsequent plasma unit partially oxidizes non-biodegradable pollutions which are finally decomposed in the oxidation catalyst section. The same company (Envisolve) describes several commercialised combinations of non-thermal plasmas with catalysts or molecular sieves for waste management facilities, paintshops and other industrial applications generating exhaust streams of up to 300,000 m3/h. In case of VOC removal different types of power supplies are used in the terms of voltage type and shape, operation frequency, supply system topology. All of the above mentioned parameters can strongly influence overall system performance and an optimum for most cases can be found. Power supply properties may influence the nature of reactor operation just like the reactor construction itself. First of all the operating frequency influences the breakdown voltage (Valdivia-Barrientos et al. 2006) according to the semi – empiric equation:

$$U_{bd} = 1.4 \left(C_d / C_g \right)^2 / \ln(f)$$

(48)

where U_{bd} – reactor breakdown voltage, C_d – dielectric barrier capacitance, C_g – gas gap capacitance, f – supply frequency. Second of all for aimed chemical process often an optimal set of supply parameters can be found yielding in maximal destruction and removal efficiency (Magureanu et al., 2007) or productivity (Buntat et al., 2009) when recalculated into SIE. Such dependencies are often hard to follow in industrial cases, where gas composition is complex and varying with time but nevertheless power supply system parameters play an important role in the VOC removal process.

Flue Gas Treatment by Means of Plasma-Enhanced Catalysis

Non-thermal plasma has been applied for the treatment of exhausts of varying sizes of diesel engines from small cars, heavy trucks and marines (Miessner et al., 2002; Bröer & Hammer, 2000; Mok & Huh, 2005; Mizuno, 2007; Cha et al., 2007; McAdams et al., 2008). Same technology has also tested for oil fired boilers (Park et al., 2008). Typically the flue gases of these sources contains 200-1,000 ppm NO_x, 10-200 ppm hydrocarbons, 200-700 ppm C_o, 2-8 % CO_2, 1-5 % H_2O and 10-18 % O_2. The exhaust gas contains also particles with varying sizes. A great deal of effort was devoted to the treatment of particulate matter in flue gas from diesel engines. In (Müller & Zahn, 2007) a reactor combining a DBD with a wall flow filter for soot reduction is described. In this system one electrode is porous and gas-permeable. The flue gas is let out through the porous electrode, which filters and holds back the soot particles. Thus soot-particles are stored on this electrode which faces to an electrode surrounded by a dielectric material. Toxic and soot-containing harmful substances are decomposed in the plasma. The accumulated soot is decomposed due to cold oxidation process initiated by active plasma species leading to constant regeneration of the filter at low temperatures during all engine operation conditions. In (Yamamoto et al., 2003) the diesel particulate filter (DPF) regeneration for real diesel engine emissions at low temperatures by means of indirect or direct non-thermal plasma treatment was demonstrated. In other studies (Chae et al., 2001, Mok & Huh, 2005) corona and DBD reactors were successfully used for the removal of smoke and particulate maters from diesel engines. For the reduction of NOx from diesel engine exhausts selective catalytic reduction is used but the catalysts do not work properly at low temperatures below (200-300°C) (Penetrante et al., 1998; Bröer & Hammer, 2000; Tonkyn et al., 2003). For improvement of the reduction efficiency at lower temperatures, plasma enhanced selective catalytic reduction (PE-SCR) has been investigated (Penetrante et al., 1998; Bröer & Hammer, 2000; Tonkyn et al., 2003; Miessner et al., 2002; Mizuno 2007; Mok & Huh, 2005; Cha et al. 2007; McAdams et al., 2008; Hammer et al., 1999). In PE-SCR the plasma serves for the oxidation (NO to higher nitride oxides and hydrocarbons to partially oxidized ones). Oxidation is needed because many NOx-reduction catalysts have a higher activity toward NO_2 and thus the removal efficiency at low temperatures is significantly enhanced (Penetrante et al., 1998; Bröer & Hammer, 2000). Further enhancement of NOx reduction on catalyst is achieved by hydrocarbon radicals generated in the plasma (Penetrante et al., 1998; Miessner et al., 2002; Tonkyn et al., 2003). In the oxidizing environment of diesel exhaust, an effective reduction of NOx to N_2 on catalyst takes place only when there are enough reducing agents (NH_3, hydrocarbons like propene). In the case of optimized burning process, there are usually not enough hydrocarbons in the engine exhaust for efficient reduction of NOx (Tonkyn et al., 2003; Miessner et al., 2002; Cha et al., 2007). Thus it is necessary to inject additional reducing agent to the exhaust gas. When NH_3 or urea is used as reducing agent, it has to be carried in a separate tank while hydrocarbons can also be obtained directly from the fuel. Presence of hydrocarbons in the plasma stage enhances the oxidation of NO to NO_2 and additionally inhibits the oxidation of SO_2 and the formation of HNO_3 (Penetrante et al., 1998). The production of HCN can be problematic when hydrocarbons are used (Tonkyn et al., 2003) while the ammonia slip and catalyst poisoning by NH_4NO_3 have to be considered when ammonia is used (Dors & Mizeraczyk, 2004).

For the removal of diesel exhaust, the additional fuel consumption due to plasma treatment should not exceed 5 % which corresponds to SIE of 15-60 J/sl for different diesel engines (Tonkyn et al., 2003; Mizuno 2007). In experiments with synthetic exhaust gas without particles, 80 % of NOx reduction has been achieved with energy input of 27-32 J/sl when hydrocarbons and NH3 were used as reducing agent and temperature was 150 °C (Bröer & Hammer, 2000; Mizuno 2007; Lee et al., 2007). V_2O_5/TiO_2 based catalysts or Co-ZSM$_5$ were used and space velocities were 15,000 and 2,000 h-1. When only hydrocarbons have been used as catalysts, the NOx reduction above 70 % has been achieved in temperature range of 170°C to 260°C with BaY zeolite (space velocity 12,000 h-1) and by using both BaY and γ- Al_2O_3 the temperature range has been extended to 500 °C (Kwak et al., 2004). For real diesel exhaust gases, the reduction efficiencies are usually smaller because of the presence of particles which reduce NO_2 back to NO (Dorai et al., 2000). Diesel exhaust gas of an Multicar M25[-10] engine (1,997 cm^3, without catalyst) having gas flow of 10 sl/min (space velocity 20,000 h^{-1}) and typically 434 p$_{pm}$ NOx was treated with plasma-enhanced catalysis where catalyst was placed downstream from the plasma reactor (Miessner et al., 2002). Catalyst alone (γ-Al2O3 + oxidation catalyst at 250°C) removed only CO and hydrocarbons (10^{-50} ppm in the exhaust). The NOx removal by plasma (54 J/sl) and catalyst was only 7% while the injection of additional propene (1.2 sl/h or 2,000 ppm) increased the NOx reduction to 56 %. Further increase of energy density did not improve NOx removal. For another diesel engine exhaust the NOx reduction of 73 % was achieved at energy densities 43 J/sl and at low temperature of 150 °C (Mizuno 2007). The gas flow rate was of 6 sl/min (space velocity of 36,000 h-1) while inlet gas had about 313 ppm of NOx and 881 ppm acetylene. In this experiment, Pt-Al2O3 pellets were used as catalyst inside the backed bed plasma reactor. For both systems, the estimated additional fuel consumption due to plasma generation was 4-5 % (Miessner et al., 2002; Mizuno 2007). In the second experiment, the removal of particulate matter was determined to be 95 % (5-7 mg/m3).

NH$_3$ as reducing agent was used for a commercial diesel engine from a used truck (Mok & Huh, 2005). Part of the diesel exhaust (10 sl/min) at no load condition with 180 ppm NOx and around 0.6 mg/m^3 particulate matters was introduced to the reactor system where monolithic V$_2$O$_5$/TiO$_2$ catalyst was placed downstream of the reactor. The effect of plasma SIE was tested in the temperature range of 100 to 200°C with the ratio of NH^3 and inlet NOx concentration set to 0.9. Plasma had strongest effect at 150 °C where the NOx reduction increased from 45 % to 80 % at input energy density of 25 J/sl whereas further increase in input energy did not improve the reduction. At 200 °C, the reduction was above 65 % already without plasma. Same system was also used for the removal of PM and it was possible to remove 50 % and 80 % of PM at SIE of 20 and 40 J/sl respectively (Mok and Huh, 2005). Up to 85 % of NOx reduction with 2 % of fuel penalty has also been achieved with similar dielectric-barrier discharge/urea-SCR hybrid system applied to VW Passat TDI engine exhaust (cold start and urban driving condition) (Hammer, 2002). An earlier experiment with Hatz 1D$_3$0 engine resulted in more than 75 % of NOx reduction with 17 J/sl and at catalyst temperature 170 °C (Hammer et al., 1999). Plasma-enhanced catalysis has also used to improve the cleaning of marine diesel exhaust at low temperatures below 200 °C where commercial NH$_3$ SCR catalysts do not work properly (Cha et al., 2007). In this study, 1/10th of the exhaust (100 Nm3/h) from 300 hp Yanmar engine was directed to hybrid plasma-catalyst system and the engine load was 25 % (550 ppm NOx, 116 ppm C$_3$H$_6$). The plasma reactor operated properly even after more than 1000 hr of work in highly humid and sooting conditions. The NOx reduction efficiency on catalyst (space velocity of 450,000 h^{-1}) increased from 20 to 80 % at 100 °C and from 55 to 90% at 200 °C at energy density of 40 J/sl with additional C$_3$H$_6$ above 1.5 times the NOx concentration injected to the exhaust and 550 ppm NH$_3$ injected after the plasma reactor. The estimated power consumption of plasma device for the warming period of the engine (500 Nm^3/h) was 5-6 kW and this corresponds for about 2 % of the engine power. The plasma reactor reduced also 45 % of the particulate maters (Cha et al., 2007). For larger NOx concentrations of 1,200 ppm, simulated marine diesel exhaust experiments have been carried out (McAdams et al., 2008). At 250°C and with Ag/Al$_2$O$_3$ catalyst, it was possible to obtain 50% of NOx reduction at energy densities of 60 J/sl (with the C$_3$H$_6$:NOx ratio of 2). At 350°C, above 90% of NOx reduction was measured at same energy density values. The catalysts were sulphur tolerant up to the concentrations of 1 %. The fuel penalty of 10 % was estimated for the type of engines simulated in the experiment.

When hydrocarbons are used as reducing agents, some of the NO_2 is reduced back to NO on the catalysts and this limits the maximum achievable reduction efficiency (Tonkyn et al., 2003). Use of multiple stages of plasma reactors and catalysts can overcome this limitation and increase of the reduction up to 90 % has been demonstrated in simulated exhaust gases (Tonkyn et al., 2003). Hybrid plasma-catalysis reactor with modular design was recently also tested for removal of NOx in oil-fired boiler (Park et al., 2008). The reactor consisted of four consequent plasma/catalyst modules where catalysts could be either TiO_2 or Pd/ZrO_2. The hybrid system with two first catalyst modules from Pd/ZrO_2 and last two from TiO_2 allowed to obtain the best results giving 74 % of NOx reduction with stoichiometric amount of C_3H_6 at 150 °C and space velocity of 3,300 h-1. Initial NOx concentration was 500 ppm. In addition to NOx reductions, several examples for large- and full-scale demonstration installations for flue-gas cleaning of SOx, dioxin and some VOCs are given in (H.H. Kim 2004; Mizuno 2007). Most recent review of research on catalytic processes enhanced by nonthermal plasma are presented in (Van Durme et al., 2008). Pulsed DC-driven FHC discharges with aerodynamic stabilization were used for conversion of NOx in air mixtures. The discharge works as 100% oxidation catalyst, converting NO to NO_2, without any additives (Barankova & Bardos 2010). The electrode material plays an important role in the plasma chemical kinetics as it brings about its own material constants, e.g. work function, secondary electron emission coefficient and catalytic activity. Due to an optimized geometry and efficient transfer of power to the electrons in the system, the power consumption for gas conversion is extremely low. Typical specific energy densities within the processing window are around 5 J/sl, i.e. 0.14 kWh/100 Nm^3.

INJECTION METHODS AND SCRUBBING-COMBINED PLASMA PROCESSES

Alternatively to the direct treatment (gas is passing the plasma reactor completely) indirect, remote or so-called injection methods are possible. In this case clean gas will be treated by the plasma and then admixed to the flue gas. The most well known example is the low temperature oxidation (LTO) of NOx by ozone injection. The idea of LTO is to oxidize relatively insoluble NOx to higher oxides such us N_2O_5 that are highly soluble and can easily be removed in wet scrubbers (Jarvis et al., 2003; Ferrell, 2000). In non-thermal plasmas in oxygen, ozone formation starts by dissociation of O2 via electron impact (49). Resulting oxygen atoms form ozone in three-body collisions (50, M is a third partner), while ozone production is balanced by the decomposition reaction (51) and thermal dissociation (52) at steady state conditions.

$$e^- + O_2 \rightarrow 2\, O(^3P,\, ^1D) \tag{49}$$

$$O(^3P,\, ^1D) + O_2 + M \rightarrow O_3 + M \tag{50}$$

$$O(^3P) + O_3 \rightarrow 2\, O_2 \tag{51}$$

$$O(^3P) + O_3 \rightarrow O_2 + O \tag{52}$$

For example, exhaust NO can be oxidized by O_3 to form NO_2, NO_3 and, subsequently, N_2O_5 (53-55). Reaction (54) is the slowest reaction in this chain. Nitric pentoxide (N_2O_5, anhydride of nitric acid) can be efficiently removed from the exhaust by a washing bottle or scrubbing forming nitric acid (HNO_3, 56). In humid exhaust gases, HNO_3 may be formed in the exhaust gas itself. It can be used as chemical feedstock or it can be neutralized and used e.g. as fertilizer, similar as in the EBFGT process.

$$O_3 + NO \rightarrow NO_2 + O_2 \tag{53}$$

$$O_3 + NO_2 \rightarrow NO_3 + O_2 \tag{54}$$

$$NO_3 + NO_2 \rightarrow N_2O_5$$

$$O_3 + 2\, NO_2 \rightarrow N_2O_5 + O_2 \tag{55}$$

$$N_2O_5(g) + H_2O \rightarrow 2\, HNO_3(aq) \tag{56}$$

The deNOx efficiency was found to be maximum at 100 °C and the addition of small water droplets improves the NOx oxidation rate (Stamate et al., 2010). Advantages of LTO NOx is that the plasma discharge is kept clean and the removal rate of NO is higher than direct oxidation methods where the reverse reactions occur to reform NO and NO2 by the O radical (Yoshioka et al., 2003; Eliasson & Kogelschatz, 1991 as cited in Stamate et al., 2010). A commercial system applying ozone injection is available under the trademark LoTox (Low Temperature Oxidation). The process works within the Electro-dynamic Venturi (EDV) wet scrubbing system in order to achieve a combined reduction of PM, SOx and NOx of stationary emission sources, especially refinery applications (Confuorto & Sexton, 2005). Ozone is generated on site and on demand and injected after the dry ESP directly into the wet scrubber. N_2O_5 is converted to HNO_3 and finally neutralized by the scrubbers alkali reagent to $NaNO_3$. Other pollutants such as SO_2 and HCl are removed in the wet scrubbing process simultaneously. There exist a number of commercial installations in the USA and in Asia on different emission sources. NOx removal higher than 90 % has been reported. The removal of mercury was demonstrated, too. Several refinery installations have demonstrated LoTox performance and reliability on an applicable scale, the process is available from DuPont BELCO Clean Air Technologies. There are several advantages combining plasma treatment of gases with scrubbing processes. Gutsol et al. reports on a wet or spray pulsed corona discharges studied for the VOC-removal from paper mill exhaust gases (Fridman, 2008). In case of spay corona water is injected to the corona discharge like a shower, while in wet corona a thin water film rinse on the outer wall electrode. In such arrangement soluble VOCs adsorb on the water droplets or film while non-soluble VOCs can be converted to soluble compounds (e.g. peroxides and peroxide radicals) by means of plasma treatment and subsequently scrubbed within the same arrangement. This results in much lower energy requirements. Furthermore, plasmastimulated oxidation continuous after adsorption resulting in a larger adsorbing capacity of the water and thus water consumption. However such process is only applicable where already large amounts of polluted water are generated and which requires effective water cleaning.

The ECO (Electro-Catalytic Oxidation) process is another example for a commercialized plasma-assisted depollution process combined with scrubbing (Boyle, 2005). The process is designed for installation downstream of a dry ESP or fabric filter (ash removal). The flue gas is directly exposed to DBD and oxidizes pollutants to soluble or capturable compounds (e.g. NO to NO_2; SO_2 to SO_3; Hg to mercury oxide HgO) and form particulate matter and aerosol mist. SO_2, NO_2 and HgO are removed in a subsequent absorber vessel (two-loop scrubber). Ammonia is added to the scrubber to maintain the pH of the solution for keeping high SO_2 scrubbing rate. NO_2 formed in the ECO reactor is scrubbed by sulphite ions, which are formed by SO_2. Finally $(NH_4)2SO_4$ and NH_4NO_3 are formed as well. Several preliminary designs for coal-fired electric utility applications ranging from $175 - 1,000$ MW has been developed and long time performance and reliability test were successfully completed. The process is available by the company Powerspan Corporation and has recently combined with post-combustion CO_2 capture technology.

CONCLUSION

To a great extend non-thermal plasma processes were demonstrated for commercial pollution control applications having following peculiarities: The decomposition of contaminants without heating of the gas can be achieved, while a wide range of pollutants (gases and particulate matter/aerosols) can be treated. Organic particles can be decomposed due to oxidation at low temperatures. The best efficiency is reached in low contaminated gases making them well suited for deodorization issues, too. An advantage of plasmas for gas depollution is that the energy consumption of the plasma stage can be regulated easily with the pollutant mass flow by the electrical parameters. However, in all examples the plasma is one part of a complete depollution system, since plasma-chemical conversion is not selective and mainly oxidative Furthermore, energetic efficient treatment is achieved in case of low contaminated exhaust air and the formation of undesirable by-products has to be taken into account. Plasma processes combined with other treatment processes give synergetic effects. The addition of ammonia-based substances as reducing agents for plasma generated higher nitrogen oxides can be considered as state-of-the-art since it is already used in several processes (EBFGT, LoTox, ECO). Theses process has been successfully demonstrated on an industrial scale, e.g. for the flue gas treatment of coal fired power plants. New developments of EBFGT technology concern the treatment of flue gases from high sulphur oil fired boilers and the removal of (poly)chlorinated VOCs like dioxins from municipal solid waste incinerators. In this context the range of removable contaminants will be extended. Several VOCs, dioxins and mercury are under investigation with promising results at bench scale. The combination of nonthermal plasma with catalysts, absorbing agents or scrubbing techniques are promising approaches. Hybrid systems and especially plasma-driven catalysis will be one of the major prospects for future developments. Therefore the interaction of plasmas with catalysts has to be investigated more detailed and a profound understanding of the development, physics and chemistry in polluted gases is desired. In this context more efforts on the understanding of the physics of filamentary plasmas consisting of microdischarges is necessary. Furthermore the power supply system parameters play an important role in the removal process and novel topologies with high potential for further improvement are under development.

ACKNOWLEDGMENT

This chapter is dedicated to Prof. Jen-Shih Chang (McMaster University, Ontario/Canada), who passed away 2011. His work is an outstanding contribution to the present knowledge on plasma science and its application to environmental problems. The contribution was prepared within the transnational project "PlasTEP - Dissemination and fostering of plasma based technological innovation for environment protection in the Baltic Sea region "part-financed by the European Union (European Regional Development Fund)." The authors like to express their gratitude to all involved partners and colleagues within this project, in particular Alexander Schwock, Justyna Jaskowiak and Jane Schmidt from the Technology Centre of Western Pomerania in Greifswald for support.

REFERENCES

1. Baird, J.K.; Miller, G.P. & Li N. (1999). The G value in plasma and radiation chemistry. Journal of Applied Physics, Vol. 68, 7, pp. 3661 – 3668, ISSN: 0021-8979

2. Baránková, H. & Bárdos, L. (2002). Fused hollow cathode cold atmospheric plasma source for gas treatment, Catalysis Today, Vol. 72, pp. 237–241, ISSN 0920-5861

3. Baránková, H. & Bárdos, L. (2003). Hollow cathode cold atmospheric plasma sources with monoatomic and molecular gases, Surface Coating Technology, Vol. 649, pp. 163 –164

4. Baránková, H. & Bárdos, L. (2010). Effect of the electrode material on the atmospheric plasma conversion of NO in air mixtures, Vacuum, Vol. 84, pp. 1385-1388

5. Basfar, A.A.; Fageeha, O.I.; Kunnummal, N.; Chmielewski, A.G; Pawelec, A.; Licki, J. & Zimek, Z. (2008). Electron beam flue gas treatment (EBFGT) technology for simultaneous removal of SO2 and NOx from combustion of liquid fuels, FUEL, Vol. 87, 8-9, pp. 1446-1452, ISSN 0016-2361

6. Becker, K.H.; Kogelschatz, U.; Schoenbach, K.H. & Barker, R.J. (2005). Series in Plasma Physics: Non-Equilibrium Air Plasmas at Atmospheric Pressure, Institute of Physics Publishing Ltd, Bristol and Philadelphia, USA, ISBN 0-7503-0962-8

7. Boyle, P. D. (2005). Multi-Pollutant Control Technology for Coal-Fired Power Plants, Proceeding of the Clean Coal and Power Conference, Washington, DC, November 21-22, 2005; available at http://www. powerspan.com/technology/eco_overview.shtml   Brandenburg,   R.; Wagner, H.-E.; Morozov, A.M. & Kozlov, K.V. (2005). Axial and radial development of the microdischarges of barrier discharge in N2/O2 mixtures at atmospheric pressure. Journal Physics D: Applied Physics, Vol. 38, 11, pp. 1649-1657,ISSN

8. Bröer, S. & Hammer, T. (2000). Selective Catalytic Reduction of Nitrogen Oxides by Combining a Non Thermal Plasma and a V2O5-WO3/TiO2 Catalyst. Applied Catalysis B: Environmental, Vol. 28, pp. 101-111, ISSN 0926-3373

9. Buntat, Z.; Smith, I. R. & Razali, N. A. M. (2009). Ozone generation using atmospheric pressure glow discharge in air. Journal Physics D: Applied Physics, Vol. 42, 235202(5pp)

10. Casanueva, R.; Azcondo, F. J. & Bracho S. (2004). Series–parallel resonant converter for an EDM power supply. Journal of Materials Processing Technology Vol. 149, pp. 172–177, ISSN 0924-0136

11. Cha, M.S.; Song, Y.-H.; Lee, J.-O. & Kim, S.J. (2007). NOx and Soot Reduction Using Dielectric Barrier Discharge and NH3 Selective Catalytic Reduction in Diesel Exhaust. InternationalJournal on Plasma Environmental Science and Technology, Vol. 1,pp. 28-33, ISSN 1881-8692

12. Chae, J.O.; Hwang, J.W.; Jung, J.Y.; Han, J.H.; Hwang, H.J.; Kim, S. & Demidiouk, V.I. (2001). Reduction of the particulate and nitric oxide from the diesel engine using a plasmachemical hybrid system. Physics Plasmas, Vol. 8, pp. ISSN 1403-1410, ISSN 1070-664X

13. Chmielewski, A. G.; Zimek, Z.; Panta, P. & Drabik, W. (1995). The double window for electron beam injection into the flue gas process vessel. Radiation Physics and Chemistry, Vol. 45, No. 6, pp. 1029-1033, ISSN 0969-806X

14. Chmielewski, A.G., (1995a) Technological development of eb flue gas treatment based on physics and chemistry of the process. Radiation Physics and Chemistry, Vol. 46, No. 46, pp. 1057-1062, ISSN 0969-806X

15. Chmielewski, A.G.; Licki, J.; Pawelec, A.; Tyminski, B. & Zimek, Z. (2004). Operational experience of the industrial plant for electron beam flue gas treatment. Radiation Physics and Chemistry, Vol. 71, pp. 439–442, ISSN 0969-806X

16. Chang, J. S. (2008). Physics and chemistry of plasma pollution control technology. Plasma Sources Science and Technology, Vol. 17, 045004 (6 pp), ISSN 0963-0252

17. Cobine, J.D. & Wilbur, D.A. (1951). The electronic torch and related high frequencyphenomena. Journal Applied Physics, Vol. 22, No. 6, pp. 835-841, ISSN 0021-8979

18. Confuorto, N. & Sexton, J. (2007). Wet Scrubbing Based NOx Control Using LoTOx™ Technology - First Commercial FCC Start-up Experience. Proceedings of NPRA 2007 Environmental Conference, Austin/Texas, September 24-25, 2007

19. Department of Energie [DOE] (2010) Accelerators for America's Future, Washington, DC http://www.acceleratorsamerica.org/report/index.html

20. Dorai, R.; Hassouni, K. & Kushner, M.J. (2000) Interaction between soot particles and NOx during dielectric barrier discharge plasma remediation of simulated diesel exhaust. Journal Applied Physics, Vol. 88, pp. 6060-6071, ISSN 0021-8979

21. Dors, M. & Mizeraczyk, J., (2004). NOx removal from a flue gas in a corona dischargecatalyst hybrid system. Catalysis Today, Vol. 89, pp. 127-133, ISSN 0920-5861

22. Edinger, R. (2008). Reduction of SOx and NOx emissions by electron beam flue gas treatment. available at http://www.ebfgt.com/file_library/ userfiles/file/COM2008_WINNIPEG_20080529.pdf

23. Ferrell, R. (2000). Controlling NOx emissions: A cooler alternative - An LTO system usesozone injection to reduce NOx emissions. Pollution Engineering, Vol. 32, 4, pp. 50-52,ISSN 0032-3640

24. Fitch R. A. (1968). Electrical Pulse Generators. US Patent US 3,366,799. 30 Jan. 1968 Francke, K. P.; Rudolph, R. & Miessner, H. (2003). Design and operating characteristics of a simple and reliable DBD reactor for use with atmospheric air. Plasma Chemistry and Plasma Processing. Vol. 23, No. 1, ISSN 0272-4324

25. Fridman, A. (2008). Plasma Chemistry, Cambridge University Press, Cambridge, USA, ISBN-13 978-0-521-84735-3

26. Fujita, H. & Akagi, H. (1999). Control and performance of a pulse-density-modulated seriesresonantinverter for corona discharge processes. IEEE Transactions on IndustryApplications. Vol. 35, 3, pp. 621 – 627, ISSN 0093-9994

27. Grundmann, J.; Müller, S. & Zahn, R.-J. (2007). Extraction of Ions from Dielectric BarrierDischarge Configurations. Plasma Processesses and Polymers, Vol. 4, S1004–S1008, ISSN 1612-8850

28. Hammer, T. (1999). Applications of plasma technology in environemtal techniques. Contributions to Plasma Physics, Vol. 39, No. 5, pp. 441-462, ISSN 0863-1042

29. Hammer, T.; Kishimoto, T.; Miessner, H. & Rudolph, R. (1999). Plasma Enhanced Selective Catalytic Reduction: Kinetics of Nox-Removal and Byproduct Formation. SAE Technical Paper 1999-01-3632, ISSN 0148-7191

30. Hammer, T. (2002). Non-thermal plasma application to the abatement of noxious emissions in automotive exhaust gases. Plasma Sources Science and Technology, Vol. 11, A196–A201, ISSN 0963-0252

31. Hirota, K.; Hakoda, T.; Taguchi, M.; Takigami, M.; Kim H. & Kojima, T. (2003) Application of electron beam for the reduction of PCDD/F emission from municipal solid waste incinerators. Environmental Science and Technology, Vol. 37, No. 14, pp. 3164-3170,ISSN 0013-936X

32. Holzer, F.; Kopinke, F.D. & Roland, U. (2005). Influence of ferroelectric

materials and catalysts on the performance of non-thermal plasma (NTP) for the removal of air pollutants. Plasma Chemistry and Plasma Processing, Vol. 25, 6, pp. 595-611, ISSN0272-4324

33. Jarvis, J.B.; Naresh, A.T.D. & Suchak, J. (2003). LoTOx process flexibility and multi-pollutant control capability, Presented at the Combined Power Plant Air Pollutant Control MegaSymposium, Washington, DC, May 2003, ISBN 0923204547, 9780923204549

34. Jasinski, M.; Mizeraczyk, J.; Zakrzewski, Z.; Ohkubo, T., & Chang, J.S. (2002). CFC-11 Destruction by Microwave Torch Generated Atmospheric-Pressure Nitrogen Discharges. Journal Physics D: Applied Physics, Vol. 35, No. 18., pp. 2274-2280, ISSN 0022-3727

35. Jasinski, M.; Dors, M. & Mizeraczyk, J. (2009). Destruction of Freon HFC-134a using a nozzleless microwave plasma source. Plasma Chemistry and Plasma Processessing,, Vol. 29, No. 5 , pp. 363-372, ISSN 0272-4324,

36. Kim, H.-H. (2004). Nonthermal plasma processing for air-pollution control: A historical review, current issues, and future prospects. Plasma Processes and Polymers, Vol. 1, pp. 91-110, ISSN 1612-8850

37. Kim, J.C.; Kim, K.H.; Al Armendariz & Al-Sheikhly, M. (2010). Electron Beam Irradiation for Mercury Oxidation and Mercury Emissions Control. Journal of EnvironmentalEngineering, pp. 554 – 559, ISSN 0733-9372

38. Kogelschatz, U. (2002). Filamentary, patterned, and diffuse barrier discharges. IEEE Transactions on Plasma Science, Vol. 30, 4, pp. 1400-1408, ISSN 0093-3813

39. Kogelschatz, U. (2004). Atmospheric-pressure plasma technology. Plasma Physics and Controlled Fusion, Vol. 46, 12B, pp. B63-B75, ISSN 0741-3335

40. Kostov, K.G.; Honda, R. Y.; Alves, L.M.S. & Kayama, M.E. (2009). Characteristics of dielectric barrier discharge reactor for material treatment. Brazilian Journal of Physics, Vol. 39, No.2, ISSN 0374-4922, 0103-9733

41. Kwak, J.H.; Szanyi, J. & Peden C.H.F. (2004). Non-thermal plasma-assisted NOx reduction over alkali and alkaline earth ion exhanged Y, FAU zeolites. Catalysis Today, Vol. 89,pp. 135-141, ISSN 0899-8388

42. Langner, M.H. (n.d.). Image brochure of airtec competence GmbH, Recke, Germany; available at http://www.plasmanorm.de Lee, J.O.; Song, Y.-H.; Cha, M.S. & Kim, S.J. (2007). Effects of Hydrocarbons and Water

43. Vapor on NOx Using V2O5-WO3/TiO2 Catalyst Reduction in Combination with Nonthermal Plasma, Industrial & Engineering Chemistry Research, Vol. 46, No. 17, pp.5570-5575, ISSN 0888-5885

44. Licki, J.; Chmielewski, A. G.; Iller, E.; Zimek, Z.; Mazurek, J. & Sobolewski, L. (2003). Electron beam flue-gas treatment for multicomponent air-pollution control. AppliedEnergy, Vol. 75, pp. 145-154, ISSN 0306-2619

45. Magureanu, M.; Mandache, N.B.; Parvulescu, V.I.; Subrahmanyam, Ch.; Renken, A. & KiwiMinsker,L. (2007). Improved performance of non-thermal plasma reactor during decomposition of trichloroethylene: Optimization of the reactor geometry and introduction of catalytic electrode. Applied Catalysis B: Environmental, Vol. 74, pp.270–277, ISSN 0926-3373

46. Marx, E. (1928). Verfahren zur Schlagpruefung von Isolatoren und anderen elektrischenVorrichtungen. Patentschrift Nr. 455933. 13 Feb. 1928

47. McAdams, R.; Beech, P. & Shawcross, J.T. (2008). Low temperature plasma assisted catalytic reduction of NOx in simulated marine diesel exhaust. Plasma Chemistry and Plasma Processing, Vol. 28, pp. 159-171, ISSN 0272-4324

48. Miessner, H.; Francke, K.-P. & Rudolph, R. (2002) Plasma-enhanced HC-SCR of NOx in the presence of excess oxygen. Applied Catalysis B: Environmental, Vol. 36, pp. 53-62, ISSN 0926-3373

49. Mizuno, A. (2007). Industrial applications of atmospheric non-thermal plasma in environmental remediation. Plasma Physics and Controlled Fusion, 49, A1-A15, ISSN 0741-3335

50. Moisan, M.; Sauve, G.; Zakrzewski, Z. & Hubert, J. (1994). An atmospheric pressure waveguide-fed microwave plasma torch: the TIA design. Plasma Sources Science and Technology, Vol. 3, No. 4, (November 1994), pp. 584-592, ISSN 0963-0252

51. Moisan, M.; Zakrzewski, Z. & Rostaing, J.C. (2001). Waveguide-based single and multiplenozzle plasma torches: the TIAGO concept. Plasma Sources Science and Technology, Vol. 10, No. 3, (August 2001), pp. 387-394, ISSN 0963-0252

52. Mok, Y.S. & Huh, Y.J. (2005). Simultaneous Removal of Nitrogen Oxides and Particulate Matters from Diesel Engine Exhaust Using Dielectric Barrier Discharge and Catalysis Hybrid System. Plasma Chemistry and Plasma Processing, Vol. 25, pp. 625-639, ISSN 0272-4324

53. Mok, Y. S.; Lee, S.-B.; Oh, J.-H.; Ra K.-S. & Sung B.-H. (2008).

Abatement of Trichloromethane by Using Nonthermal Plasma Reactors. Journal of Plasma Chemistry and Plasma Processing. Vol. 28, No. 6, ISSN 0272-4324

54. Müller, S.; Zahn, R.-J.; Grundmann, J. & Langner, M. (2006). Plasma Treatment of Aerosolsand Odours. Proceedings 3rd International Workshop on Microplasmas IWM3, Greifswald, Germany, May 9 - 11, 2006, p. 121

55. Müller, S. & Zahn, R.-J. (2007). Air Pollution Control by Non-Thermal Plasma. Contributions to Plasma Physics, Vol. 47, No. 7, pp. 520 – 529, ISSN 0863-1042

56. Ohkubo, T.; Kanazawa, S.; Nomoto, Y.; Chang, J. S. & Adachi, T. (1996). Time dependence of NOx removal rate by a corona radical shower system. IEEE Transaction on Industry Applications, Vol. 32, pp. 1058–1062, ISSN 0093-9994

57. Park, B. R. & Deshwal, S.H., Moon (2008) NOx removal from the flue gas of oil-fired boiler using a multistage plasma-catalyst hybrid system, Fuel Processing and Technology, Vol. 89, pp. 540-548, ISSN 0016-2361, 0378-3820

58. Park, J.Y.; Tomicic, I.; Round, G.F. & Chang J.S. (1999). Simultaneous removal of NOx and SO2 from NO-SO2-CO2-N-2-O-2 gas mixtures by corona radical shower systems. Journal Physics D: Applied Physics, Volume: 32, 9, pp. 1006-1011, ISSN: 0022-3727

59. Penetrante, B. M. & Schultheis, S. E. (1993). Non-Thermal Plasma Techniques for Pollution Control, NATO ASI Series, Vol. G34, Springer, Part A and B, Berlin, Germany, ISBN0258-2023

60. Penetrante B.M.; Hsiao, M.C,; Merritt, B.T.; Vogtlin, G.E.; Wallman, P.H.; Neiger, M.; Wolf O.; Hammer, T. & Broer, S. (1996). Pulsed corona and dielectric-barrier discharge processing of NO in N2. Applied Physics Letters, Vol. 68, 26, pp. 3719-3721, ISSN:0003-6951

61. Penetrante, B.M.; Brusasco, R.M.; Merritt, B.T.; Pitz, W.J.; Vogtlin, G.E.; Kung, M.C.; Kung,H.H.; Wan, C.Z. & Voss, K.E. (1998). Plasma-Assisted Catalytic Reduction of NOx.SAE Technical Paper 982508 (October 1998)

62. Rafflenbeul, R. (1998). Nicht-thermische Plasmaanlagen (NTP) zur Luftreinhaltung in derAbfallwirtschaft. Müll und Abfall 1/1998, pp. 38-44 (in german), ISSN 0027-2957

63. Rafflenbeul, R. (2008). Geringe Kosten für gering konzentrierte Abluft. Wasser, Luft undBoden (wlb) 6/2008, pp. 36-41, (in german); available at http://www.envisolve.com,ISSN 1421-8615

64. Sasoh, A.; Kikuchi K. & Sakai, T. (2007). Spatio-temporal filament behaviour in a dielectricbarrier discharge plasma actuator. Journal Physics D: Applied Physics, Vol. 40, pp.4181–4184, ISSN 1361-6463

65. Stamate, E.; Chen, W.; Jørgensen, L.; Jensen, T.K., Fateev, A. & Michelsen, P.K. (2010). IR and UV gas adsorption measurements during NOx reduction on an industrial gas fired power plant. Fuel, Vol. 89, 5, pp. 978-985, ISSN 0016-2361

66. Sun, Y.X.; Chmielewski, A.G. & Bulka, S. (2006). Influence of base gas mixture on decomposition of 1,4-dichlorobenzene in an electron beam generated plasma reactor. Plasma Chemistry and Plasma Processing, Vol. 26, 4, pp. 347-359, ISSN 0272-4324

67. Tonkyn, R.G.; Barlow, S.E. & Hoard, J.W. (2003). Reduction of NOx in synthetic dieselexhaust via two-step plasma-catalysis treatment. Applied Catalysis B: Environmental, Vol. 40, p. 207, ISSN 0926-3373

68. Tokunaga, O. & Suzuki, N. (1984). Radiation chemical reactions in NOx and SO2 removals from flue gas. Radiation Physics and Chemistry, Vol. 24, No. 1, pp. 145-165, ISSN0020-7055, 0146-5724, 0969-806X

69. Uhm, H.S.; Hong, Y.C. & Shin, D.H. (2006). A microwave plasma torch and its applications.Plasma Sources Science and Technology, Vol. 15, No. 2, (May 2006), pp. S26-S34, ISSN 0963-0252

70. Van Durme, J.; Dewulf, J.; Leys, Ch. & Van Langenhove, H. (2008). Combining non-thermalplasma with heterogeneous catalysis in waste gas treatment: A Review. Applied Catalysis B: Environmental, Vol. 78, No. 3-4, pp. 324-333, ISSN 0926-3373

71. Wagner, H.-E.; Brandenburg , R.; Kozlov, K.V.; Sonnenfeld, A.; Michel, P. & Behnke, J.F. (2003). The barrier discharge: basic properties and applications to surface treatment Vacuum, Vol. 71, pp. 417–436, ISSN: 0042-207X

72. Validivia-Barrientos, R.; Pacheco-Sotelo, J.; Pacheco-Pacheco, M.; Benitez-Read, J. S. & Lopez-Callejas R. (2006). Analysis and electrical modelling of a cylindrical DBD configuration at different operating frequencies. Plasma Sources Science and Technology, Vol. 15, pp. 237-245, ISSN 0963-0252

73. Yamamoto, T.; Ramanathan, K.; Lawless, P.A.; Ensor, D.S.; Newsome, J.R.; Plaks, N. & Ramsey, G.H. (1992). Control of volatile organic-compounds by an ac energized ferroelectric pellet reactor and a pulse corona reactor. IEEE Trans. on Industry Applications, Vol. 28, 3, pp. 528-534, ISSN: 0093-9994

74. Yamamoto, M. & Murayama, S. (1967). UHF torch discharge as an excitation source. Spectrochimia Acta A, Vol. 23, No. 4, pp. 773-776, ISSN 0370-8322, 0584-8539, 1386-1425

75. Yamamoto, T.; Okubo, M.; Kuroki, T. & Miyairi, Y. (2003). Nonthermal plasma regeneration of diesel particulate filter. SAE Technical Paper 2003-01-1182, SAE International, ISSN 0148-7191

Chapter 12

GENERATION AND DISPERSION OF TOTAL SUSPENDED PARTICULATE MATTER DUE TO MINING ACTIVITIES IN AN INDIAN OPENCAST COAL PROJECT

Ratnesh Trivedi, M. K. Chakraborty and B. K. Tewary

Scientists, Central Institute of Mining and Fuel Research, India

INTRODUCTION

The knowledge of ambient air quality plays an important role in assessing the environmental scenario of the region. The ambient air quality status in the vicinity of the mining activities forms an indispensable part of the Environmental Impact Assessment Studies. The quality of ambient air depends upon the concentrations of specific contaminants, the emission sources and meteorological conditions. The mining activities contribute to the problem of air pollution directly or indirectly (Trichy ,1996, Corti and Senatore, 2000, Baldauf et al., 2001 and Collins et al., 2001). Coal dust is the major pollutant in the air of open cast coal mining areas. (Kumar et al., 1994, Vallack and Shillito, 1998. and CIMFR, 1998) The primary source of fugitive dust at fully operational surface mine may include overburden (OB) removal, blasting, mineral haulage, mechanical handling operations, minerals stockpiles and site restoration (Appleton et. al. 2006). Major air pollutants due to opencast mining are total suspended particulate matter and respirable particulate matter whereas concentration of SO_2 and NOX is negligible (Sinha and Banerjee, 1997, CIMFR, 1998, Banerjee, 2006, and Trivedi et. al., 2009). Transportation of materials is the major source of TSPM generation in the mining areas. The vehicle and haul road intersection has been identified as the most critical source producing as much as 70% of total dust emitted from surface coal mines (Muleski and Cowherd, 1987, Sinha and Banerjee, 1997, Ghose and Majee, 2002), while it was accounted to be 80-90% of the PM_{10} emission (Cole and Zapert, 1995). Maximal concentrations of particulate matter are generally occurred during winter and minimal in the rainy season.(Ghose and Majee, 2000, Tayanc, 2000, Reddy and Ruj, 2003). However, in certain urban areas maximal concentrations of particulate matters are also observed in summer season (Crabbe et al., 2000, Almbauer et al., 2001, Triantafyllou et al., 2002 and Triantafyllou, 2003). The dispersion of particulate matter follows the annual predominant wind direction of an area (Corti and Senatore, 2000, Baldauf et al., 2001 and Pandey et. al., 2008). Such a large amount of dust generated cause safety and health hazards such as poor visibility, failure of mining equipment, increased maintenance cost etc which ultimately lowers the productivity. A prolonged exposure to air borne dust may cause to damage of lung tissues of the miners which may further lead to pneumoconiosis or black lung disease. The maximum tissue damage is caused by the dust of 5 microns lesser sizes since such particles reach the alveoli of the lung (Peavey et. al, 1985). These air pollutants reduce air quality and this ultimately affects people, flora and fauna in and around mining areas (Crabbe et al., 2000, Wheeler et al., 2000). Implementation of effective air quality control measures by the mining company are needed and green belts development can be devised wherever necessary (Kapoor and Gupta, 1984, Sharma and Roy, 1997, Shannigrahi and Sharma, 2000, Chaulya, 2004). In the present study, an attempt has been made to generate ambient air quality data, micrometeorological data, source-wise emission inventory data for an Indian coal mine namely Padampur Opencast Coal Project (O.C.P.) of Western Coalfields Ltd. (W.C.L.), India. The status of TSPM and PM10 concentration in ambient air has been monitored through a well defined at monitoring network. In the light of micro-meteorological data such as wind speed, stability class etc, dispersion coefficients of the dust for vertical as well as horizontal direction have been estimated. A correlation between TSPM and PM_{10} concentration has been sought out. Emission inventory data for all point, area and line sources of TSPM at Padampur OCP have been generated for the determination of emission rates. Air Pollution modeling has also been attempted using Fugitive Dust Model (FDM) developed by United States Environment protection Agency (USEPA). FDM has been used for the validation of the study by comparing predicted and observed values. FDM is a computerized Gaussian Plume Air Quality Model, specifically designed for the estimation of the concentration and deposition impacts from fugitive dust sources. FDM employs an advance transfer particle deposition algorithm. (USEPA, 1995). The model gives hourly average, long term concentration and deposition of particulate matters at all user selected receptor locations. FDM represents the behavior of particles in the atmosphere most accurately. Since terrain features are not included in FDM, it can be used only for local scale.

MATERIALS AND METHOD

Field Settings at OCP

As mentioned earlier that Padampur OCP is selected for the study purpose. Padampur OCP is located at Chandrapur district in Maharastra State of India. The Project is located between latitudes 20^0 2' N to 20^0 3' N and Longitudes 79^0 17' E to 79^0 19' E and is covered by Survey of India Toposheet No 55 P/8. Geologically the area forms the central part of eastern limb of Regional anticline structure of Wardha Valley Coalfield of Western Coalfields Limited (WCL). The area is undulating with few isolated ridges of Kamthi Sandstone. The area covers two separate and adjoining geological blocks namely Padampur and Motaghat blocks. The net geological reserve of Padampur OCP is about 43.5 Million Tones. The annual production is 1 Million Tones with an average stripping ratio of 3.7 m³/tones. The coal produced from the mine is of non-coaking type with 'D' and 'E' grade. The shovel dumper combination is being used to excavate the overburden as well as coal. The shovels of 4.6 m3 bucket capacity and dumpers are of 35 Ton capacities have been deployed in the field. Backfilling is also practiced simultaneously with the production of coal. Micrometeorological data collection is an indispensable part of any air pollution study. The data collected during air quality survey are used for proper interpretation of existing ambient air quality status. The ambient air quality monitoring was carried out through reconnaissance followed by air quality surveillance program and micrometeorological study of the area. A weather monitoring station and SODAR have been installed at study site. The weather monitoring station measures ambient air temperature in degree centigrade, wind speed in km. per hour and wind direction in degrees from north. It also measures relative humidity, barometric pressure, and total rainfall. Site specific relevant parameters like mixing height and stability class have been accurately measured by SODAR. The amount of turbulence in the ambient air has a major effect upon the rise and dispersion of air pollutant plumes. The amount of turbulence is categorized into "stability classes". The most commonly used categories are the Pasquill stability classes A, B, C, D, E, and F. Class A denotes the most unstable or most turbulent conditions and Class F denotes the most stable or least turbulent conditions. The most common procedure for estimating the dispersion coefficients was introduced by Pasquill (1961), modified by Gifford (1961) and adopted by the U.S. Public Health Service (Turner, 1970). Meteorological data has been collected from the nearest Indian Meteorological Department (IMD) station at Nagpur, India. The climate of the area is tropical. Summer is well defined from April to June, followed by rainy season from July to September and winter from December to February. May is the hottest month with temperature rising to a maximum of around 48°C. December is the coldest month when the temperature falls down to about 10°C. The mean annual rainfall is around 1250 mm. Wind direction is generally from North and Northwest, with velocities up to 6-7 Km./hour during monsoon and about 3-4 Km./hour in winter. Relative humidity varies from 74-83% during August and September and is about 15-20% during summer. Wind rose diagram during the study period is illustrated in Fig.1.

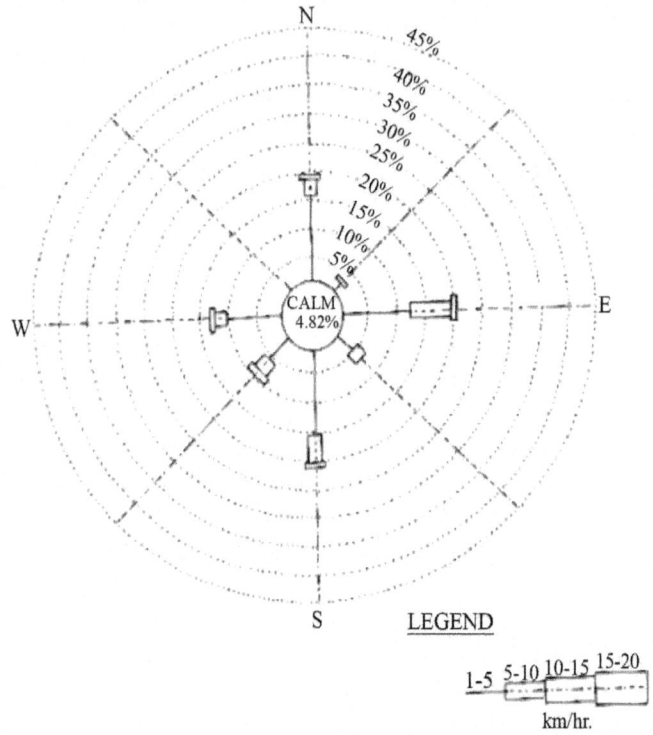

Figure. 1: Wind Rose Diagram of Padampur OCP

Ambient Air Quality Monitoring

The ambient air quality status in the impact zone was assessed through a network of ambient air quality monitoring locations. The studies on air environment include identification of specific air pollutants for assessing the impacts of proposed mining projects including other activities. Air quality monitoring was carried out in winter season. Among the ambient air quality parameters, Total Suspended Particulate Matter (T.S.P.M.) and Respirable Particulate Matter (PM$_{10}$.) have been measured at 8 hours interval for 24 hours using the High Volume Sampler with Respiratory Particulate Matter measurement arrangement with the standard methods as shown in Table 1. Other air qualities parameters are not considered because of their concentrations are found much below the threshold value in the study area. The existing status of air environment was assessed through a systematic air quality surveillance program in which five ambient air quality stations have been selected to know the air quality of the area. The measured data of TSPM and PM_{10} are shown in Table 2 along with the arithmetic mean and standard deviation (S.D.) of the measured data.

Table 1: Air Pollutant Analysis Methods: Coal Mine Standards

Para meter	Time weighted Avg.	Concentration in Ambient Air	Method	Instruments
TSPM	Annual	430 µg/ m³	IS-5182	High Volume Sampler with PM_{10}
	24 hours	600 µg/ m³	Part XIV	Measurement arrangement (Av. Flow rate not < 1.1 m³/min)
PM_{10}	Annual	215 µg/ m³	IS-5182	High Volume Sampler with PM_{10}
	24 hours	300 µg/ m³	Part XIV	Measurement arrangement (Av. Flow rate not < 1.1 m³/min)

(Source: Central Pollution Control Board Notification, 1994)

Table 2: Table Showing Ambient Air Quality of Padmapur OCP

Sl. No.	Sample Site	TSPM (µg/ m³)		PM_{10} (µg/ m³)		PM_{10} (Percent of TSPM
		mean	S.D.	mean	S.D.	
1	Filter Plant	294.30	21.67	120.34	18.90	40.89
2	SAM Office	631.72	58.44	120.03	21.55	19.00
3	Kitadi Village	390.30	32.20	103.03	25.90	26.39
4	Padampur Village	654.38	54.55	130.88	12.15	20.00
5	Manager Office Sec-IV	1078.10	75.56	226.40	35.60	20.99

Source-Wise Emission Inventory Details

Emission inventory details have been collected by installing two High Volume Samplers at down wind sides and, to know the back ground concentration, one High Volume Sampler at up wind side of the TSPM source. High Volume Samplers are placed at a distance nearly 100m from the source Emission data have been generated for various mining activities such as overburden loading, coal loading, haul road transportation, unloading of overburden, unloading of coal, stock yard, exposed overburden dumps, coal handling plant, exposed pit face and workshop. Blasting being an instantaneous source was monitored separately which is not included in the present study. The modified Pasquill and Gifford formula for ground level emission has been used to calculate the emission rate.

$$C(x,0) = \frac{Q}{\Pi u\, \sigma y\, \sigma z}$$

Where, C(x,0) =DN max - UP C (x, 0), Difference in pollutant concentration , μg/.m; DN max, maximum concentration in down wind direction; UP, back ground concentration in up wind direction; Q, Pollutant emission rate, μg/s; Π, 3.14159; u, Mean wind speed, m/s; σy, Standard deviation of horizontal plume concentration, evaluated in terms of downwind distance x, m (as shown in Fig. 2); σz, Standard deviation of vertical plumes concentration, evaluated in terms of downwind distance x, m (as shown in Fig. 3).

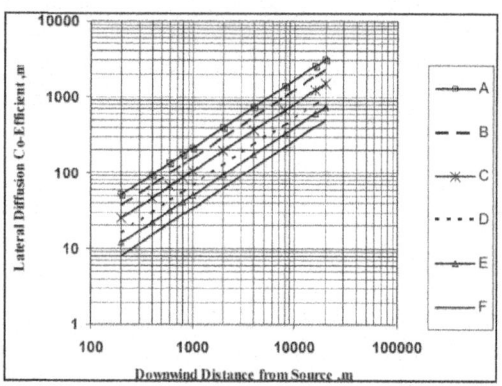

Figure. 2: Lateral Diffusion Co-Efficient Vs Downwind Distance from Source (Source : Turner, 1970).

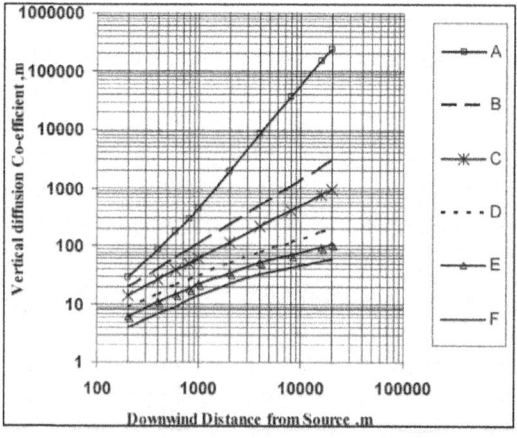

Figure. 3: Vertical diffusion Co-efficient Vs Downwind Distance from Source (Source : Turner, 1970).

Source-wise emission inventory has been shown in Table 3. Source-wise emission properties such as moisture content silt content etc. measured from the samples collected during field study have been placed in Table 4. Secondary data affecting TSPM emission such as frequency of drilling, vehicle movement on the haul road and transport road etc have also been collected as shown in Table 4.

Table 3: Source-Wise TSPM Emission Inventory at Padmapur OCP

TSPM source	TSPM Concentration ($\mu g/m^3$)				Wind velocity, m/s	Diffusion coefficient		Emission Rate	
	DN Min	DN Max	UP	DN_{max}-UP		σy, m	σz, m	Unit	Value
Drilling	1340	1758	1233	525	2.1	14	8	g/s	0.3877
Overburden Loading	1234	1660	1108	552	2.4	14	8	g/s	0.4659
Coal Loading	1648	2092	1377	715	2.1	14	8	g/s	0.5218
Haul Road	1963	2498	1336	1162	3.1	14	8	g/ms	0.0127
Transport Road	2015	2605	1387	1218	3.1	14	8	g/ms	0.0132
Overburden Unloading	1195	1605	942	663	1.9	18	12	g/s	0.8544
Coal Unloading	1438	1897	1135	762	2.0	14	8	g/s	0.5360
Exposed Overburden dump	1030	1387	1002	385	2.5	24	16	g/m²s	0.0000363
Stockyard	1482	1872	1027	845	1.8	14	8	g/m²s	0.0001981
Workshop	1062	1478	1040	438	1.7	25	15	g/m²s	0.0000878
Exposed pit surface	1015	1357	985	372	1.0	15	32	g/m²s	0.0000160
Overall mine	469	713	365	348	2.4	95	60	g/m²s	0.0000108

Table 4: Source Wise TSPM Emission Properties

TSPM sources	Source type	Moisture content, %	Silt content, %	Emission rate		Remarks
				Unit	Value	
Drilling	Point	7.4	38.0	g/s	0.443	Hole dia 160 mm;12 hole/day
Overburden Loading	Point	7.6	13.6	g/s	0.4867	drop height 1.4 m; frequency 23 no/hr
Coal Loading	Point	8.1	10.9	g/s	0.5783	drop height 0.9 m; frequency 23 no/hr
Haul Road	Line	12.4	34.5	g/ms	0.0144	Frequency 18 no/hr; average speed 2.6m/sec;
Transport Road	Line	9.8%	30.0	g/ms	0.0146	Frequency 27 no/hr; Average speed 10 m/sec.
Unloading of Overburden	Point	7.2	14.2	g/s	1.2740	Frequency 10 no/hr ;drop height 14.3m.
Unloading of coal	Point	8.0	11.2	g/s	0.7333	Frequency 7 no/hr; drop height 2.5 m
Exposed Overburden dump	Area	7.4	8.2	g/m²s	0.00004	dump area 0.029 sq. km
Stock yard	Area	6.0	12.5	g/m²s	0.00024	Unloading freq. 3 No/hr; loading freq. 12.0 No/hr
Work shop	Area	12.4	31.8	g/m²s	0.0001	Area 5000 sq. m
Exposed pit	Area	8.1	7.8	g/m²s	0.00002	Exposed area 0.03 sq.km.

RESULTS AND DISCUSSION

Air pollution modeling has been exercised with the help of Fugitive Dust Model (FDM). The input parameters include source types, dust concentration near sources, hourly meteorological data such as wind speed and direction, temperature atmospheric stability and receptor locations. All data depicted the average value for the study period. Emission source has been demarcated in three categories of sources like point, line and area sources using mine plan. All these sources have been numbered for preparation of data sheet. Emission rate has been assigned to each activity as per the field measurement data in the mine. From the modeling exercise, TSPM concentrations at certain receptor locations have been predicted. The receptor locations have been selected such that these are exactly same of one where ambient air quality measurement was carried out. The predicted values at receptor locations have been added to regional background levels to get the total predicted TSPM concentration. Regional background data are the average of the monitored data in no activity zone. The predicted and observed TSPM concentrations at receptor locations for different mines are listed in Table 5 and Table 6.

Field observations of ambient air quality of Padampur OCP have been placed in the Table 2. The Ambient Air quality at the five sites of Padmapur is well within the limit except Manager Office sec IV The higher value of TSPM and PM_{10} at Manager Office sec IV may be contributed by the presence of main transport road and other industries nearby. The 24-hr average of TSPM concentrations ranged from 294.3 to 1078.1 µg m^{-3} in industrial area i.e. mining area and from 390.3 to 654.38 µg m^{-3} in residential area respectively. The 24-hr average of PM_{10} concentrations ranged from 120.34 to 226.4 µg m−3 in industrial area and from 103.03 to 130.88 µg m^{-3} in residential area respectively as shown in Table 2. On average the PM10 in the ambient air constituted 19.00 % to 40.89 % of the TSPM in mining area and 20.00 % to 26.39 % of the TSPM in residential area. The concentration of particulate matters vary with the meteorological parameters and a relation also exist between TSPM and PM_{10} (Tayanc, 2000, Jones et al., 2002, Triantafyllou et al., 2002, Triantafyllou, 2003, Chaulya, 2004). The case under study also reveals that there exists a relationship between TSPM and PM_{10} concentrations. Linear regression correlation coefficient (R^2) between TSPM with PM_{10} has been found to be 0.8116 as shown in Fig. 4. With the help of FDM, TSPM concentration at five monitoring stations has been predicted. The variation between measured and predicted values, as shown in Table 5 and fig. 6 , may be due to non-accountability of emission from various other sources like non-mining area activities, domestic use of fuels, transportation network nearby thermal power plant , cement plant etc. The value of coefficient of correlation between observed values of TSPM Concentration and predicted values by FDM have been calculated to be 0.969

Source-wise emission inventory data placed in Table 3. Stability classes have been found to be B, C & D. It is clearly evident from the Table 3 that among the point sources namely drilling, overburden loading, overburden unloading and coal unloading highest value of emission rates (g/s) has been found in case of unloading of overburden. Among the area sources namely exposed OB dump, stockyard, workshop, exposed pit surface, highest values of emission rate (gm/m²/s) has been found in case of exposed OB dump. Among the line sources, emission rates have been in case of haul found and transport road to be 0.0127 gm per meter per second and 0.0132 gm per meter per second respectively. In terms of overall TSPM pollution line sources contribute more than other sources because of their lengths and nature of mining operations. This very fact again confirms that the vehicle and haul road intersection is the major source of dust in opencast mines (Muleski and Cowherd, 1987, Sinha and Banerjee, 1997, Ghose and Majee, 2002) Emission rate for whole mine is found 0.0000108 gm per sq. meter per second.

With the help of FDM, TSPM concentration has been predicted at various distances in down wind direction as shown in Table 6. As far as rate of fall in concentration of TSPM with the distance from the source is concerned, an exponential fall in the TSPM concentration with the distance from the source has been observed which can be clearly seen in the Fig. 5. Maximal concentrations of TSPM and PM_{10} have been found to occur within the mine. Again the dust generated due to mining activities does not contribute to ambient air quality in surrounding areas beyond 500 meters in normal meteorological condition as shown in the Table 6. Thus the result matches with the findings of the Several researchers (Hanna et al., 1982, Chaulya et al., 2001 and Jones et al., 2002, Chaulya, 2004) that maximal concentrations of TSPM and PM_{10} are found in a mining area and the concentrations are gradually diminished with increase in distance due to transportation, deposition and dispersion of particles. The value of coefficient of correlation between observed values of TSPM Concentration and predicted values by FDM have been calculated to be 0.9957. From the emission study, it is quite clear that haul road and transport road were the major contributor to the pollution load of ambient air quality. Therefore proper dust suppression arrangement is to be made. The prevailing practice of water sprinkling does not seem to be adequate. Therefore, the installation of continuous atomized spraying system for haul roads should be used. Exposed overburden dump is another major contributor of pollution load. These dumps not only contribute to air pollution by way of wind erosion but also spread the dump it self. Therefore, judicious, plantation on these dumps is highly recommended. These plantations will not only stabilize the dump but also attenuate the dust emission. Biological reclamation of overburden dumps and wastelands is also essential. Effective control measures at the coal handling plant, excavation area and overburden dumps should also be implemented to mitigate the TSPM emissions at source. Green belt development around the mining area is highly recommended. The capacity of plants to reduce air pollution is well known (Kapoor and Gupta, 1984, Sharma and Roy, 1997, Shannigrahi and Sharma, 2000, Chaulya, 2004). A Few plant species can be grown around highly polluted areas where dust (TSPM) is the main pollutant as given Table 7. These species not only reduces air pollutants but also retards water and soil pollution.

Table 5: Comparison between observed and predicted values of TSPM Concentration

S. No.	Sample site	Observed value (μg/m³)	Predicted value (μg/m³)
1	Filter plant	294.30	250
2	SAM Office	631.72	540
3	Kitadi Village	390.30	325
4	Padmapur Village	654.38	564
5	Manager Office Sec-IV	1078.10	910

Table 6: Predicted values of TSPM Concentration along down Wind Direction

TSPM Source	Predicted values (μg/m³)					
	At source	100 m	200 m	300 m	400 m	500 m
Padampur mine as a whole	698	505	375	260	165	125

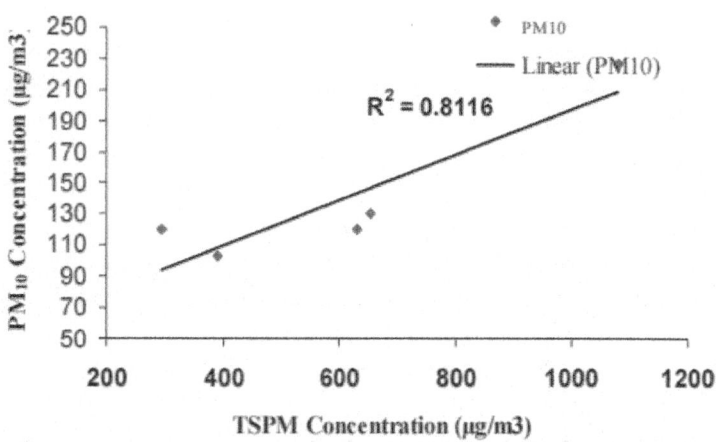

Figure. 4: Correlation between TSPM and PM$_{10}$ Concentration.

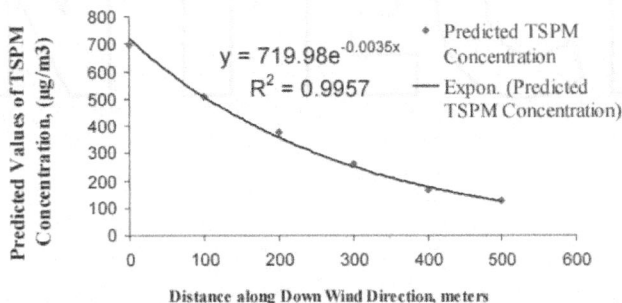

Figure. 5: Relation of TSPM Concetration with Distance from OCP.

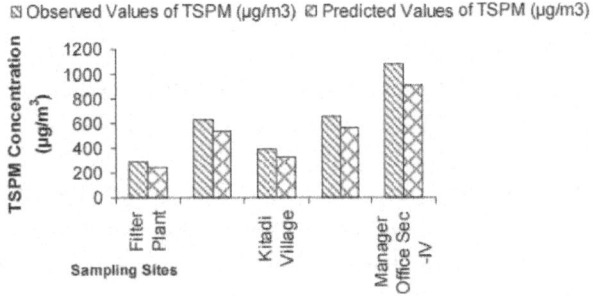

Figure. 6: Comparision between Observedvalues and Predicted Values of TSPM.

Table 7: Recommended pollution retarding plant species for green belt development

Species Name	Family	Local Name of Plants	Evergreen (E) or deciduous
Butea monsperma	Moraceae	Palas	Deciduous
Spathodea companulata	Bignoniaceae	Sapeta	Evergreen
Fiscus infectoria	Moraceae	Pakur	Evergreen
Cassia fistula	Caesalpiniaceae	Amaltas	Deciduous
Anthocephalus cadamba	Rubiaceae	Kadam	Deciduous
Cassia siamea	Caesalpiniaceae	Minjari	Deciduous

CONCLUSIONS

TSPM and PM10 are the major sources of emission from various opencast coal mining activities. The predicted values of TSPM using FDM are 70 percent to 94 percent of observed values. The difference between observed values and predicted values of TSPM indicates that there are non-mining sources of emission viz. domestic transportation network near by mine sites and other industries etc. Fugitive Dust Model (FDM) has been found to be most suitable for modeling of dispersion pattern of fugitive dust at Padampur

Opencast Coalmine Project of W.C.L. PM_{10} is the main focus of concern for human health. Correlation between PM_{10} and TSPM would help in predicting the PM10 concentration by knowing the concentration of TSPM for a similar mining site. Maximal concentration of TSPM is found in a mining area and the concentrations falls exponentially with increase in distance due to transportation, deposition and dispersion of particles.

Of the various sources of TSPM pollution, line sources contribute more than other sources because of their lengths and nature of mining operations. Among the line sources, emission rates have been in case of haul found and transport road to be 0.0127 gm per meter per second and 0.0132 gm per meter per second respectively. Emission rate for whole mine is found 0.0000108 gm per sq. meter per second. Various management strategies are evaluated for reduction of dust emission at the source and design of green belt with few recommended species is also very effective tool to mitigate air pollution. Proper dust suppression arrangement is to be made including installation of continuous atomized spraying system for haul roads and transport roads. As exposed overburden dump is another major contributor of pollution load, judicious, plantation on these dumps is highly recommended. However, for achieving the effective result to bring down the air pollution level in the mining area a constructive measure at political level is also highly essential. This would lead to an eco-friendly mining and better habitat for all those living in the area.

ACKNOWLEDGEMENTS

Authors are grateful to the Director, Central Institute of Mining and Fuel Research (CIMFR), Dhanbad, India for giving permission to publish this article. Authors are also thankful to M/s Western Coalfields Limited, Nagpur for sponsoring this study and providing necessary facilities.

REFERENES

1.   Almbauer, R.A., Piringer, M., Baumann, K., Oettle D., & Sturm P.J. (2001). Analysis of the   daily variations of winter time air pollution concentrations in the city of Graz,

2.   Austria., Environmental Monitoring and Assessment, Vol. 65, pp. 79–87.

3.   Appleton, T.J., Kingman, S.W., Lowndes I.S., & Silvester, S.A. (2006). The development of a modeling strategy for the simulation of fugitive dust emissions from in-pit quarrying activites: a UK case study, International Journal of Mining, Reclamation and Environment, Vol. 20, P. 57-82.

4.   Baldauf, R.W., Lane D.D., & Marote, G.A. (2001). Ambient air quality monitoring network design for assessing human health impacts from exposures to air-borne contaminants, Environmental Monitoring and Assessment, Vol. 66, pp. 63–76.

5.   Banerjee, S.P (2006). TSP emission factors for different mining activities for air quality impact prediction as collated from different sources, Minetech, Vol. 27, pp. 3-18.

6.   CPCB, Central Pollution Control Board Notification, India, (1994).

7.   Chaulya, S.K., Chakraborty, M.K., & Singh R.S.(2001). Air pollution modelling for a proposed limestone quarry. Water, Air, and Soil Pollution Vol. 126, pp. 171–191.

8.   Chaulya, S. K. (2004). Assessment and management of air quality for an opencast coalmining area, Journal of Environmental Management, Vol. 70, No. 1, pp. 1-14

9.   CIMFR. Central Institute of Mining and Fuel Research (erstwhile Central Mining Research Institute) Report (1998). Determination of Emission Factor for Various OpencastMining Activities, GAP/9/EMG/MOEF/97, Dhanbad, India

10.   Collins, M.J., Williams P.L, & MacIntosh, D.L (2001). Ambient air quality at the site of a former manufactured gas plant. Environmental Monitoring and Assessment Vol. 68,pp. 137–152.

11.   Corti, A. & Senatore, A. (2000). Project of an air quality monitoring network for industrialsite in Italy. Environmental Monitoring and Assessment, Vol. 65, pp. 109–117

12.   Crabbe, H., Beaumont R. & Norton, D. (2000). Assessment of air quality, emissions and management in a local urban environment. Environmental Monitoring and Assessment, Vol. 65 , pp. 435–442.

13.   Cole, C.F. & Zapert, J.C. (1995).Air Quality Dispersion Model Validation at Three Stone Quarries, Washington DC, National Stone Association.

14.   Ghose, M.K. & Majee, J. (2000). Assessment of dust generation due to opencast coal mining—an Indian case study. Environmental Monitoring and Assessment, Vol. 61,pp. 255–263.

15.   Ghose, M.K. & Majee, S.R. (2000). Assessment of the impact on the air environment due to opencast coal mining -an Indian case study, Atmospheric Environment ,Vol. 34, pp.2791-2796

16.   Gilford, F.A. (1961). Uses of Routine Meteorological Observations for estimating atmospheric Dispersion, Nuclear Safe, Vol. 2

17. Grundnig, P.W., Höflinger, W., Mauschitz, G., Liu, Z., Zhang G., & Wang, Z. (2006).

18. Influence of air humidity on the suppression of fugitive dust by using a waterspraying system, China Particuology, Vol. 4, No. 5, pp. 229-233.

19. Hanna, S.R., Briggs, G.A. &. Hosker, R.P. (1982). Handbook on Atmospheric Diffusion,DOE/TIC-11223, US Department of Energy, Technical Information Center Jones, T. Blackmore, P. Leach, M. Matt, B.K. Sexton K. &. Richards, R. (2002). Characterisation of airborne particles collected within and proximal to an opencast coalmine: SouthWales. UK Environmental Monitoring and Assessment, Vol. 75, pp. 293–312.

20. Kapoor, R.K. &. Gupta, V.K., (1984). A pollution attenuation coefficient concept for optimization of green belt. Atmospheric Environment, Vol. 18, pp. 1107–1117.

21. Karaca, M., Tayanc M. &. Toros, H. (1995). The effects of urbanization on climate of Istanbul and Ankara: a first study. Atmospheric Environment, Vol. , pp. 3411–3429.

22. Kumar, C.S.S., Kumar, P., Deshpande, V.P., &. Badrinath S.D. (1994). Fugitive dust emission estimation and validation of air quality model in bauxite mines, Proceedings of International Conference on Environmental Issues in Minerals and Energy Industry, IME Publications, New Delhi, India, pp. 77–81.

23. Muleski G.E. &. Cowherd, C. (1987). Evaluation of the effectiveness of Chemical dust Suppressants on Unpaved Roads, EPA/600/2-87.102. U.S. Environmental Protection Agency, Research Triangle Park N.C., pp-81.

24. Pandey, S.K., Tripathi, B.D. &. Mishra, V.K. (2008). Dust deposition in a sub-tropical opencast coalmine area, India, Journal of Environmental Management, Vol. 86, No. 1,pp. 132-138.

25. Pasquill, F. (1962). Atmospheric Diffusion, Van Nostrand Co. Ltd. Londan

26. Peavy, H.S., Rowe, D.R. &. Obanoglous, G. Tech (1985). Environmental Engineering, Megraw Hill, New York, pp. 668-670.

27. Reddy, G.S. &. Ruj, B. (2003). Ambient air quality status in Raniganj–Asansol area, India. Environmental Monitoring and Assessment, Vol. 189, pp. 153–163.

28. Roney J. A. &. White, B. R. (2006). Estimating fugitive dust emission rates using an environmental boundary layer wind tunnel, Atmospheric Environment, Vol. 40, pp. 7668-7685

29. Shannigrahi, A.S. &. Sharma, R.C. (2000). Environmental factors in green belt developmentan overview. Indian Journal of Environmental Protection , Vol. 20, pp. 602–607.

30. Sharma, S.C. &. Roy, R.K. (1997). Green belt—an effective means of mitigating industrial pollution. Indian Journal of Environmental Protection Vol. 17, pp. 724–727.

31. Sinha, S. &. Banerjee, S.P. (1997). Characterisation of haul road in Indian open cast iron oremine. Atmospheric Environment, Vol. 31, pp. 2809–2814.

32. Tayanc, M. (2000) An assessment of spatial and temporal variation of sulphur dioxide levels over Istanbul, Turkey. Environmental Pollution, Vol. 107, pp. 61–69.

33. Tichy, J. (1996). Impact of atmospheric deposition on the status of planted Norway spacestands: a comparative study between sites in Southern Sweden and the North Eastern Czech Republic. Environmental Pollution, Vol. 93, pp. 303–312

34. Triantafyllou, A.G. (2003). Levels and trends of suspended partcles around large lignitepower station. Environmental Monitoring and Assessment, Vol. 89, pp. 15–34.

35. Triantafyllou, A.G., Kyros E.S. &. Evagelopoulos, V.G. (2002). Respirable particulate matter at an urban and nearby industrial location: concentrations and variability, synoptic weather conditions during high pollution episodes. Journal of Air and Waste Management Association, Vol. 52, pp. 287–296.

36. Trivedi, R. Chakraborty M. K., &. Tiwary, B.K. (2009). Dust Dispersion Modeling Using Fugitive Dust Model at an Opencast Coal Project of Western Coalfields Limited, India, Journal of Scientific and Industrial Research, Vol. 68, pp71-78

37. Turner, D.B. (1970).Workbook of atmospheric Dispersion Estimates, U.S.E.P.A.,Washington, DC. USEPA, United States Environmental Protection Agency (1995). User's guide for the fugitive dust model (FDM), vol. 1, User Instructions, Region 10, 1200 sixth Avenue, Seattle, Washington, USA.

38. Vallack, H.W. &. Shillito, D.E. (1998). Suggested guidelines for deposited ambient dust, Atmospheric Environment, Vol. 32, No. 16, pp. 2737–2744.

39. Wheeler, A.J., Williams, I. Beaumont, R.A., &. Manilton, R.S. (2000). Characterisation of particulate matter sampled during a study of

children's personal exposure to air borne particulate matter in a UK urban environment. Environmental Monitoring and Assessment, Vol. 65, pp. 69–77.

CITATION

CHAPTER 1

Frederick W. Lipfert, An Assessment of Air Pollution Exposure Information for Health Studies, doi:10.3390/atmos6111736.

CHAPTER 2

Rohde RA, Muller RA (2015) Air Pollution in China: Mapping of Concentrations and Sources. PLoS ONE 10(8): e0135749. doi:10.1371/journal.pone.0135749.

CHAPTER 3

Diana Linhares, Patrícia Ventura Garcia, Fátima Viveiros, Teresa Ferreira, and Armindo dos Santos Rodrigues, "Air Pollution by Hydrothermal Volcanism and Human Pulmonary Function," BioMed Research International, vol. 2015, Article ID 326794, 9 pages, 2015. doi:10.1155/2015/326794.

CHAPTER 4

Laura E. Venegas, Nicolas A. Mazzeo and Andrea L. Pineda Rojas (2011). Evaluation of an Emission Inventory and Air Pollution in the Metropolitan Area of Buenos Aires, Air Quality-Models and Applications, Prof. Dragana Popovic (Ed.), ISBN: 978-953-307-307-1, InTech, DOI: 10.5772/18767.

CHAPTER 5

Al-Hassen, S. , Sultan, A. , Ateek, A. , Al-Saad, H. , Mahdi, S. and Alhello, A. (2015) Spatial Analysis on the Concentrations of Air Pollutants in Basra Province (Southern Iraq). Open Journal of Air Pollution, 4, 139-148. doi: 10.4236/ojap.2015.43013.

CHAPTER 6

Wang J-F, Hu M-G, Xu C-D, Christakos G, Zhao Y (2013) Estimation of Citywide Air Pollution in Beijing. PLoS ONE 8(1): e53400. doi:10.1371/journal.pone.0053400

CHAPTER 7

Gervasio A. Degrazia, Andrea U. Timm, Virnei S. Moreira and Debora R. Roberti (2011). Meandering Dispersion Model Applied to Air Pollution, Air Quality-Models and Applications, Prof. Dragana Popovic (Ed.), ISBN: 978-953-307-307-1, InTech, DOI: 10.5772/16892.

CHAPTER 8

Eleni Metaxa (2011). Air Pollution and Cultural Heritage: Searching for "The Relation Between Cause and Effect", Monitoring, Control and Effects of Air Pollution, Prof. Andrzej G. Chmielewski (Ed.), ISBN: 978-953-307-526-6, InTech, DOI: 10.5772/17704.

CHAPTER 9

Mohamed Kamal Khallaf (2011). Effect of Air Pollution on Archaeological Buildings in Cairo, Monitoring, Control and Effects of Air Pollution, Prof. Andrzej G. Chmielewski (Ed.), ISBN: 978-953-307-526-6, InTech, DOI: 10.5772/16748.

CHAPTER 10

Javier Reyes, Francisco Corvo, Yolanda Espinosa-Morales, Brisvey Dzul, Tezozomoc Perez, Cecilia Valdes, Daniel Aguilar and Patricia Quintana (2011). Influence of Air Pollution on Degradation of Historic Buildings at the Urban Tropical Atmosphere of San Francisco de Campeche City, México, Monitoring, Control and Effects of Air Pollution, Prof. Andrzej G. Chmielewski (Ed.), ISBN: 978-953-307-526-6, InTech, DOI: 10.5772/18739.

CHAPTER 11

Ronny Brandenburg, Hana Barankova, Ladislav Bardos, Andrzej G. Chmielewski, Miroslaw Dors, Helge Grosch, Marcin Hołub, Indrek Jõgi, Matti Laan, Jerzy Mizeraczyk, Andrzej Pawelec and Eugen Stamate (2011). Plasma-Based Depollution of Exhausts: Principles, State of the Art and Future Prospects, Monitoring, Control and Effects of Air Pollution, Prof. Andrzej G. Chmielewski (Ed.), ISBN: 978-953-307-526-6, InTech, DOI: 10.5772/20351.

CHAPTER 12

Ratnesh Trivedi, M. K. Chakraborty and B. K. Tewary (2011). Generation and Dispersion of Total Suspended Particulate Matter Due to Mining Activities in an Indian Opencast Coal Project, Monitoring, Control and Effects of Air Pollution, Prof. Andrzej G. Chmielewski (Ed.), ISBN: 978-953-307-526-6, InTech, DOI: 10.5772/20828.

INDEX

A

Air pollution 41, 42, 43, 44, 47, 48, 49, 50, 51, 52, 53, 54, 55, 56, 57, 58, 93, 175, 196
Air pollution modeling 270
American Thoracic Society (ATS) 44

B

best linear unbiased estimates (BLUE) 112

C

Carbon monoxide (CO) 61, 64, 85, 98
Central Institute of Mining and Fuel Research (CIMFR) 274
Chemical formula 103
Chemical processes 239
Chemistry, plasma 230
City of Buenos Aires (CBA) 62, 72
Conventional epidemiological 4
Cultural heritage 144, 146

D

Diffuse degassing structure (DDS) 44

E

Environmental tobacco smoke (ETS) 1, 10

F

Fundamental parameters 129, 131

G

Geographical distribution 105
Geographic Information System (GIS) 111
Greater Buenos Aires (GBA) 62, 72

H

Historic buildings 201, 204, 206, 209, 217, 218
Hydrocarbon (HCs) 103
Hydrogen sulfide 102

I

Idaho Engineering Laboratory (INEL) 131
Indian Meteorological Department (IMD) 265
Islamic Cairo 173

L

Landing-take-off (LTO) 65
Laser-induced breakdown spectroscopy

(LIBS) 148
Light microscopy (LM) 148

M

Masonry cleaning 186
Metropolitan Area of Buenos Aires
    (MABA) 61, 62, 72, 85
Microwave plasma source (MPS) 232

N

National Health and Nutrition Examina-
    tion Survey (NHANES III) 46
National Oceanic and Atmos-pheric
    Administration (NOAA) 131
Natural environment 143
Natural fires 208
Nitric oxide (NO) 9

O

Odds ratio (OR) 47

P

Particulate matter (PM) 1, 110
Particulate matter (PM2.5) 6
Plasma pollution 231
Plasma torches 231, 232, 259
Polarizing Microscope (PM) 179
Pollution concentrations 8, 10, 12, 13
Pulsed corona discharges (PCD). 235

R

Reversed-Flow Inverse Gas Chromatog-

raphy (RF-IGC) 146, 151

S

Scanning electron microscope (SEM)
    179
Scanning Electron Microscope [SEM]
    178
Single Point Areal Estimation (SPA) 112
Stone materials 201
Studied extensively 9
Studies involving 3

T

Total suspended particles (TSP) 211
Turbulent velocity 126, 128, 130, 131,
    141

U

United States Environment protection
    Agency (USEPA) 264

V

Velocity fluctuations 125, 126, 128

W

Western Coalfields Limited (WCL) 265
World Health Organization (WHO) 5,
    119

X

X-ray diffraction (XRD) 178